Lecture Notes in Networks and Systems

Volume 1453

The series "Lecture Notes in Networks and Systems" publishes the latest developments in Networks and Systems—quickly, informally and with high quality. Original research reported in proceedings and post-proceedings represents the core of LNNS.

Volumes published in LNNS embrace all aspects and subfields of, as well as new challenges in, Networks and Systems.

The series contains proceedings and edited volumes in systems and networks, spanning the areas of Cyber-Physical Systems, Autonomous Systems, Sensor Networks, Control Systems, Energy Systems, Automotive Systems, Biological Systems, Vehicular Networking and Connected Vehicles, Aerospace Systems, Automation, Manufacturing, Smart Grids, Nonlinear Systems, Power Systems, Robotics, Social Systems, Economic Systems and other. Of particular value to both the contributors and the readership are the short publication timeframe and the world-wide distribution and exposure which enable both a wide and rapid dissemination of research output.

The series covers the theory, applications, and perspectives on the state of the art and future developments relevant to systems and networks, decision making, control, complex processes and related areas, as embedded in the fields of interdisciplinary and applied sciences, engineering, computer science, physics, economics, social, and life sciences, as well as the paradigms and methodologies behind them.

Indexed by SCOPUS, EI Compendex, INSPEC, WTI Frankfurt eG, zbMATH, SCImago.

All books published in the series are submitted for consideration in Web of Science.

For proposals from Asia please contact Aninda Bose (aninda.bose@springer.com).

Chaminda E. R. Hewage ·
Mohammad Haseeb Zafar · Nishtha Kesswani
Editors

AI Applications in Cyber Security and Privacy of Communication Networks

Proceedings of 10th International Conference on Cyber Security, Privacy in Communication Networks (ICCS 2024)

Springer

Editors
Chaminda E. R. Hewage
Cardiff School of Management
Cardiff Metropolitan University
Cardiff, UK

Mohammad Haseeb Zafar
Cardiff Metropolitan University
Cardiff, UK

Nishtha Kesswani
Department of Computer Science
Central University of Rajasthan
Kishangarh, Rajasthan, India

ISSN 2367-3370 ISSN 2367-3389 (electronic)
Lecture Notes in Networks and Systems
ISBN 978-981-96-7399-5 ISBN 978-981-96-7400-8 (eBook)
https://doi.org/10.1007/978-981-96-7400-8

This Springer imprint is published by the registered company Springer Nature Singapore Pte Ltd.
The registered company address is: 152 Beach Road, #21-01/04 Gateway East, Singapore 189721, Singapore

Organization

Program Committee Chairs

Bentotahewa, Vibhushinie
Hewage, Chaminda, Cardiff Metropolitan University, Cardiff School of Technologies, Cardiff, UK
SSE, IAA, IAASSE, LA, India
Tahir, Sabeen, CST, Computer Science, Cardiff, UK
Zafar, Mohammad, Cardiff Metropolitan University, Computing, Cardiff, UK

Program Committee Members

Bentley, Barry
Bentotahewa, Vibhushinie
Hewage, Chaminda, Cardiff Metropolitan University, Cardiff School of Technologies, Cardiff, UK
Jenkins, Paul
Nawaf, Liqaa
SSE, IAA, IAASSE, LA, India
Singh, Chandrani, Sinhgad Institute of Management, MCA, Pune, India
Tahir, Sabeen, CST, Computer Science, Cardiff, UK
Tahir Bakhsh, Sheikh
Tomlinson, Matthew, Cardiff Metropolitan University, Cardiff School of Technologies, Cardiff, UK
Uddin, Jasim
Ullah, Rehmat
Zafar, Mohammad, Cardiff Metropolitan University, Computing, Cardiff, UK

Reviewers

Alotaibi, Reem
Baig, Imran
Bentley, Barry
Bentotahewa, Vibhushinie
Damaj, Issam
Hewage, Chaminda, Cardiff Metropolitan University, Cardiff School of Technologies, Cardiff, UK
Jenkins, Paul
Khalil, Ruhul Amin, United Arab Emirates University, Engineering Requirement Unit, Al-Ain, United Arab Emirates
Khan, Imtiaz Hussain
Nawaf, Liqaa
SSE, IAA, IAASSE, LA, India
Singh, Chandrani, Sinhgad Institute of Management, MCA, Pune, India
Tahir, Sabeen, CST, Computer Science, Cardiff, UK
Tahir Bakhsh, Sheikh
Tomlinson, Matthew, Cardiff Metropolitan University, Cardiff School of Technologies, Cardiff, UK
Uddin, Jasim
Ullah, Rehmat
Zafar, Mohammad, Cardiff Metropolitan University, Computing, Cardiff, UK

Preface

The *AI Applications in Cyber Security and Privacy of Communication Networks: Proceedings of 10th International Conference on Cyber Security, Privacy in Communication Networks (ICCS 2024)*, held at Cardiff Metropolitan University, UK, on December 9–10, 2024, presents a significant compilation of cutting-edge research and innovations in cybersecurity and privacy for communication networks. Organized in collaboration with California State University, San Bernardino (CSUSB), and the International Association of Academicians (IAASSE), this conference served as a global forum for experts, researchers, and industry leaders to address the challenges posed by increasingly sophisticated cyber threats and the growing demand for secure, efficient communication infrastructures.

As digital landscapes continue to evolve with technologies like 5G, IoT, and cloud computing, ensuring robust security and privacy in communication networks has become a top priority. This event explored the transformative role of artificial intelligence (AI), machine learning, and other advanced technologies in enhancing network security, threat detection, data privacy, and risk management. The proceedings reflect a diverse range of topics, from AI-based intrusion detection systems to innovations in secure data transmission, offering invaluable insights for advancing cybersecurity strategies and solutions in today's interconnected world.

Cardiff, UK — Chaminda E. R. Hewage
Cardiff, UK — Mohammad Haseeb Zafar
Kishangarh, India — Nishtha Kesswani

Contents

About the Editors

Chaminda E. R. Hewage currently serves as the Associate Dean of Research in the Cardiff School of Technologies at Cardiff Metropolitan University, UK. He obtained a B.Sc. Engineering (First Class Honours Degree) in Electrical and Information Engineering from the University of Ruhuna (Sri Lanka) in 2004, followed by a Ph.D. in Multimedia Communications from the University of Surrey (UK) in 2009. Recognized for his outstanding undergraduate performance, he was awarded the Gold Medal for Engineering Excellence in 2014 by the University of Ruhuna. Beginning his professional career as a Telecommunication Engineer at Sri Lanka Telecom (Pvt.) Ltd. in 2004, Hewage received the Overseas Research Scholarships (ORS) award from the Higher Education Funding Council of England in 2005 to pursue his Ph.D. at the University of Surrey. Additionally, he obtained a Postgraduate Certification in HE Teaching and Learning from Kingston University—London in 2014 and is a Fellow of the Higher Education Academy (HEA), UK.

Dr. Mohammad Haseeb Zafar is an esteemed Researcher and Educator, currently serves as the Director of the Cybersecurity and Information Networks Centre (CINC) at Cardiff Metropolitan University (CMU), Cardiff. In this role, he oversees all activities related to cybersecurity and information networks research and education at the center. Dr. Zafar is also a Professor in the Faculty of Electrical and Computer Engineering at the University of Engineering and Technology (UET), Peshawar, Pakistan. Previously, he held leadership roles, including Secretary of the Board of Advanced Studies and Research at UET and Professor at King Abdulaziz University (KAU), Jeddah, Kingdom of Saudi Arabia, where he led the Postgraduate Committee and secured research grants on IoT and 5G Networks. With over 100 research publications, Dr. Zafar's research interests span cybersecurity, communication networks, and digital inclusion's impact on society. He actively contributes to the academic community through conference organization, such as ICCS 2023 in Cardiff, and technical program committee memberships.

Nishtha Kesswani received the M.Tech. degree from the Malaviya National Institute of Technology and the Ph.D. degree in Computer Science from the University of Rajasthan. From 2014 to 2015, she was a Postdoctoral Researcher with California State University at San Bernardino, USA. She has a vivid teaching experience at several reputed universities, including California State University at San Bernardino and the University of Ljubljana, Slovenia. She has visited more than 15 countries and delivered invited talks at several conferences and workshops. She is currently with the Central University of Rajasthan, India. Her current research interests include wireless networks and Internet of Things. She has received prestigious awards, including the UGC Raman Postdoctoral Fellowship tenable in USA and the Young Teacher Award. She has edited a number of journals special issues as a guest editor. She is on the reviewer board of several reputed journals, including the *IEEE IoT*, the *IEEE Transactions on Emerging Topics in Computing*, and *Applied Soft Computing* (Elsevier).

Next-Gen Recruitment: Harnessing Face Recognition and Biometrics for Smart Hiring

K. Pal Amutha, D. Ethirajan, N. Mehanathen, and K. Ramkumar

Abstract This paper presents an approach to enhance the security and integrity of the recruitment process by integrating face recognition, biometric verification, and replacing traditional paper-based exams with computer-based assessments. The objective is to automate candidate identification, verification, and secure data handling throughout the recruitment process. Traditional systems face challenges in maintaining accurate candidate records, such as application forms and supporting documents. Paper-based exams also present several key issues, including errors in manual grading, logistical difficulties in handling and storing exam papers, and security risks like cheating and tampering. Additionally, manual result processing is slow, prone to human error, and can cause delays in releasing results. These existing systems are vulnerable to security risks, such as potential loss or manipulation of records and are inefficient in handling large volumes of data. To address these challenges, we propose an automated system that simplifies and optimizes the entire process, from candidate registration to result announcement. The system is designed to securely collect and manage candidate information at every stage. Furthermore, it incorporates face recognition for candidate identification and biometric verification to ensure identity confirmation throughout the recruitment process.

Keywords Face recognition · Biometric verification · Computer-based exam · MTCNN · Next-gen recruitment

K. Pal Amutha (✉) · D. Ethirajan · N. Mehanathen · K. Ramkumar
Centre for Development of Advanced Computing, Chennai, India
e-mail: palamuthak@cdac.in

D. Ethirajan
e-mail: ethirajand@cdac.in

N. Mehanathen
e-mail: mehanathenn@cdac.in

K. Ramkumar
e-mail: ramkumark@cdac.in

C. E. R. Hewage et al. (eds.), *AI Applications in Cyber Security and Privacy of Communication Networks*, Lecture Notes in Networks and Systems 1453,
https://doi.org/10.1007/978-981-96-7400-8_1

1 Introduction

The primary focus of this system is to optimize the recruitment process by incorporating face recognition, biometric verification, and a secure computer-based exam platform. This system aims to streamline the candidate registration process and ensure the integrity of the recruitment cycle, from initial registration through to the final announcement of results. Traditional systems face significant challenges, such as the inability to consistently maintain candidate information throughout the recruitment process. This includes managing candidate details, fee records, and essential documents like experience certificates, educational qualifications, and images, even when dealing with large volumes of data. Furthermore, manual candidate verification frequently leads to inaccuracies and is time-consuming. To address these challenges, an automated system has been implemented to streamline the entire recruitment process, from candidate registration to the announcement of final results. This system enhances both security and efficiency, reducing errors and accelerating the verification process. By integrating face recognition for candidate identification and biometric verification to confirm the identity of the exam taker, the system ensures a secure and streamlined recruitment process. The core intention of this project is to securely collect and manage candidate data for subsequent stages.

Our contribution to this project is as follows: (1) secure collection of candidate information to collect candidate data and ensuring data security at all times; (2) identity verification to authenticate the candidate's identity; (3) secure data transmission and storage is to ensure secure transmission and protection of data throughout the recruitment process, safeguarding it from unauthorized access at all stages; (4) candidate verification to ensure accurate identification and continuous verification in all stages; and (5) result processing and announcement to announce results securely, maintaining the confidentiality and integrity of all candidate data.

The remaining of the paper is organized as follows: Sect. 2 discusses the literature review in this area. The system architecture and overview are described in Sect. 3. The entire system processing and implementation are presented in Sect. 4. Conclusion and future work are described in Sect. 5.

2 Literature Review

G. T. Mardiani et al. proposed a system and used MOORA to provide recommendations on which employees should be recruited according to company needs [1]. Charithma Jayasekara et al. present a system that enhances the recruitment process by merging AI advancements. It automates traditional methods while aiding HR experts in identifying the ideal candidates aligned with company goals [2]. JIANJUN MIAO1 et al. formulated the recruitment quota allocation problem using the Markov decision process (MDP) approach where the effect of the information of potential applicants is innovatively taken into account [3]. Liu Youping and Shen Jiangang proposed a system to screen resumes automatically to embody the necessary requirements of employers. Deep neural network model is used to calculate the matching degree of the talent information that has passed the preliminary screening, short description [4]. Thaya Madhavi, and Avulakunta Kaveri proposed a system where AI enhances the expediting of recruitment processes, particularly in customer service and applicant communication, leading to quicker response times and improved candidate engagement [5].

G. Mageshvaran et al. build a recruitment system that provides intuitive interface, such as enrollment, photo capture, training, and automated attendance marking. The facial recognition algorithm's efficiency greatly simplifies the attendance management procedures [6]. Thuat NGUYEN-KHANH et al. proposed to improve the transparency between candidates and companies and ensure candidates' privacy using blockchain, federated learning, and facial recognition [7]. S. Durai et al. proposed a system that can instantly identify and detect the faces of offenders. By utilizing the MTCNN algorithm, the model hopes to facilitate quick and accurate identification, improving public safety by proactively identifying those with criminal histories and bolstering social security in general [8]. Marko Arsenovic et al. proposed a novel approach of face augmentation for the purpose of extending the dataset which could lead to achieving higher accuracy on smaller datasets of the original images [9]. Jung Hee Lee et al. aim to deriving the priority of factors affecting the utilization of AI recruitment system using analytic hierarchy process [10]. Adil Benabou et al. present AI's capacity to automate administrative operations, offer HR support, enable predictive analysis, and ensure compliance oversight, HR professionals are presented with a unique opportunity to refocus their efforts on strategic imperatives and foster deeper employee connections [11].

3 Overview and Architectures

3.1 System Design

Proposed recruitment system is designed to automate the recruitment process until the result announcement. The system comprises three main components: 1. Registration Portal Design: This component facilitates the collection of candidate data for further processing. 2. Integration of Face Recognition Technology: This technology enables candidate identification during Stage I and Stage II. 3. Biometric Data Capture and Verification: This feature captures biometric data and verifies the candidate in subsequent stages.

Registration Portal Design. The registration portal is a key component of the recruitment system. It is offering an intuitive and responsive interface for candidates to input personal details, supporting documents, images, and capture live images. Once payment is completed, the portal ensures the secure transmission and storage of data. It primarily collects four live images for candidate identification in subsequent stages. Additionally, the portal provides user assistance documentation to facilitate easy use.

Integration of Face Recognition Technology. System retrieves candidate information from the dataset collected during registration, including their name, address, photo, and application number, based on the candidate's hall ticket number. The system identifies the candidate using face recognition by capturing a live image and comparing it with stored images, as required at each stage for candidate identification.

Biometric Data Capture. Biometric information is collected during the first level of the computer-based exam, with fingerprint data being captured based on the recruiter's preference (e.g., which thumb or finger). System provides the process of capturing biometric data and verifying candidate images during Stage I and Stage II. Once collected, this data is securely stored in a database for future process. The purpose of capturing biometric data from candidates is to accurately verify their identity for subsequent stages.

3.2 System Workflow

The recruitment system is composed of several integrated software components, each specifically designed to automate different aspects of the hiring process, significantly reducing the need for manual intervention. By leveraging these automated tools, the system streamlines complex tasks such as candidate registration, data collection, exam scheduling, biometric verification, interview management, and result processing. These components operate together to ensure that each step, from the initial registration to the final selection of candidates, is efficiently executed without unnecessary delays or human error. Figure 1 shows that overall flow diagram for next-gen recruitment system.

Registration via Web Portal. The recruitment process begins with candidate registration through a dedicated web portal, which serves as the entry point for all applicants as shown in Fig. 2. Upon accessing the portal, candidates are required to provide essential preliminary information, such as personal details, contact information, relevant qualifications and their preferred written exam location, using the User Registration section. The Supporting Documents section allows candidates to upload proof of documents, such as qualification certificates, photo identity, and age proof. The Capture Live Photo section collects four live images of the candidate. After completing the Capture Live Photo section, the Payment section is activated, allowing candidates to make the required payment. This confirms their registration and generates a unique ID for further processing.

Pre-Exam Portal. The system transitions to the Pre-Exam Portal stage once the Registration Process is complete. This stage includes the center allocation, hall ticket generation, data grouping, and intimate user section. The Center Allocation Section generates the exam venue, date, time, and slot for each candidate based

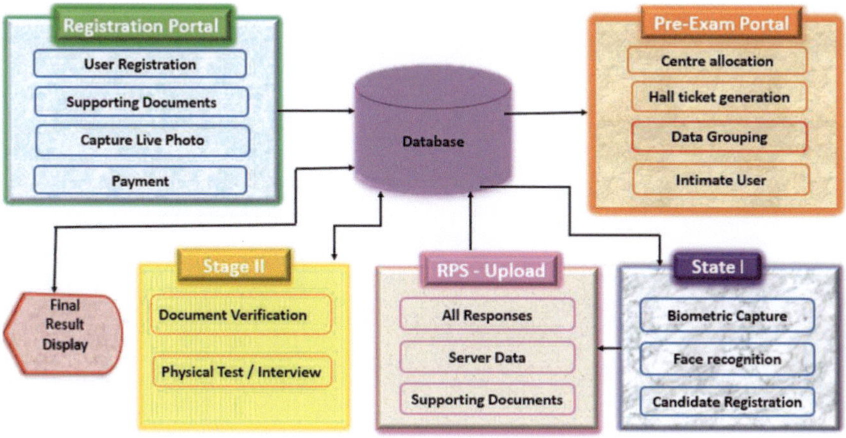

Fig. 1 Overall flow diagram

on their preferred written exam location, which is collected during the Registration Process. The Hall Ticket Generation Section prepares the admit card document, which includes the exam location, date, time, slot, instructions for the candidate, reporting time, etc. The Data Grouping Section generates an encrypted compressed file containing the candidate's image and admit card, organized by slot, time, date, and location. This file serves as the input for candidate verification whenever verification is required. Once the hall tickets are generated, the Intimate User Section notifies the candidates of their exam date, time, and venue through predefined communication channels, such as email or SMS. This automated approach ensures timely and reliable communication, reducing the risk of miscommunication.

Stage I Process. A feature of our platform facilitates candidate authorization by incorporating face recognition technology. This application serves as a registration platform for candidates enrolling to take an exam. During enrollment, candidates are required to provide a live image, and the application verifies their photo identity against five stored images collected during the registration process in the Photo Capture section, as shown in Fig. 3. The application displays candidate information based on the hall ticket. Additionally, the system collects another live image and biometric information to proceed to the next stage. This application includes the key features such as biometric capture, face recognition, and candidate registration. Candidates must successfully complete each feature to be eligible to proceed and write the exam. Finally, it generates a compressed file containing candidate images, biometric data, attendance information (present/absent status), gender data (male/female), and other relevant details. This file helps as one of the inputs for the RPS uploading process.

Result Processing System (RPS) Process. We have developed a bootable operating system specifically designed with robust security features. It functions as a server,

Fig. 2 Registration portal

primarily connecting to client machines (used by candidates for the written exam) and enabling their bootable capabilities. This server system establishes a secure network connection, ensuring the system remains isolated and inaccessible from external sources. Additionally, it provides monitoring features for the client machines to prevent malpractice during the exam. Once the exam is completed, the server collects the answer reports from all present candidates and prepares the system for the next exam slot. The system now uploads essential information to the centralized server, including answer reports, candidate attendance data generated during Stage I, and

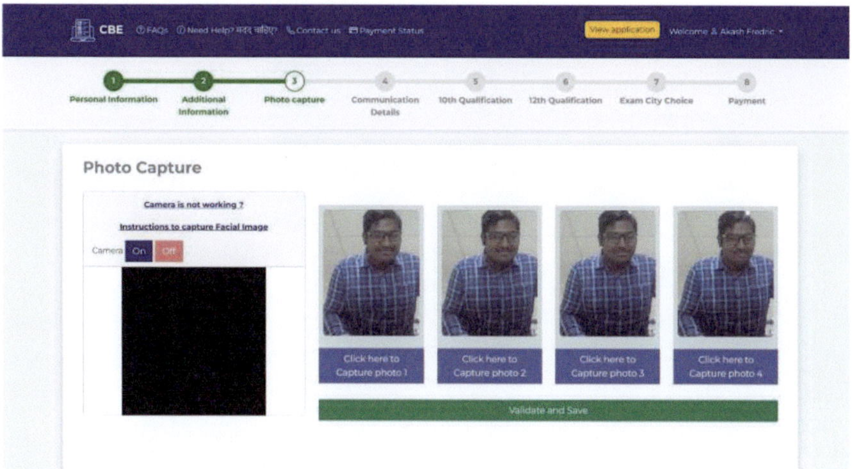

Fig. 3 Capture four live images

any other supporting documents collected during Stage I registration, through the RPS portal.

Stage II Process. The Stage II process begins after the shortlisted candidate list is published based on the recruiter's preferences. This process may involve another written exam or a personal interview or physical test. The cycle can be repeated with some conditional changes, such as admit card generation, candidate notifications, and data grouping, but only for shortlisted candidates. The application shown in Fig. 4 facilitates the registration of shortlisted candidates and authorizes them using face recognition. It captures live biometric data and compares it with the stored biometric data collected during the Stage I process. Finally, it generates a compressed file containing the candidate's Stage II registration data, which serves as one of the input files for RPS upload for further action.

Final Result Process. After all interviews are concluded, the system processes the results and, using dedicated result announcement software, automatically publishes the final outcomes based on the recruiter's requirements. This step marks the conclusion of the recruitment process by informing candidates of the final hiring decisions without the need for manual intervention. By automating the result announcement, the system ensures timely communication to all candidates, enhancing efficiency, transparency, and consistency in delivering the final decisions.

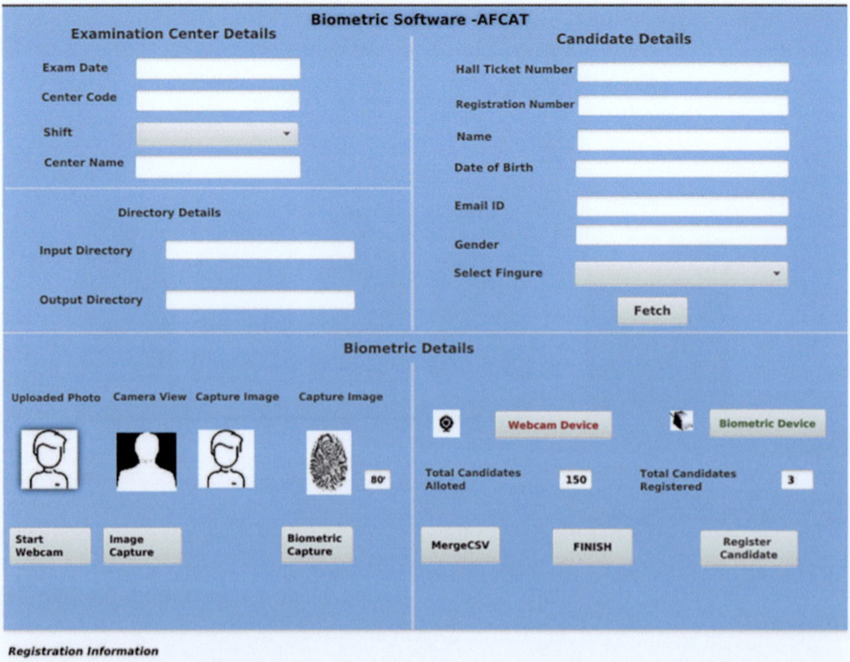

Fig. 4 Application for Stage I and stage II registration

4 Implementation

This system securely collects and manages candidate data throughout multiple stages of the recruitment process which involves several key components. Each stage, from initial registration to the final result announcement, must ensure data security, accuracy in identification, and seamless transition between stages. Below is a detailed breakdown of how this process can be implemented.

4.1 Candidate Registration and Data Collection

Candidates will be able to enter their personal information, upload necessary documents, and submit a clear photograph with the help of online registration portal. The system will ensure secure data transfer and storage by employing encryption methods. Each candidate will be assigned a unique ID (UID) that will be used throughout the recruitment process to link all subsequent stages, including biometrics and exam results. A user-friendly portal is designed to make it easy for candidate to navigate and access. The portal typically offers an intuitive interface with straightforward instructions, eliminating the need for advanced technical skills. The backend process collects and manages candidate data, securely storing it in database servers by incorporating advanced data encryption algorithms.

4.2 Biometric Data Capture and Live Image Verification

The face recognition process using MTCNN (Multi-Task Cascaded Convolutional Networks) begins by detecting faces in the input image. MTCNN generates bounding box coordinates around detected faces and identifies key facial landmarks for further processing. The detected faces are aligned and resized to a standard size, ensuring consistency for subsequent stages. MTCNN is adaptable for a range of applications that require images with changing face sizes since it can handle faces of different sizes and resolutions inside an image. In addition to detecting faces, MTCNN can also give facial landmarks that aid in aligning and normalizing the faces that are discovered [12]. The recognized face information is then stored, accurately recording the candidate's identity for subsequent stages. In general, to address face recognition using a prediction model to locate a person's face, cropping the face from the image based on the identified coordinates, and converting the cropped face into vector values. Reduce the size of the image to 224 * 224 and feed it into the model as input. Convert to vector values using face-net [13], to extract key facial features for comparison.

The system captures the candidate's live image and biometric data whenever they appear for a computer-based exam, personal interview, or physical test. Upon receiving the live data, the system compares it with the stored data, calculates the match probability, and retrieves the stored image if it meets the required threshold for confirming the registration process. If the match probability does not meet the threshold, manual verification of the candidate is needed for further assessment. During an interview process the interviewer will scan the student face and at the backend the process of facial recognition will run which will convert the facial values to face vector and this face vector will be compared to already present face vectors. If the similarity will cross the threshold value, then the details of the person with whom the face vector has been matched will get displayed in front of the user and the user then can access resume, GitHub profile, LinkedIn profile, mark sheets, certificates, etc. with the button present LinkedIn profile, mark sheets, and certificates [14].

4.3 Comprehensive System Workflow

In our system, biometric data, live images, and exam responses are encrypted and securely transmitted in real time to a centralized database, ensuring that sensitive information remains confidential and its integrity is maintained. At each stage of the recruitment process, candidates are required to submit live images and biometric data for re-verification, which is compared with the data previously collected during the exam to ensure consistency and accuracy. The results of these verification checks are systematically logged for auditing purposes. This system also integrates role-based access control (RBAC), which restricts access to sensitive data based on user roles, ensuring that only authorized individuals can manage and view critical information. This security approach protects against unauthorized access and potential breaches, maintaining the system's overall reliability and security. Trust and managing risks in AI require a combination of technical and ethical as well as a commitment to continuous improvement and accountability. Nowadays, organizations must actively work toward building and maintaining trust while mitigating potential risks [15].

After completing all stages, including the exam, interviews, and background checks, the system consolidates and analyzes the candidate data, making final decisions based on the recruiter's criteria. Candidates are promptly notified of their selection status via email or SMS, ensuring that they receive timely communication regarding their application outcome. Our contribution to this system includes

1. **Secure Collection of Candidate Information**: Collect candidate information securely, including images and biometric data, while enforcing privacy protections and ensuring confidentiality.
2. **Identity Verification**: Comparing the live captured image with the stored image to verify the candidate's identity.
3. **Secure Data Transmission and Storage**: Securing and organizing data effectively to ensure its safe handling and accessibility when needed.
4. **Candidate Verification**: Recognize and validate candidate identities at every stage of the recruitment process.
5. **Result Processing and Announcement**: Managing and announcing results securely while ensuring the protection of candidate data, including biometric and personal details, to maintain integrity and confidentiality at every stage of the process.

5 Conclusion and Future Work

This system securely collects and manages candidate data throughout multiple stages of the recruitment. Using this system, we have successfully conducted over 30 exam processes. Initially, we encountered a face recognition accuracy rate of 75%. However, by enhancing our system with a larger and more diverse dataset for training the model, and optimizing the background settings (e.g., using a white background), we can improve the overall accuracy and robustness of the system. Face identification requires invariance to pose, illumination, expression, background, and other factors [16]. Now system has improved accuracy to 89%. Challenges still remain, particularly in addressing trivial variations in live images, such as background differences, head angles, and eye movements during image capture. To address these challenges and improve candidate verification, biometric verification features have been integrated into the system. By combining facial recognition with biometric data, the system enhances the accuracy of candidate identification during interviews and physical tests. This integration ensures that the verification process is more reliable and reduces the chances of errors or fraud. This dual approach helps ensure more reliable candidate verification and enhances the overall effectiveness of the recruitment process.

Our future research focuses on improving exam integrity through the introduction of video proctoring. This system will actively monitor candidates throughout the exam, ensuring they remain focused and preventing unauthorized interactions, either with other candidates or the invigilator. The goal is to minimize distractions and potential for malpractice, thus maintaining a fair and secure testing environment. Additionally, the system will track variations in mouse movements and other behavior indicators to identify irregularities, further enhancing the reliability of the exam process. By closely observing candidate behavior, this approach aims to uphold the integrity of the exam while safeguarding against cheating or other forms of dishonesty.

Acknowledgements We sincerely thank everyone involved in this project. The team's dedication was essential in creating a secure candidate verification and exam management system. Special thanks to everyone who contributed to improving face recognition accuracy and integrated biometric verification. We also appreciate the valuable input from stakeholders, which contributed to the project's success.

References

1. Mardiani GT, Susanto R, Harihayati T (2023) Employee recruitment analysis using the MOORA method. In: 2023 international conference on informatics engineering, science & technology (INCITEST). IEEE, Indonesia
2. Jayasekara C, Senarathne I, Wickramasinghe A, Jayathilaka N, Thelijjagoda S, Silva HD, Giguruwa N, Abeysinghe N (2023) Artificial intelligence agent to identify the correct human resources. In: 2023 5th international conference on advancements in computing (ICAC). IEEE, Sri Lanka
3. Miao J, Liu J, Feng R, Zhang Y, Zhu H, Bai AL (2022) Optimization of the recruitment quota allocation in intra-organizational networks. Digital Object Identifier. https://doi.org/10.1109/ACCESS.2022.3230717, IEEE Access (2022)
4. Youping L, Jiangang S (2022) Talent recruitment platform for large-scale group enterprises based on deep learning. In: 2022 7th international conference on cloud computing and big data. IEEE, China
5. Madhavi T, Kaveri A (2024) The impact of artificial intelligence in recruitment and selection processes in IT companies. In: 2024 16th international conference on electronics, computers and artificial intelligence (ECAI). IEEE, Romania
6. Mageshvaran G, Sowmya VJ (2024) A machine learning-based attendance management system. In: 2024 IEEE international conference on information technology, electronics and intelligent communication systems (ICITEICS). IEEE, India
7. Nguyen-Khanh T, Ngo-Phuc T, Van-Thien L (2022) Building a recruitment system based on blockchain and federated learning. In: 2022 international conference on computing and communication technologies (RIVF). IEEE, Vietnam

8. Durai S, Sujithra T, Satyam BV, Keshetty SN, Sagar CNS, Charan AS (2024) Real time facial recognition-based criminal identification using MTCNN. In: Proceedings of 2nd international conference on sustainable computing and smart systems (ICSCSS 2024). IEEE, India
9. Arsenovic M, Sladojevic S, Anderla A, Stefanovic D (2017) Face time—deep learning based face recognition attendance system. In: SISY 2017 IEEE 15th international symposium on intelligent systems and informatics. IEEE, Serbia
10. Lee JH, Kim JH, Kim YH, Song YM (2021) A study on priorities for utilization of AI recruitment system, In: 21st ACIS international winter conference on software engineering, artificial intelligence, networking and parallel/distributed computing (SNPD-Winter). IEEE, Vietnam
11. Benabou A, Touhami F, Demraoui L (2024) Artificial intelligence and the future of human resource management. In: 2024 international conference on intelligent systems and computer vision (ISCV). IEEE, Morocco
12. Tariyal S, Chauhan R, Bijalwan Y, Rawat R, Gupta R (2024) A comparative study of MTCNN, Viola-Jones, SSD and YOLO face detection algorithms. In: 2024 international conference on intelligent and innovative technologies in computing, electrical and electronics (IITCEE). IEEE, India
13. Thipbumrung R, Aonpong P (2024) Automated face selection and censoring on image and video using artificial intelligent. In: 2024 21st international joint conference on computer science and software engineering (JCSSE). IEEE, Thailand
14. Shivhare R, Guru Prasad MS, Singh RK, Neha K, Dwivedi D, Ram M (2023) Automated student's details access system using face recognition. In: 2023 international conference on computer science and emerging technologies (CSET). IEEE, India
15. Yu S, Carroll F, Bentley BL (2024) Trust and risk: psybersecurity in the AI era, 1st edn. CRC Press, eBook ISBN9781032664859
16. Rozner A, Battash B, Lindenbaum O, Wolf L (2024) Efficient verification-based face identification. In: 2024 18th international conference on automatic face and gesture recognition (FG). IEEE, Turkey

Implementing Deep Learning to Detect Malicious URLs

Rhodri Thomas, Sabeen Tahir, Sheikh Tahir Bakhsh, and Reem Alotaibi

Abstract The GoPhish Chrome extension aims to increase web security by providing users with an easy-to-use tool for identifying malicious URLs. With a focus on simplicity, the extension lets users start URL scans with a context menu interaction. This sets off a machine learning system that analyzes the input URL and produces a confidence score that indicates how malignant it is. The main functionality is underpinned by a structured manifest file outlining required rights and JavaScript components handling data processing and user interaction. For effective client-side execution, the machine learning model, which was initially created in Keras, is transformed into TensorFlow.js format.execution. After thorough evaluation and comprehensive testing, it shows an accuracy rate of 74.626% when classifying real-world URLs, this is below the targeted standard of 95% for dependable security applications. This research emphasizes the necessity of additional model optimization to improve its prediction efficacy in real-world situations, demonstrating the continuous challenges associated with implementing machine learning solutions in browser extensions for cybersecurity.

Keywords GoPhish · Machine learning · Browser extensions · Cybersecurity

R. Thomas · S. Tahir (✉) · S. T. Bakhsh
Department of Computer Science, Cardiff School of Technologies, Cardiff Metropolitan University, Cardiff, UK
e-mail: stahir@cardiffmet.ac.uk

R. Thomas
e-mail: rthomas6@cardiffmet.ac.uk

S. T. Bakhsh
e-mail: sbakhsh@cardiffmet.ac.uk

R. Alotaibi
Department of Information Technology, Faculty of Computing and Information Technology, King Abdulaziz University, Jeddah, Saudi Arabia
e-mail: ralotibi@kau.edu.sa

C. E. R. Hewage et al. (eds.), *AI Applications in Cyber Security and Privacy of Communication Networks*, Lecture Notes in Networks and Systems 1453,
https://doi.org/10.1007/978-981-96-7400-8_2

1 Introduction

The human factor is the most commonly exploited attack vector in any IT infrastructure or system. This vulnerability is especially susceptible to social engineering tactics, the most common of which is phishing. Phishing attacks take advantage of psychological tricks to trick people into revealing private information. According to authors [1], there are two primary forms of phishing, each with unique techniques:

1. Technical phishing: This type of phishing involves tricking customers into revealing personal information utilizing advanced technological techniques, such as malware-infected URLs.
2. Social engineering phishing: It occurs when cybercriminals construct enticing emails or communications to trick the recipient into willingly providing personal information.

There are serious risks associated with both approaches for regular Internet users. Phishing scams will continue to be a major cybersecurity concern until extensive countermeasures are implemented, such as improved security standards and user awareness training. According to the Office for National Statistics [2], phishing attacks have been identified as one of the most persistent threats facing both the public and private sectors, highlighting the ongoing need for vigilance and proactive defense mechanisms.

In the field of Internet security, identifying malicious URLs is a critical challenge, as attackers continuously evolve their techniques to exploit vulnerabilities. One of the most common methods for detecting malicious URLs is the use of blacklist sites. These blacklists compile URLs that have been previously identified as malicious through various techniques such as web scraping, user reporting, and site analysis heuristics [3]. Web scraping makes it possible to automatically collect data from websites in order to detect possible dangers, whereas user reporting depends on people or organizations reporting questionable URLs. Heuristic analysis is the process of looking for indications of malicious intent in URLs by analyzing their properties and behavior.

While these methods have proven effective, they come with significant limitations. Blacklists are usually reactive in nature, which means that before a URL can be added to the list, it has to be found and reported. Because of this, there is a lag in time between the creation of a malicious URL and its detection, making people exposed. Moreover, the dynamic and ever-changing nature of malicious URLs often using techniques such as URL shorteners, obfuscation, and frequent domain changes makes it difficult for blacklist sites to keep up with emerging threats. As a result, new malicious URLs may remain undetected until they cause harm. In conclusion, we offer a thorough analysis that highlights our results, evaluates the research's consequences, and makes recommendations for additional studies.

2 Literature Review

Marwa Al Saedi and Flayh [4] conducted a thorough analysis of the main features leveraged by previous studies in identifying malicious URLs. Their work emphasizes the importance of extracting specific URL components, which serve as indicators of malicious behavior. These components may include elements such as domain age, domain registration details, subdomain length, and the presence of suspicious characters in the URL structure. They argue that these indicators provide critical insights into URL legitimacy, forming the foundation for many machine learning models aimed at phishing detection. In addition to URL feature extraction, Marwa Al Saedi and Flayh [6] highlight the necessity of page content analysis, a method some prior approaches have implemented. Page content analysis investigates the structure of HTML, embedded scripts, and visible text to identify malicious intent or behavior. While this form of analysis may not align precisely with the goals of their paper, they argue that it remains a critical phase for increasing future detection systems. This approach adds another layer of security beyond URL-based detection by considering the actual contents of the webpage that the URL points to, potentially identifying threats that URL-based methods might miss.

The identification of fundamental difficulties in phishing website detection is a significant contribution of their work [5]. They offer a thorough rundown of problems such as attackers' constantly changing strategies, the scarcity of high-quality datasets, the high incidence of false positives, and the resource-intensiveness of current detection methods. These challenges highlight the demand for more effective, scalable detection methods that are flexible enough to change with the ever-changing and dynamic landscape of phishing attempts. One of the major focuses of their study is addressing the problem of resource-intensive solutions. They propose the development of a lightweight, client-side program that leverages TensorFlow.JS, a JavaScript-based machine learning library, to handle all predictions locally on the user's device. This method would bypass the need for resource-heavy server-side processing and avoid the latency issues associated with using slower model languages such as Python. By moving the computation to the client side, this solution aims to significantly reduce the computational load and improve real-time detection performance without relying on external servers, making it a scalable and effective alternative for phishing website detection.

The authors [6] proposed a sophisticated method for identifying URLs using Long Short-Term Memory (LSTM) networks, which are a specialized type of Recurrent Neural Network (RNN) and Artificial Neural Network (ANN). This innovative approach leveraged similar data sources as many other methodologies in the domain, and the data was meticulously processed through the extraction of relevant features from the URLs themselves, such as their structure, length, and the presence of certain keywords. The researchers implemented two models, gradually increasing the volume of data used in testing to evaluate the models' accuracy across larger datasets. Their results demonstrated that, on a large dataset of 456,300 records, the ANN performed excellently with an accuracy of 95.57%, while the LSTM model fared better, with an astounding accuracy of 96.89%. This emphasizes how crucial it is to use more datasets to improve machine learning models' accuracy and resilience.

3 System Design Basis Architecture

The basic design of the architecture started with the idea of a simple framework where the application would wait for input from the user, process the URL, and then output a prediction result. Establishing a clear user interface (UI) that promotes seamless interaction is essential in the early design phase. To improve user experience and make sure the program satisfies their needs, user feedback tools will also be incorporated.

The implementation of a machine learning model into a Chrome extension requires several areas of development. To monitor user interactions and the extension's performance, the design should also provide strong error recording and monitoring tools. To make sure the extension can handle a variety of user inputs and functions consistently across diverse web environments, testing will be essential. The ML model will be developed via supervised learning techniques. The study will also cover data preprocessing techniques to enhance model performance, including normalization, encoding categorical variables, and handling missing values. This system will demonstrate the functions of built-in ML libraries such as Keras and TensorFlow.js and will not create an ML model from scratch. The extension itself should be simple and lightweight so as not to interfere with the process of the TF.JS model that it contains and the user's browser experience. Performance optimization of the extension is essential because a slow extension might irritate users and reduce their interest. To locate bottlenecks and maximize resource utilization, the design will incorporate performance-monitoring tools. The extension only needs to manipulate the DOM to extract the URL from a context menu click.

The detailed flow of GoPhish described in Fig. 1 gives more detail into the inner workings of the project. It describes the flow of the data from module to module, including how the ML model interacts with the data extracted from the DOM of the webpage. Understanding the way data is used and altered at each processing stage—which might lead to possible optimizations and improvements—will depend heavily on this interaction flow. Additionally, thorough illustrations that show the connections between modules will facilitate the communication of the architecture to team members and stakeholders.

3.1 Data Source

Data required for the machine learning algorithm requires some key features for it to be fair, complete, and unbiased. The data will need to come from a reliable source. There needs to be a large and diverse set collected and the data will need to be fair and unbiased. Reliability is a major factor in considering what data is

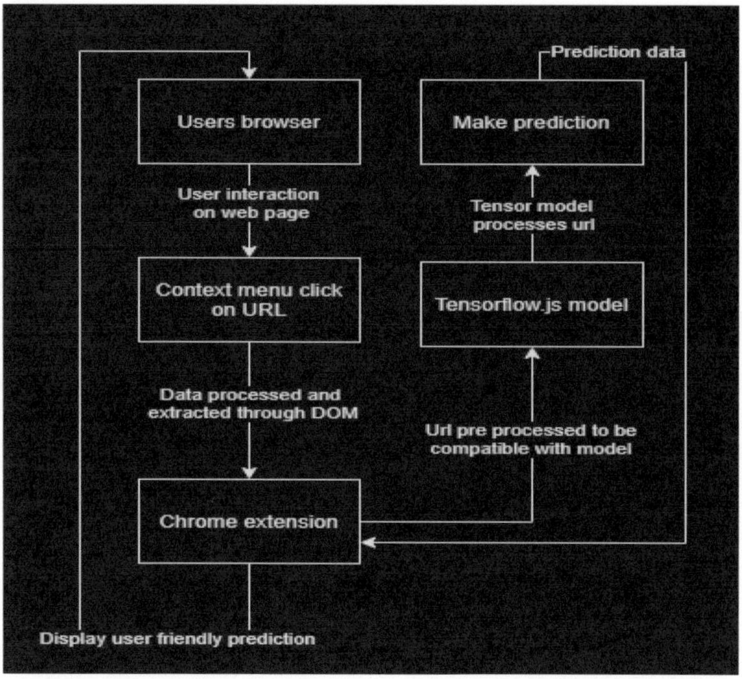

Fig. 1 Detailed flow of GoPhish

required for the model. This is due to the nature of the data itself, being a collection of malicious URLs, they need to be verified as malicious URLs, otherwise, the model will be trained on fake data and will not produce accurate results on unseen data after training and testing. For this reason, the data will be collected from Kaggle. The data is readily available and already labeled ready to extract the feature set from. Further data can be extracted from other sources such as URLHaus where data can be manually labeled and cleaned for future training. The dataset I am using is an amalgamation of several other data sources that have been merged, retaining the URLs and their labels.

In terms of reliability, data sources from Kaggle are rated by users on the site who extract them and utilize them for other projects. They are given ratings such as usability and cleanliness. They are also labeled with the usage rights. All data used in this project are under the CCO 1.0 Deed which means that the user has dedicated the dataset to the public domain. The dataset that is being used for GoPhish has been rated highly by users and has been given a usability score of 10.0/10.0.

3.2 Data Pre-processing and Cleaning

The dataset to be used in the Keras model will have two columns. The URL column contains the long URL that will be processed by the machine learning algorithm and the second column is the type or "label" column that will be used to identify the nature of the URL itself. The dataset originally had four labels, benign, defacement, malicious, and malware. To make the data more uniform and more compatible with the machine learning model, I have changed the labels to only be a 1 or a 0. 1 meaning the URL is malicious and 0 meaning that the URL is benign. Simplifying the label makes it easier to process in the model. Additionally, when considering the original labels, you would not consider anything other than benign to be safe. Therefore, a basic assumption would be to label everything not benign as malicious (or 1(positive)).

4 Algorithm Development

Developing the algorithm required several steps. First there had to be an identification of the best method for testing and deployment. For ease of use and testing, Python would be the choice language for the model. In this case, Jupyter Notebook was chosen as the IDE. This was because it is easy to implement and test machine learning models, and the testing of the models can be done locally using personal hardware. Other options included Collaboratory from Google. Where this environment is a suitable method for testing and deployment, the free version had limitations such as limited processing power and limited processing times. I had used this environment initially and found that testing the model early on when it was slower and less accurate resulted in various errors and slow processing speeds. There was also limited processing times for the free version which further impeded on this report's findings. To avoid these errors, the decision to process and test locally was easy. To test and develop in Jupyter notebook there needed to be some libraries imported via command lines such as TensorFlow, TF.JS, and Keras.

A simple feedforward sequential model will be used in conjunction with Keras. With these ML libraries, it makes it easy to implement techniques and processing power. First importing the libraries and tools required. Keras libraries are used to call specific learning models and the tokenizer. Scikit-learn will be used to pre-process and split the data for the sequential model. There will also be libraries required for saving the model and converting to required format for TF.JS. After testing the model with various hyperparameters, it was discovered that using the sigmoid activation function was the most effective model for learning on the dataset as the data requires binary classification.

Table 1 Impact of LSTM embedded layer sizes on model performance

LSTM embedded layer	Result
32	Higher overfitting of the data
64	Found to provide the best fit on the dataset
128	Overfitting of the data found

4.1 Model Parameters

It was discovered that the best parameters for accuracy and reduction in loss involved the sigmoid activation function. This was because other activation functions such as SoftMax did not improve the accuracy but did reduce the loss.

Other parameters included the amount of LSTM layers included were 64. This resulted in a higher accuracy for the model, also included in this layer was the dropout and recurrent dropout. Both parameters were set to 0.2. These metrics saw better testing on the test dataset and yielded lower loss rates on the testing dataset, Table 1 shows the impact of LSTM.

The optimizer Adam was chosen with a learning rate of 0.1. Combined with the loss metric of binary cross-entropy, this proved to be a sufficient loss metric and training parameter for the model. For the evaluation of the test dataset, 10 epochs and a batch size of 2500 proved to be the best parameters for the testing. This testing yielded an accuracy of 0.9767 and a loss of 0.0679. This proved to be the best accuracy and loss metric received from the training and testing of the model. Therefore, these parameters are going to be used for the final model going forward.

Finding the correct combination of Epochs to batch size for the testing resulted in some interesting data. The results fluctuated considerably when tests were run but eventually became stable after reducing the epochs and the batch sizes. This may be because of the limitations of the hardware that the tests were run on. When training and testing the dataset, the machine CPU was at 100% usage constantly. This may have had an impact on the testing of the data as the machine was not able to handle the processing power required for the dataset size and parameters' set. After tweaking the parameters, the metrics became stable, and a good level of accuracy and loss was found, the model accuracy and loss are listed in Table 2.

5 Chrome Extension Development

Extension framework and development, the extension aims to be simple and easy to use. Thus, the extension has a simple, yet effective dataflow. The user can install the extension from the Google extension store. The nature of GoPhish being a Chrome extension is so that they are able to access the features on any web page and be able to utilize the features at any time. After installing, the user can input a URL to the extension that will then in turn feed the URL to the ML algorithm that will output

Table 2 Effect of epochs, learning rate on model accuracy, and loss

Epoch	Batch size	Learning rate	Result (accuracy/loss)
1000	500	0.01	0.3/0.7
1000	50	0.05	0.4/0.7
100	5000	0.02	0.6/0.5
100	500	0.01	0.5/0.6
10	5000	0.02	0.7/0.3
10	2000	0.1	0.8/0.05
10	2500	0.1	0.9/0.06

a confidence score on the likelihood that the input URL is malicious. To initiate the action of "scanning" the URL, GoPhish uses the features of context menus for the user to interact with. This context menu makes it more intuitive to use as the user only has to right-click a specific URL and select "Scan now". This will then in turn send the URL object to the model for analyzing.

There are only a few files required for the extension to be functional. The main one being the "manifest.JSON". This file describes the permissions and parameters of the extension itself. This file is required to be on the top level of the extension file structure. Within this file, there are parameters that are useful to analyze as they grant permissions and describe the structure of the extension. The .JSON file, as described earlier, aims to define the rules, permissions, and parameters for the extension to function. The main section includes the version, manifest version, name, and permissions.

The most important being the manifest version and permissions. The manifest version is important as this defines the scope and general functionality with updated features and permissions set out by Google and the extension environment. The permissions section itself is aimed to define the permissions of the extension itself on the users' browser. For example, the permissions seen here are "tabs", "activeTab", and "contextMenus". The tabs' permissions allow for access to the current tab of the user's browser and allow for DOM manipulation which is important for GoPhish as the application needs access to the users' DOM to retrieve any URLs the user would like to select. The context menu permission allows the extension to access the user's context menu right-click option, thus allowing GoPhish to provide an easy option for the user to select and "scan" a URL on the webpage.

6 Implementation of Algorithm and Testing of GoPhish

To implement the algorithm, the Keras model developed needs to be converted to a format that can be used by tensorflow.js. The converted files are then imported to the GoPhish extension file structure which in turn can then be used to run the model on the client side. The converted files are in the form of a .JSON file that describes

Fig. 2 Using command-line controls to create the .JSON files in collaboratory

```
!cd /tmp/goPhish/
!tensorflowjs_converter
--input_format=keras
--output_format=tfjs_layers_model
/tmp/goPhish/saved_goPhish.h5 /tmp/tfjs_goPhish
```

the parameters of the model and what weights are attributed to the saved model. To convert the files, the model and all its parameters need to be ported to Google Collaboratory. Here, the model can be converted due to the server-side operating system in Linux. Collaboratory uses Google servers to run code, like Jupyter but instead of running locally. Collaboratory allows for command-line instructions to be run on the server-side file structure, allowing for I/O commands and the creation of the .JSON files required for GoPhish to be converted to a tensorflow.js format. Figure 2 illustrates the process of using command-line controls to generate .JSON files.

First, the correct file location needs to be navigated to in order to open and save the files in the proper location. The location for the files to be saved in the collab server-side file structure is /tmp/GoPhish. This location is created as a temporary location for the model files. Once the location has been navigated to, then the converter and its parameters can be used. The [!tensorflowjs_converter] is used here, with the input and output parameters specified to what the desired files will be. The input format will be Keras as the ML model is a Keras model. The output format will be a tensorflow.js layers model. This layers model can be used in the extension and run on the client side without any heavy server-side processing power needed.

The input file will be the Keras saved model which is in the .h5 format. This file is then converted to the TFJS model in a .JSON file format in the pre-defined file location. These files that are stored in /tmp/GoPhish will be the files used in the extension for the ML model predictions. Deploying and testing GoPhish can be hosted and conducted on chrome extension developer. Chrome allows for "unpacked" extensions to be deployed on the browser for debugging and testing. All the files that are included in the software are packaged together and deployed in the Chrome extension environment. Figure 3 provides an example of how the Chrome Extension Developer environment displays GoPhish.

Deploying an unpackaged extension is an effective way to test the whole system and also an insight into how the users will interact with the extension. The developer mode also allows error codes in the file structure and so gives a good environment for any developer to make an effective extension. Once the extension is deployed, it is instantly available to be interacted with and integrated to the Chrome browser. Here is where all the testing for the final solution will be conducted.

When testing the prediction metric, the threshold for classifying if a URL is malicious or not will be set at 0.75. This is a suitable threshold for URL classification as it suggests proving beyond doubt that the URL is not benign. As tested in the model output section of this report, seemingly benign URLs score around 0.05~. To simulate a realistic user browser experience, the ratio of benign to malicious would be around

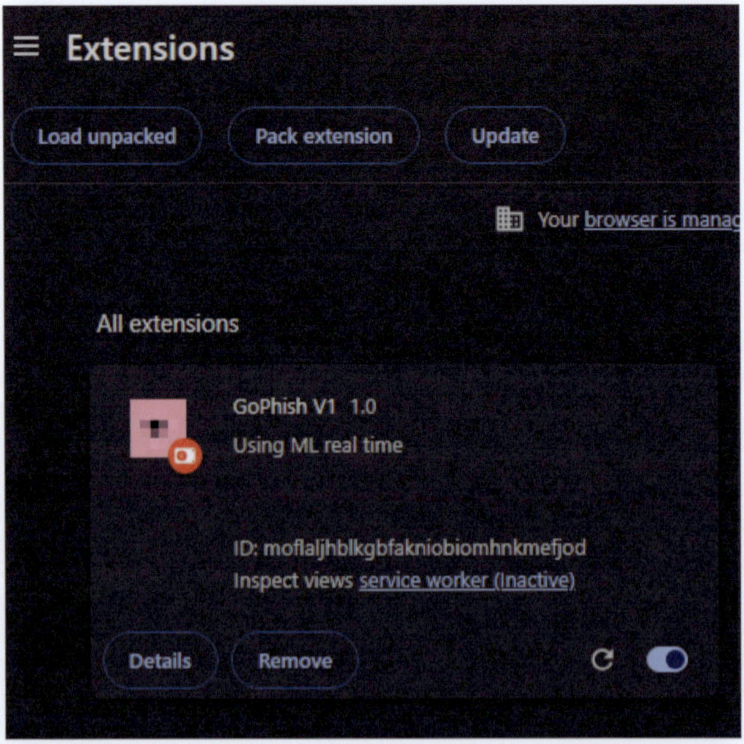

Fig. 3 Example of how the Chrome extension developer environment displays GoPhish

113:67 (benign to malicious). This has been calculated from the estimated number of websites and number of malicious URLs recorded in 2022–2023 [1]. For this reason, this report will test GoPhish with 180 total URLs to make it a realistic and thorough test. The model accuracy on the predictions will be recorded.

The results from the testing on real-world URLs yield some interesting results. The accuracy of successfully predicting malicious URLs is 74.626%. This figure, where successful in some regards, is not suitable for a production application. Where GoPhish can make successful predictions, there needs to be an accuracy of >95% to be considered secure and reliable enough for users to browse the Internet safely. This testing of accuracy differs from the Keras model for several possible reasons. Firstly, the testing data fed into the Keras sequential model differs from the real-world URLs input in this testing. The real-world inputs are raw data extracted through the DOM and do not reflect the dataset URLs provided.

7 Conclusion

The goal of GoPhish is to protect users in their day-to-day browsing behaviors. The aim of GoPhish being deployed on the user's web browser is so that the application can target 100% of phishing attempts made on the user. Building the ML algorithm to have an accurate and high certainty allows for a confident (but not 100%) accuracy when predicting if a URL is malicious or not. Where the application aims to not be intrusive on the user's browsing ability and only scans the URLs the user chooses, GoPhish can only have a limited effectiveness. This is one area that the application can be improved on. Instead of waiting for the user to select a URL to feed into the application, the GoPhish extension can be enabled to scan the whole web page in the background and feed all URLs into the ML model for processing.

The results from the ML model itself have proven effective and accurate. However, there can be further improvements made on the model itself. This can be through improving the dataset, collecting more up-to-date and relevant datasets. There can also be improvements made to the dataset features. The results from the testing of the GoPhish application on actual URLs proved some shortcomings. This accuracy can be further improved by possibly making the Keras model more accurate and efficient.

Acknowledgements This research was supported by the UK-Saudi Challenge Fund 2024, British Council, under the project titled "Data Fusion Dynamics: A Collaborative UK-Saudi Initiative in Cybersecurity and Artificial Intelligence".

References

1. Atkins B, Huang W (2013) A study of social engineering in online frauds. Open J Soc Sci 1(3):23–32. https://doi.org/10.4236/jss.2013.13004
2. Office for National Statistics: Phishing attacks—who is most at risk? Office for National Statistics (2022). https://www.ons.gov.uk/peoplepopulationandcommunity/crimeandjustice/articles/phishingattackswhoismostatrisk/2022-09-26
3. Yan CH, Liu F, Farooqi S, Zhu S, Liu D, Ma J, Fan J, Tao T (2023) Recent trends in internet threats: common industries impersonated in phishing attacks, web skimmer analysis and more. Unit 42 (2023). https://unit42.paloaltonetworks.com/internet-threats-late-2022/#:~:text=Indicators%20of%20Compromise
4. Al Saedi M, Flayh NA (2023) Phishing website detection using machine learning: a review. Wasit J Pure Sci 2(2), 270–281. https://doi.org/10.31185/wjps.145
5. Aslam S, Aslam H, Manzoor A, Hui C, Rasool A (2024) AntiPhishStack: LSTM-based stacked generalization model for optimized phishing URLs detection. Symmetry 16(2). https://doi.org/10.20944/preprints202401.1142.v1
6. Shivangi S, Debnath P, Sajeevan K, Annapurna D (2018) Chrome extension for malicious URLs detection in social media applications using artificial neural networks and long short-term memory networks. In: 2018 international conference on advances in computing, communications and informatics (ICACCI), pp 1–6. IEEE (2018). https://doi.org/10.1109/ICACCI.2018.8554647

Collaborative Farming: Innovative Approaches to Agricultural Sustainability and Efficiency

Tushar Gaikwad, Ankush Kudale, Chandrani Singh, and Bhagwat Kaulwar

Abstract As farmers face escalating challenges such as climate change, resource scarcity, and increasing operational costs, collaborative farming and joint ventures offer promising solutions for resource optimization and cost reduction. Collaborative farming involves shared efforts in utilizing land, equipment, and labor, while joint ventures formalize these partnerships, enabling participants to share profits, risks, and resources. This paper explores successful models of collaboration across various countries, including equipment-sharing cooperatives and farming networks. It highlights the benefits of these approaches, such as economies of scale, risk distribution, and enhanced innovation, fostering sustainable agricultural practices through efficient land use and resource conservation. However, challenges such as legal complexities, trust, and resource management are discussed. collaborative farming and joint ventures present viable strategies for achieving agricultural sustainability and economic stability.

Keywords Collaborative farming · Joint ventures · Agricultural sustainability · Resource sharing · Economies of scale

1 Introduction

The agricultural sector faces unprecedented challenges, including climate change, resource scarcity, rising operational costs, and market inefficiencies. These issues threaten the sustainability and economic stability of farming communities worldwide, necessitating innovative solutions. Collaborative farming has emerged as a transformative strategy to address these challenges by promoting resource optimization, risk-sharing, and enhanced operational efficiency [1, 2]. Through shared efforts

T. Gaikwad (✉) · A. Kudale · C. Singh · B. Kaulwar
Sinhgad Institute of Management, Vadgaon Bk, Pune, India
e-mail: tushar.wjrwm22@sinhgad.edu

B. Kaulwar
e-mail: bhagwatkaulwat.siom@sinhgad.edu

C. E. R. Hewage et al. (eds.), *AI Applications in Cyber Security and Privacy of Communication Networks*, Lecture Notes in Networks and Systems 1453,
https://doi.org/10.1007/978-981-96-7400-8_3

in land, labor, and equipment, collaborative farming models help farmers achieve economies of scale while fostering community resilience.

Advanced technologies, such as the Internet of Things (IoT) and blockchain, have further strengthened the potential of collaborative farming by offering tools for improved decision-making, real-time data sharing, and secure resource management. For instance, systems powered by IoT can track crop health, soil conditions, and weather patterns, enabling farmers to make informed decisions while minimizing resource waste [3, 4]. Likewise, blockchain technology enhances transparency and traceability within supply chains, building trust among all involved parties [5]. These technological innovations, when integrated with collaborative farming models, enable farmers to address long-standing issues such as market inefficiencies and resource underutilization.

This study explores the dynamics of collaborative farming by reviewing successful models from around the world, including equipment-sharing cooperatives in France and Germany, joint ventures in Brazil, and Community-Supported Agriculture (CSA) in the United States [1, 2, 6, 7]. The research also investigates the role of technical platforms, such as the Collaborative Resource Management System (CRMS), in driving efficiency and sustainability. Despite its advantages, collaborative farming faces challenges, including legal complexities, digital literacy gaps, and trust issues, which must be addressed to ensure widespread adoption and long-term success [8, 9]. Collaborative approaches help ensure the long-term success of farms while strengthening community ties, benefiting both socially and economically [10].

1.1 Concept of Collaborative Farming

Collaborative farming, based on mutual support, refers to cooperation among farmers to share resources and knowledge. Modern collaborative farming encompasses a wide range of practices, from informal sharing of equipment to formal business partnerships. The central idea is that farmers working collectively can achieve greater results by pooling their resources and efforts.

1.2 Approaches to Collaborative Farming

As shown in Fig. 1, collaborative farming adapts to diverse needs and circumstances, employing key approaches such as resource sharing, collective purchasing, and joint ventures.

- Resource Sharing: Farmers share machinery, labor, and inputs, reducing costs and improving productivity. This can be informal or structured through cooperatives.
- Collective Purchasing and Marketing: Collaboration enables farmers to negotiate better prices for inputs and secure stronger market positions.

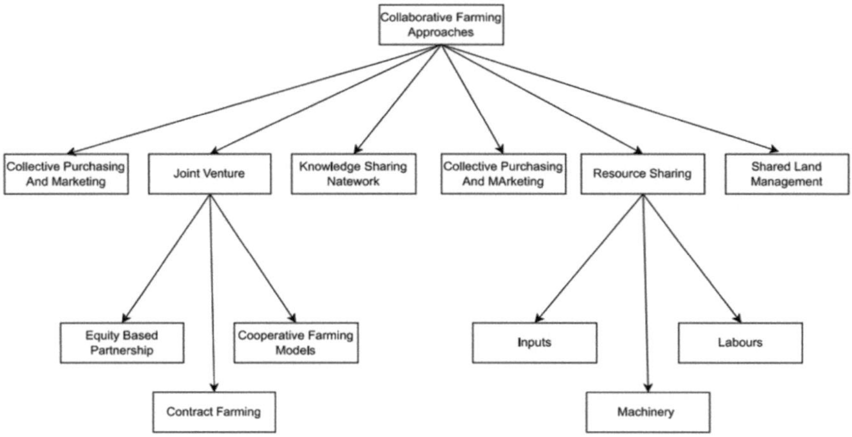

Fig. 1 Collaborative farming approaches

- Shared Land Management: Cooperative land management optimizes land use through practices like crop rotation.
- Joint Ventures: Formal agreements in which resources are pooled for specific farming enterprises [11].
- Knowledge Sharing Networks: Facilitates the exchange of information and best practices among farmers.
- Community-Supported Agriculture (CSA): A model where consumers and farmers partner to share the risks and benefits of food production [6].

These approaches aim to enhance economic resilience and operational efficiency while promoting sustainable agricultural practices.

1.3 Research Objectives

The objectives of this research are to

- Define and differentiate between collaborative farming and joint ventures.
- Examine the advantages and difficulties of these approaches.
- Examine case studies of successful implementations.
- Assess the potential impact on agricultural sustainability and efficiency.
- Provide recommendations for policies and practices to support effective collaborative farming initiatives.
- Assessing the potential impact of technological platforms, such as the Collaborative Resource Management System (CRMS), on agricultural sustainability and efficiency.

2 Literature Survey

2.1 *Collaborative Farming: A Growing Trend*

Collaborative farming is an innovative model that helps farmers share resources, reduce costs, and improve market access. By working together, they achieve better efficiency and sustainability.

Advantages:

Cost Efficiency: Farmers share equipment, labor, and land, lowering operational costs.

Risk Reduction: Spreading financial and operational risks makes farming more stable.

Market Strength: Collective bargaining improves product pricing and market access.

Challenges:

Coordination Issues: Effective collaboration requires structured agreements.

Loss of Autonomy: Farmers may feel restricted by group decisions.

Legal Barriers: Establishing formal partnerships can be complex.

2.2 *Research on Collaborative Farming*

Studies show that cooperatives in France and Germany help small farmers access advanced machinery. In the UK, farmer clusters improve cost efficiency and environmental conservation [2]. However, legal hurdles and organizational challenges persist, especially for small-scale farms.

2.3 *Global Practices and Applications*

2.3.1 Equipment Sharing

Cooperatives in France and Germany, such as CUMA, allow farmers to jointly purchase expensive machinery.

2.3.2 Land Sharing

In New Zealand and Australia, farmers collaborate on land management, improving sustainability through crop rotation and livestock sharing.

2.3.3 Community-Supported Agriculture (CSA)

In the U.S. and Europe, consumers invest in farms upfront, ensuring financial security for farmers while receiving fresh produce.

2.3.4 Joint Ventures in Brazil

Farmers collaborate with large corporations in soybean farming, gaining capital and technology but facing dependency risks.

2.3.5 Smart Farming in Eastern Europe

IoT-based networks help farmers share real-time data to optimize resources and reduce costs.

2.3.6 UK Farmer Clusters

Groups of farmers jointly manage land and promote environmental sustainability.

2.4 Literature Gaps

Existing studies focus on large-scale farms in developed nations, with limited research on small farmers and long-term sustainability [1, 8]. This paper aims to bridge that gap by analyzing how collaborative farming benefits both large and small farms, particularly in developing regions. We'll also look at how these methods could help farming be more sustainable and help farmers become more financially stable [12].

3 Methodology

This study adopts a mixed-methods approach to comprehensively explore collaborative farming practices and their impact on agricultural sustainability. The following subsections outline the research design, data collection methods, and analytical strategies.

3.1 *Research Design*

The study combines qualitative and quantitative approaches to examine the benefits, challenges, and technological aspects of collaborative farming. A case study approach is employed to gather in-depth insights from real-world examples across diverse geographical regions.

3.2 *Data Collection Methods*

- Literature Review: Scholarly articles, reports, and conference proceedings were reviewed to establish a theoretical foundation and identify research gaps.
- Surveys and Interviews: Structured surveys and semi-structured interviews were conducted with farmers, agricultural experts, and policymakers to collect primary data.

3.3 *Analytical Tools and Techniques*

- Thematic Analysis: Qualitative data from interviews were coded and analyzed to identify recurring themes, challenges, and success factors.
- Statistical Analysis: Quantitative data from surveys and secondary sources were analyzed using tools like SPSS and Excel to evaluate the impact of collaborative farming on productivity and sustainability metrics.
- Comparative Analysis: Case studies from different regions were compared to identify best practices and contextual variations in collaborative farming models.

4 Case Studies: Real-Life Examples of Collaborative Farming

This section explores various collaborative farming models worldwide, highlighting strategies that enhance efficiency and sustainability.

4.1 France and Germany

CUMA cooperatives enable farmers to share costly machinery, reducing expenses and increasing access to modern technology.

4.2 United States

Community-Supported Agriculture (CSA) fosters direct farmer-consumer engagement, ensuring financial stability and promoting sustainable food systems.

4.3 Brazil

Smallholder farmers form joint ventures with agribusinesses, gaining market access and technology while requiring safeguards against dependency risks.

4.4 India

The Kisan Network connects farmers directly with consumers through a digital platform, ensuring fair prices and reducing middlemen intervention.

4.5 Eastern Europe

Smart farming networks use IoT solutions for data-driven farming, optimizing resource use and improving sustainability.

5 Challenges in Collaborative Farming

While collaborative farming offers many advantages, several challenges can hinder its effectiveness and success. Understanding these obstacles is essential for developing strategies to overcome them.

5.1 *Infrastructure Challenges*

Poor roads, transport, and storage make it hard for farmers to deliver produce, causing losses and limiting market access. Better infrastructure is essential.

5.2 *Limited Market Access*

Small farmers often face challenges in reaching buyers due to a lack of connections and resources. Providing them with digital platforms, training on market trends, and direct sales opportunities can enhance their economic sustainability.

5.3 *Technical Knowledge Gaps*

The adoption of modern farming techniques and technologies is often limited by a lack of technical expertise. Training programs focused on smart farming, precision agriculture, and digital tools can help farmers improve productivity and efficiency.

5.4 *Legal and Policy Barriers*

Complex regulations, unclear contracts, and legal uncertainties can discourage farmers from forming partnerships. Simplified legal frameworks, policy support, and access to legal guidance can encourage collaboration and long-term success.

6 Integrating Technology into Collaborative Farming: Strategies and Solutions

As illustrated in Fig. 2, the integration of technology in farming includes smart farming tools, online marketplaces, data sharing networks, and mobile agricultural applications.

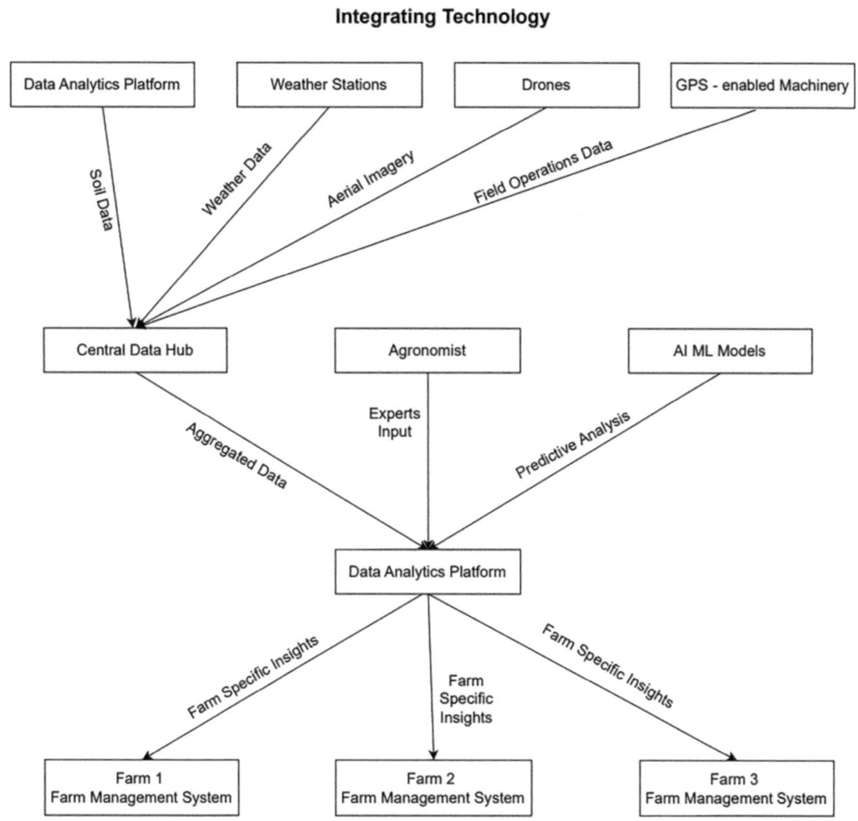

Fig. 2 Integrating technology

6.1 Adoption of Smart Farming Technologies

Smart tools like IoT sensors and drones help monitor crop health, soil moisture, and weather conditions, improving decision-making and yields. Training programs and partnerships with tech firms can enhance accessibility and usage [13].

6.2 Establishing Online Marketplaces

Digital platforms allow farmers to sell directly to consumers, eliminating middlemen and boosting profits. Developing user-friendly e-commerce platforms with training support can facilitate smooth transactions.

6.3 Data Sharing and Collaborative Networks

Farmers can improve decision-making by sharing data on crops, markets, and best practices. Online forums and training sessions can promote knowledge exchange and cooperation.

6.4 Utilization of Agricultural Apps

Mobile apps provide critical data on weather, pest control, and market trends, helping farmers make informed choices. Creating farmer-friendly apps with proper training can enhance their effectiveness.

6.5 Blockchain for Transparency and Trust

Blockchain technology plays a crucial role in supply chain transparency, as demonstrated in Fig. 3. It ensures secure, tamper-proof records, fostering trust between farmers and consumers.

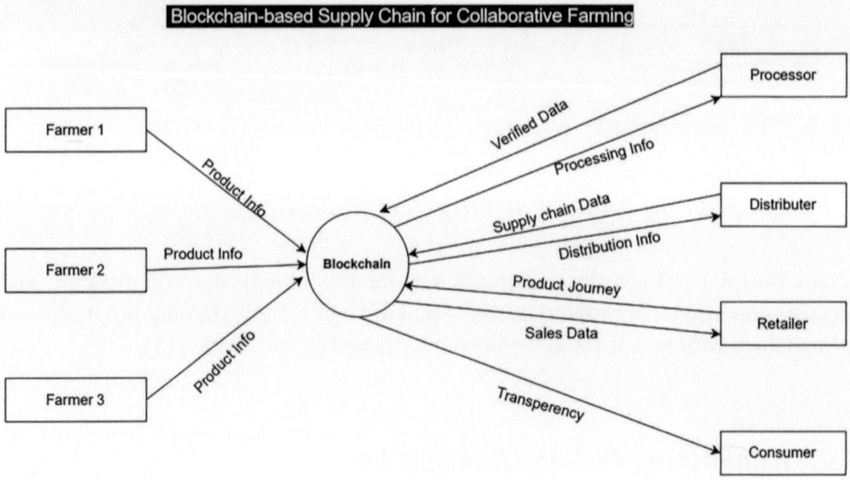

Fig. 3 Blockchain-based supply chain for collaborative farming

6.6 Government and NGO Support

Funding, training, and infrastructure improvements by governments and NGOs can boost technology adoption. Policies, grants, and awareness programs can aid farmers.

6.7 Addressing Digital Literacy

Farmers need digital skills for effective technology use. Training programs, in collaboration with tech firms and educational institutions, can bridge this gap.

6.8 Enhancing Internet Connectivity

Reliable Internet is essential for digital farming. Government and telecom partnerships can expand rural connectivity and enable access to online resources.

6.9 Ensuring Data Privacy and Security

Protecting farmers' data is crucial for trust in technology. Clear policies and strong cybersecurity measures can prevent data breaches and misuse.

6.10 Continuous Evaluation and Feedback

Regular assessment of technology in farming helps identify issues and improvements. Creating feedback channels ensures better integration and farmer support.

7 Overcoming Challenges and Future Directions

To unlock the full potential of collaborative farming and technical platforms, it is essential to address several challenges that currently hinder progress. Here are some key areas of focus.

7.1 *Digital Literacy*

Many farmers lack digital skills to use online platforms effectively. Training programs can equip them with knowledge on marketing, resource management, and technology adoption, enhancing participation in collaborative farming.

7.2 *Internet Connectivity*

Poor Internet access in rural areas limits digital farming opportunities. Investments in rural infrastructure through government and private sector collaboration can ensure reliable connectivity, enabling farmers to engage in online markets and networks.

7.3 *Data Privacy*

With increased digital adoption, farmers face data security concerns. Implementing strong cybersecurity measures, clear data policies, and farmer education on privacy rights can build trust and encourage participation.

7.4 *Future Directions*

Emerging technologies like blockchain, IoT, and big data can improve efficiency in farming. Blockchain ensures transparent transactions, IoT enables real-time monitoring, and big data analytics provides insights for better decision-making.

8 Discussion

Collaborative farming has evolved from traditional shared labor to organized, technology-driven systems that enhance efficiency and affordability. Digital platforms and mobile apps facilitate data sharing, transparency, and trust among farmers. This model also promotes sustainability by optimizing resource use and reducing waste. Successful examples include cooperative dairy farming in Europe and market-access initiatives in developing nations. However, challenges such as managing shared resources, integrating new technologies, and accessing markets remain. A balanced assessment of its benefits and limitations is crucial to shaping its future role in sustainable agriculture.

8.1 The Role of Technical Platforms

Technical platforms play a vital role in boosting collaborative farming initiatives. By facilitating better communication, enhancing knowledge sharing, improving market access, and optimizing resource management, these platforms significantly contribute to the success of farming communities.

8.1.1 Improving Communication and Coordination

Online forums help farmers exchange knowledge, discuss best practices, and seek advice, fostering collaboration. Better communication enables informed decision-making, improving efficiency in collaborative farming.

8.1.2 Enhancing Knowledge Sharing

Agricultural databases provide insights on farming techniques, market trends, and pest control. AI-powered tools offer personalized recommendations, helping farmers optimize productivity and adopt sustainable practices.

8.1.3 Facilitating Market Access

E-commerce platforms enable direct farmer-to-consumer sales, increasing profits and eliminating middlemen. Supply chain management tools improve logistics, ensuring product quality and market efficiency.

8.1.4 Improving Resource Management

As depicted in Fig. 4, a Collaborative Resource Management System (CRMS) aids in effective decision-making, optimizing resource allocation, and improving farm productivity. Weather forecasting tools assist farmers in planning planting and harvesting schedules, reducing climate risks.

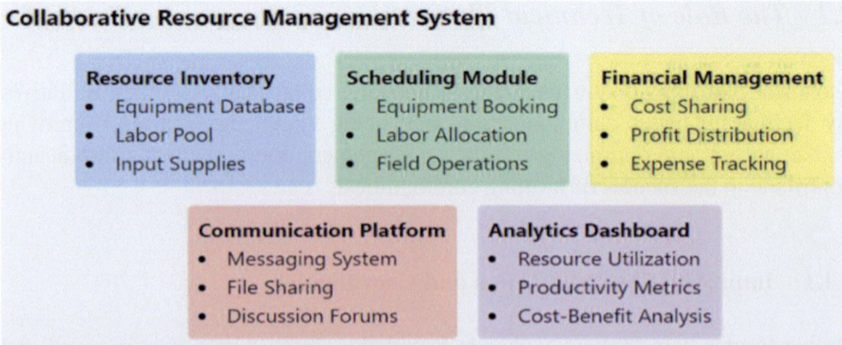

Fig. 4 Collaborative resource management system

8.2 Case Studies of Successful Technical Platforms

8.2.1 eNAM (National Agricultural Market)

eNAM connects farmers with buyers nationwide, ensuring price transparency and reducing intermediaries. This enhances market access and allows farmers to negotiate better prices, increasing income and sustainability [9].

8.2.2 Kisan Network

This platform facilitates direct farmer-to-consumer sales, providing market insights and eliminating middlemen. It helps farmers understand demand, plan production efficiently, and retain higher profits [14].

8.2.3 AgriTech Startups

AgroStar provides farmers with quality inputs and expert guidance, optimizing productivity [15].

Ninjacart streamlines the supply chain, linking farmers with retailers and restaurants, reducing waste, and ensuring fresh produce reaches consumers efficiently [16].

9 Conclusion

Collaborative farming, enhanced by technology, offers a viable solution to modern agricultural challenges. By sharing resources and knowledge, farmers can boost productivity and resilience against market and climate fluctuations. Digital platforms, farm management software, and online marketplaces facilitate better decision-making and operational efficiency. However, challenges such as inadequate infrastructure, digital literacy gaps, and data privacy concerns must be addressed. Investing in connectivity, farmer training, and data security is essential. A future-focused approach integrating collaboration and technology can create a sustainable, productive, and secure agricultural ecosystem for farmers and consumers alike.

References

1. Béné C et al (2019) Collaborative farming: opportunities and challenges. Food Policy 82:23–32. https://doi.org/10.1016/j.foodpol.2018.11.005
2. Van der Meer T (2021) Farmer clusters in the UK: a new approach to sustainable agriculture. J Agric Stud 9(2):45–62. https://doi.org/10.5296/jas.v9i2.18954
3. Khanna M, Kaur P (2022) Smart farming: Internet of Things (IoT)-based sustainable agriculture. Agriculture 12(10), Article 1745. https://doi.org/10.3390/agriculture12101745
4. Wolfert S et al (2024) Data analytics in agriculture: enhancing decision-making for sustainable practices. Sustainability 16(17), Article 7331 (2024). https://doi.org/10.3390/su16177331.com/2071-1050/16/17/7331
5. Rehberge S (2022) Collaborative farming models: creating a closed-loop system and sharing resources in a highly specialized agricultural sector. Plant Agri Biosci. https://www.plantagbiosciences.org
6. Food and Agriculture Organization of the United Nations (FAO) (2022) The role of technology in agricultural transformation. FAO Publication. https://www.fao.org/publications
7. UnCommon Farms (2023) The power of collaboration: farm teams and their impact. UnCommon Farms Blog. https://uncommonfarms.com/blog/the-power-of-collaboration-farm-teams-and-their-impact
8. International Food Policy Research Institute (IFPRI) (2020) Scaling up collaborative farming in developing countries. In: IFPRI discussion paper. https://www.ifpri.org/publication/scaling-collaborative-farming-developing-countries
9. eNAM (National Agricultural Market) (2023) Empowering farmers through technology: a review of eNAM. eNAM Website. https://www.enam.gov.in
10. Hayden J et al (2023) The role of collaborative farming in generational renewal and farm succession. Int Food Agribus Manage Rev. https://www.researchgate.net/publication/377435202
11. Newig J et al (2019) Success of collaboration for sustainable agriculture: a case study meta-analysis. Sustainability 11(2):251–267. https://doi.org/10.3390/su11020251
12. World Bank (2021) Digital agriculture: a new wave of innovation for farmers. World Bank Report. https://www.worldbank.org/en/news/feature/2021/10/11/digital-agriculture-innovation
13. Kudale A (2019) IoT application for smart agriculture. In: Proceedings of the national conference on recent trends and advances in the field of information technology. Karpagam Academy of Higher Education, 27 September 2019. ISSN: 0973-2926
14. Kisan Network (2023) Transforming agriculture through technology: Kisan Network. Kisan Network Website. https://www.kisannetwork.com

15. AgroStar (2023) Revolutionizing agricultural supply chains. AgroStar Website. https://www.agrostar.in
16. Ninjacart (2023) Connecting farmers to consumers. Ninjacart Website. https://www.ninjacart.in

Next-Generation Medical Education: Integrating Virtual Reality into Pedagogical Simulation

Sankari Sankari, P. Vignesh Raj, and S. Sakthivel

Abstract In order to aid human beings, understand better outcomes, and employ virtual reality (VR), an emerging technology is being employed extensively in the defence and innovative sectors. In our work, we have involved the development of a VR application that helps professionals effectively comprehend human anatomy. For the benefit of potential medical students and professionals, we have developed a versatile VR application that is accessible as a mobile app and a web platform. Users can advance from basic courses to complicated medical ideas with this platform, depending on their ability level and goals. Based on this concept, our endeavors use real-time VR scenarios made using Unity and Blender to deliver a real-time medical simulation and lessons learnt about mastering the intricate skills and knowledge of a medical setting.

Keywords Blender · Scenario simulation · 3D data model · Medical simulation · Unity · VR environment · Natural language processing

1 Introduction

Traditional medical education, which relies on cadaver dissections, faces challenges like limited access, ethical issues, and lack of interactive technology. VR environments are revolutionizing medical education by providing immersive learning experiences, allowing students and experts to practice procedures in controlled, risk-free settings, offering accessibility and flexibility. The research aims to use virtual reality to enhance medical education by providing a customizable, accessible, and engaging learning environment, ensuring universal accessibility of high-quality medical education, regardless of geographical constraints.

VR simulations offer a dynamic alternative to traditional cadaveric dissections, overcoming geographical barriers and fostering inclusivity in healthcare education.

S. Sankari (✉) · P. Vignesh Raj · S. Sakthivel
Department of Information Technology, Sri Sai Ram Engineering College, Chennai, India
e-mail: sankari2705@gmail.com

C. E. R. Hewage et al. (eds.), *AI Applications in Cyber Security and Privacy of Communication Networks*, Lecture Notes in Networks and Systems 1453,
https://doi.org/10.1007/978-981-96-7400-8_4

They replicate intricate medical scenarios, providing tailored learning experiences and revolutionizing medical education by equipping healthcare professionals with real-world practice skills.

2 Literature Survey

The authors have developed a systematic approach to Augmented and Virtual Reality (AR/VR) in education, aiming to enhance learning experiences, increase student interest, and provide real-world scenarios [1]. The authors propose a new medical device education strategy utilizing VR to bridge the gap between academic learning and real-world healthcare practice, enhancing proficiency, preventing malfunctions, and improving healthcare education quality [2].

In [3], IFOREAL developed a VR-assisted training system for gynecology students and trainees, enhancing user experience and perceptions by providing real-time, immersive practical knowledge through visual representation. VR is utilized in surgical education to teach teamwork and patient safety through stereoscopic 360 VR videos, recorded using a Vuze360 3D camera and edited using VR plugin tools [4].

Researchers utilized Oculus Quest2 to enhance medical education during the COVID-19 pandemic by providing hands-on learning experiences and knowledge about medical equipment handling [5]. The study proposes a VR gaming application to enhance medical students' empathy for patients with Parkinson's disease by providing hands-on experience and understanding of symptoms and patient experiences [6].

A multi-user liver anatomy prototype was proposed, incorporating 19 liver datasets, 3D surface models, 2D imagine data, diagnosis techniques, and treatment plans in augmented and virtual reality [7]. The authors emphasize the importance of using a virtual reality training platform for novices to learn about minimally invasive knee surgery regulation, incorporating medical data visualization, dynamic tissue models, and high-speed graphics [8].

The paper proposes a surgical simulation solution for medical students to enhance their understanding of surgical procedures by utilizing VR technology for real-time interactive demonstrations [9]. In [10], the Aesculapian simulation learning system is presented, which is a virtual reality tool that provides a hands-on anatomy training experience for medical students, eliminating the need for real cadavers and enhancing anatomy education in medical academies.

In [11], a virtual reality application is introduced that improves medical training by preparing students for real-world experiences in the operating room through simulations, head-up displays, and self-guided learning. The author explores the use of virtual and augmented reality in emotive virtual patients, highlighting its potential to enhance empathy and user interfaces while addressing platform-specific challenges [12]. The authors explore the use of virtual reality (VR) in medical experiment teaching, focusing on enhancing interactivity and addressing low simulation levels using interactive models and 3D VR systems to simulate human body organs and diseased areas [13].

The paper explores the use of Augmented Reality and Virtual Reality (AR/VR) technology in healthcare, comparing traditional and immersive methods. It uses UCSIM resources to conduct experiments and projects, highlighting its potential for revolutionizing healthcare practices [14]. The paper explores the use of VR in medical education, presenting a 3D First Aid VR prototype that offers immersive experiences, revolutionizing medical training methods and providing valuable resources [15].

Thus, the literature survey gives an idea about simulating the surgical procedures and usage of immersive technology on medical education for real-time interactions and demonstrations using VR technology besides the training system for students and professionals for the better understanding of various procedures.

3 Proposed System

The proposed VR medical education system operates through a logical flow, as illustrated in Fig. 1. It ensures a personalized and effective learning experience for medical students and professionals as follows:

User Database: The user database securely stores student and professional information, regulates access, and monitors individual learning progress in the VR training curriculum.

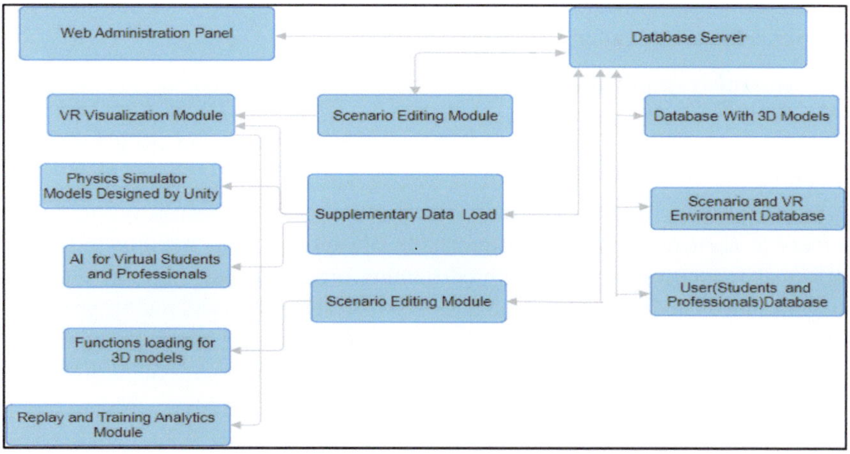

Fig. 1 Proposed system work flow

Scenario Editing Module: This module allows educators to create personalized VR training scenarios, ensuring alignment with medical fields and student requirements, thereby enhancing the learning experience.

Supplementary Data Load: This module enhances medical training by incorporating 3D models and textures, bridging content gaps and enhancing the realism and effectiveness of VR simulations.

Scenario and VR Environment Database: This database houses a vast collection of VR scenarios and environments created using the editing module, providing users with diverse training simulations for their medical education journey.

Database with 3D Models: This database stores 3D models crucial for VR simulations, including anatomical structures, medical instruments, and virtual environments, forming the basis for realistic training experiences.

Functions Loading for 3D Models: The VR training space is equipped with detailed anatomical structures and interactive elements, thanks to a dedicated retrieval function that uses optimized queries.

VR Visualization Module: This module provides an immersive learning experience through VR headsets, enhancing users' understanding and retention of medical concepts through interactive navigation and interaction.

AI for Virtual Students and Professionals: This module utilizes artificial intelligence to create interactive simulations, allowing users to interact with virtual patients or instructors, providing realistic scenarios and feedback.

Replay and Training Analytics Module: This module allows users to post-training to analyze their performance, identify strengths and areas for improvement, and refine their skills for future training sessions.

Web Administration Panel: This panel serves as the central control hub, enabling administrators to efficiently manage system operations, from user management to course creation, ensuring the smooth functioning of the VR Medical Education System.

A revolutionary VR system for medical training allows instructors to design immersive scenarios, provide customizable 3D models, and monitor individual performance metrics, enabling tailored training regimens. This innovative system integrates medical data, blurring the lines between simulation and reality, providing a secure learning environment for trainees. It uses powerful physics engines and extensive databases, allowing for realistic scenarios and personalized training. This VR system is poised to revolutionize medical education.

4 Experimental Set-Up and Results

Unity is used to create 3D environments for VR head-mounted displays, with the XR Interaction Toolkit enhancing functionality. Blender is used for asset and environment creation. The 3D Medical Simulation Designing Software is combined with scenario creation software to design specific medical scenarios. A medical simulation system using VR technology is proposed, breaking down simulations into four modules for web and run-time applications for medical professionals and students.

4.1 VR Visualization Simulation

Our first module focuses on creating 3D models of human organs and bone structures using Blender and Unity software. At first, an entire human skeleton model is designed by adding their names and functions and adding the data takes place in the database as the later part. The model is designed, added to a database, exported to Unity for animations and textures for a realistic simulation.

The VR application and simulation 3D model are developed in Unity using Steam VR plugin, Unity XR interaction toolkit, and VRTK packages, incorporating textures and materials from Blender software for realistic simulation.

Figures 2 and 3 show a 3D simulation model of the human skeleton, loaded via C# scripting and animated for movement. Each organ and structure is meticulously designed in Blender and scripted in Unity for integration into the VR simulation application. This allows for realistic movements and interactions, deepening understanding of specific human body organs and structures. All components are integrated into the database for a streamlined application building process.

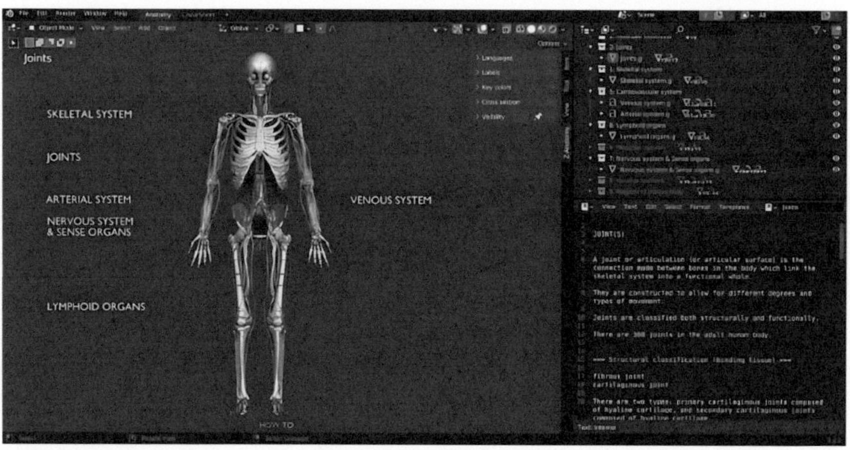

Fig. 2 Skeleton model representation in blender

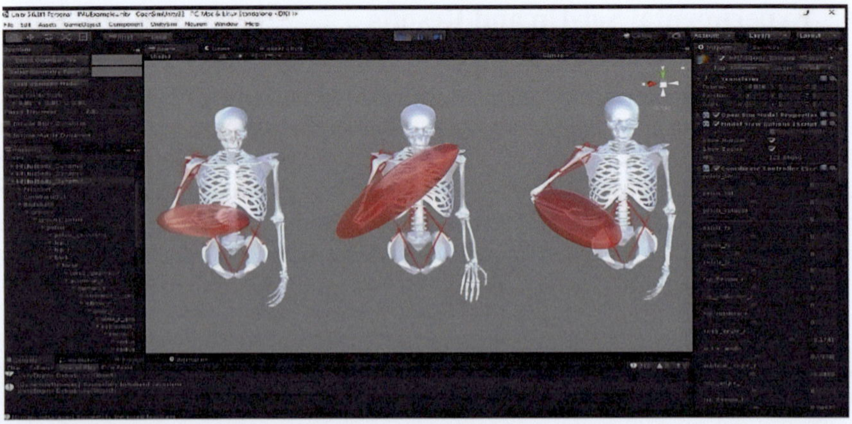

Fig. 3 Animation of the skeleton model for simulation

4.2 Scenario Simulation and Model Editing

The second module focuses on creating a medical education-specific virtual environment using Blender and Unity software. This realistic environment enhances understanding through real-time observation, allowing educators and learners to observe anatomical models and medical equipment, enhancing the learning experience.

4.3 Supplementary Data Load by Adding Voice Assistant

The third module of VR application development incorporates a 3D model and virtual scenarios, enhancing user and educator experience. Voice assistant is also integrated for organ function explanations, improving education through simulation. The VR application features a voice assistant and real-time conversation using natural language processing, enhancing functionality and addressing simple queries related to simulation or medical concepts.

TIMIT Dataset: This dataset is an instrumental for phonetic analysis and training. It provides a comprehensive set of acoustic–phonetic transcriptions, allowing the voice assistant to accurately interpret diverse pronunciations and improve user interaction in a medical education context.

LibriSpeech Dataset: A robust dataset of English speech derived from audiobooks, LibriSpeech is used to train conversational models. This enables the voice assistant to process natural language inputs effectively, facilitating smooth and context-aware interactions during VR simulations.

Medical-Specific Corpora: Custom corpora fine-tuned on medical terminology and contextual scenarios ensure that the voice assistant understands and responds accurately to complex medical queries. These corpora include terminologies related to anatomy, clinical procedures, and diagnostic workflows, enhancing the system's domain-specific proficiency.

The voice assistant uses a combination of datasets to recognize user commands, answer questions, and guide learners through VR-based simulations, providing an interactive and engaging learning environment for medical education. The system includes 3D models and immersive simulations, allowing users and educators to explore concepts and making VR technology accessible for medical training and advanced professional education.

Figure 4 showcases the script for voice command processing with NLP using Google Dialogflow and triggers appropriate actions based on user input.

```
using UnityEngine.UI;
using Google.Cloud.Dialogflow.V2;

public class VoiceCommandProcessor : MonoBehaviour
{
    public Text responseText;
    private SessionsClient sessionClient;
    private string sessionId = "session-id";

    void Start()
    {
        sessionClient = SessionsClient.Create();
    }

    public void ProcessVoiceCommand(string userInput)
    {
        var queryInput = new QueryInput
        {
            Text = new TextInput { Text = userInput, LanguageCode = "en"
}
        };

        var response = sessionClient.DetectIntent(new DetectIntentRequest
        {
            Session = new SessionName("project-id",
sessionId).ToString(),
            QueryInput = queryInput
        });

        DisplayResponse(response.QueryResult.FulfillmentText);
    }

    private void DisplayResponse(string response)
    {
        responseText.text = response;
        if (response.Contains("help"))
        {
            ProvideAssistance();
        }
    }

    private void ProvideAssistance()
    {
        Debug.Log("Providing assistance to the user.");
        // Logic to assist the user based on their request
    }
}
```

Fig. 4 C# script for voice command processing with NLP

4.4 Making VR Application and Adding 3D Model Data Simulation in the Database

The final module focuses on developing a medical simulation application using 3D models and C# scripts, incorporating AI-based assistants for effective communication, and exporting a database of models and techniques from Unity and Blender.

Natural Language Processing (NLP): Google Dialog flow uses Natural Language Processing (NLP) to enable users to interact with virtual patients through voice commands, converting spoken language into text and analyzing user input for appropriate responses.

Behavioral AI: Behavioral AI governs how virtual patients respond to user actions which includes dynamic responses for virtual patients reacting appropriately based on user interactions and scenario adaptation to adjust patient behavior according to user performance.

Adaptive Learning Algorithms: These algorithms customize training experiences based on user performance data for performance tracking to monitor user interactions to identify strengths and weaknesses and to customize scenarios dynamically based on analytics.

The final phase of our work involves integrating 3D models into a virtual reality program, evaluating the simulation, and conducting testing to identify any issues or bugs. After the testing phase, the proposed work is complete, with the final simulation output and haptic feedback for VR head mounts and controllers for each anatomical model and surgical simulation in the virtual world.

4.5 User Testing and Validation

The user testing aimed to assess the VR application's effectiveness in improving medical education, focusing on user engagement, learning outcomes, and satisfaction. 30 individuals, including 15 medical students and 15 healthcare professionals aged 20–40, with varying VR experience levels participated. The evaluation metrics and testing procedures are showcased in Table 1. The results indicated significant improvements in learning outcomes and high user satisfaction as shown in Table 2.

Qualitative Feedback—Participants expressed positive experiences as follows:

- The immersive nature of VR made learning anatomy much more engaging.
- The simulation enhanced my understanding of surgical procedures.

Validation of Realism—To validate anatomical accuracy, expert reviews confirmed the realism of the models used in simulations. Users reported high satisfaction with the lifelike representations.

Table 1 Evaluation metrics and testing procedure of VR environment

S. no.	Evaluation metrics	Testing procedure
1	**Knowledge retention**—Assessed through pre- and post-tests	Pre-test knowledge assessment
2	**User engagement**: Measured using a Likert scale (1–5) on immersion, interactivity, and satisfaction	VR interaction for 30 min
3	**Qualitative feedback**: Collected through open-ended survey	Post-test assessment and completion of a user feedback survey

Table 2 Comparison of VR environment with the traditional system

Metric	Pre-test score (mean)	Post-test score (mean)	Improvement (%)
Knowledge retention	65%	85%	30%
User engagement (satisfaction)	3.5/5	N/A	N/A

Table 3 Benchmarking performance of the proposed system

Metric	Proposed system	Benchmark VR system
Latency (ms)	20	25
Rendering speed (fps)	60	50
Anatomical accuracy (%)	95	90

Benchmarking Performance—The VR system's performance was benchmarked against existing solutions as given in Table 3.

4.6 Experimental Setup, Participant Demographics, and Testing Results

A VR system was tested using Oculus Quest 2 headsets, Intel i7 processors, NVIDIA RTX 3060 GPUs, and 16 GB RAM. Participants were medical students and healthcare professionals. The system was found to be highly realism, accurate, and immersive. It exceeded industry benchmarks, achieving latency below 20 ms and frame rates above 60 FPS. Future improvements include multi-user environments, haptic feedback, and advanced physiological models to bridge the gap between virtual training and real-world medical education. Table 4 showcases the feature comparison of the proposed VR system with the existing system.

Table 4 Feature comparison with existing VR systems

Feature	Existing system	Proposed system
AI-driven interactions	Limited or not available	Includes interactive virtual patients
Accessibility	Primarily PC based	Mobile and web platforms available
Realistic 3D models	Standard 3D models, limited realism	High-resolution models with animation
Customization for users	Minimal customization options	Educators can create tailored scenarios

5 Conclusion and Future Scope

Virtual reality technology can significantly enhance medical education and training by providing hands-on experience in a controlled environment, enhancing diagnostic and treatment skills. It can also create immersive learning experiences, making medical simulations more engaging and effective. The use of VR in medical education can enhance the quality of training for future healthcare professionals and students. Further improvements can include more realistic anatomy and physiological responses, enabling doctors to practice complex surgical procedures in a safe and controlled environment.

References

1. Danaristo J et al (2022) Systematic literature review of augmented reality and virtual reality development in education. In: 2022 international conference on informatics, multimedia, cyber and information system (ICIMCIS), pp 552–556
2. Ozdemir H et al (2020) Virtual reality-based medical device education. medical technologies congress (TIPTEKNO), pp 1–4
3. Chang CW et al (2019) The introduction of a novel virtual reality training system for gynecology learning and its user experience research. In: IEEE access, vol 7, pp 43637–43653
4. Todsen T, Melchiors J, Wennerwaldt K (2018) Use of virtual reality to teach team work and patient safety in surgical education. In: 2018 IEEE conference on virtual reality and 3D user interfaces (VR), Tuebingen/Reutlingen, Germany, pp 1–1
5. Muntahir MI et al (2022) Implementation of immersive technology on medical education. In: 2022 international electronics symposium (IES), Surabaya, Indonesia, pp 651–657
6. Li YJ et al (2021) Parkinson's disease simulation in virtual reality for empathy training in medical education. In: 2021 IEEE conference on virtual reality and 3D user interfaces abstracts and workshops (VRW), Lisbon, Portugal, pp 56–59
7. Schott D et al (2021) A VR/AR environment for multi-user liver anatomy education. In: 2021 IEEE virtual reality and 3D user interfaces (VR), Lisboa, Portugal, pp 296–305
8. Li R, Lou Y (2019) Research and development of the regulation training system of minimally invasive knee surgery based on VR technology. In: 14th international conference on computer science & education (ICCSE), Toronto, ON, Canada, pp 161–164
9. Zhou L, Sato R (2020) Development and application of a surgical process simulation system using VR technology. In: IEEE 9th global conference on consumer electronics (GCCE), Kobe, Japan, pp 655–657

10. Kubur A, Korkmaz I (2022) A virtual reality based hands-on simulation learning system for anatomy education in medical schools. In: 2022 medical technologies congress (TIPTEKNO), Antalya, Turkey, pp 1–4
11. Gruenewald et al (2021) Virtual reality training application to prepare medical student's for their first operating room experience. In: 2021 IEEE international conference on artificial intelligence and virtual reality (AIVR), pp 201–204
12. Zielke MA et al (2017) Developing virtual patients with VR/AR for a natural user interface in medical teaching. In: IEEE 5th international conference on serious games and applications for health (SeGAH), pp 1–8
13. Wang Q (2022) Application of virtual reality technology in experimental teaching of medical specialty. In: 3rd international conference on internet technology and educational information (ITEI), pp 219–222
14. Ajmera H, Gonen B (2020) International conference on computing, networking and communications (ICNC), pp 51–55
15. AL-Hiyari NN, Jusoh SS (2021) Healthcare training application: 3D first aid virtual reality. In: International conference on data science, E-learning and information systems, pp 107–116

Improving Honey Adulteration Detection with Feature Selection and Resampling

Sumayyea Salahuddin, Mohammad Haseeb Zafar, Maryam Mahsal Khan, Nasru Minallah, Shah Haseeb Ahmad Khan, and Esha Rizwan

Abstract Pure and reliable honey is essential. This research addresses honey adulteration detection using hyperspectral imagery, feature selection, resampling, and machine learning. Hyperspectral data from the New Zealand honey dataset, representing various honey types and adulteration levels, was analyzed using SVM, LR, and RF. The dataset was split 80–20% for training and testing, with fivefold cross-validation on the training set. Four models were evaluated: baseline, RFE-based feature selection, SMOTE for class imbalance, and SMOTE with PCA for dimensionality reduction. RF performed best, achieving a 0.996 mean accuracy, 0.997 test accuracy, and F1 scores of 0.986 (pure) and 0.998 (adulterated) for the RFE model with 75 key features. This study offers an accurate, efficient solution for honey adulteration detection, enhancing quality assessment and consumer trust.

Keywords Honey adulteration detection · Hyperspectral imagery · Machine learning · Recursive feature elimination · Synthetic minority over-sampling technique · Principal component analysis

1 Introduction

Detecting honey adulteration is crucial for consumer health and confidence. Adulteration with cheaper substances, such as sugar syrups, undermines the authenticity and poses health risks [1]. Studies emphasize the need for reliable detection methods, both chemical and non-chemical. For example, Guellis et al. [2] showed that

S. Salahuddin (✉) · N. Minallah · S. H. A. Khan · E. Rizwan
Department of Computer Systems Engineering, University of Engineering and Technology, Peshawar, Pakistan
e-mail: sumayyea@uetpeshawar.edu.pk

M. H. Zafar
Cardiff School of Technologies, Cardiff Metropolitan University, Cardiff, Wales, UK

M. M. Khan
Department of Computer Science, CECOS University of IT & Emerging Sciences, Peshawar, Pakistan

C. E. R. Hewage et al. (eds.), *AI Applications in Cyber Security and Privacy of Communication Networks*, Lecture Notes in Networks and Systems 1453,
https://doi.org/10.1007/978-981-96-7400-8_5

the combination of electrochemical and spectrophotometric techniques can distinguish pure honey from adulterated honey, although these lab-based methods have limitations in accurately quantifying adulteration.

Considering these factors, a non-destructive and effective hyperspectral imaging technique can be used for the detection of adulterated honey. It extracts detailed information about the composition of honey by using a wide range of spectral wavelengths and generates a rich dataset. Hyperspectral imaging has recently been developed to detect the botanical origins and quality of pure honey [3]. It has demonstrated high accuracy in detecting adulteration [4].

This research uses hyperspectral imagery with machine learning techniques, feature selection, and resampling to detect honey adulteration and provide accurate results. Its implications include a flexible approach to safeguarding honey integrity, ensuring quality, boosting consumer trust and health, and enabling efficient, non-destructive detection. In addition, it offers a scalable and adaptive model that significantly contributes to the research community.

Our main contributions are as follows:

1. Developing a baseline model for binary honey adulteration detection using Support Vector Machine (SVM), Logistic Regression (LR), and Random Forest (RF) without feature selection or resampling.
2. Implementing a feature-based model with Recursive Feature Elimination (RFE) to assess the impact of key feature selection on classification accuracy.
3. Creating a balanced dataset model using the Synthetic Minority Over-sampling Technique (SMOTE) to address class imbalance and improve model robustness.
4. Designing a feature reduction model that applies Principal Component Analysis (PCA) after SMOTE to reduce dimensionality and evaluate its effect on classification performance.
5. Comparing and analyzing the performance of all models to identify the most effective approach for honey adulteration detection.

2 Literature Review

In recent years, researchers have worked on honey adulteration detection employing a range of techniques, including chromatography, DNA-based, and NMR technologies [5], but these techniques continue to face challenges due to their complexity, cost, and viability. More research is needed to improve the accuracy of detection using various models. Tomczyk et al. [6] studied the effects of 0–30% sugar syrup adulteration on honey's thermal properties, physicochemical parameters, antioxidant content, and enzymatic activity. They observed significant reductions in enzymatic and antioxidant activities ($p < 0.05$) and changes in viscosity, while DSC revealed that syrup type and dosage notably affected crystallization, viscosity, and heat flow rate.

Hyperspectral imagery detects adulterated honey with over 95% accuracy for binary and multi-class classification, providing a reliable, low-cost solution to honey fraud [7]. This method outperforms previous approaches on a broad public dataset. Additionally, a colorimetric sensor array with an electronic tongue and multivariate analysis classified honey by botanical origin (Astragalus, Ziziphus, Polyfloral) and production region in Iran [8]. Classification with PLS-DA achieved 100% training and 95% cross-validation accuracy, while PCA effectively identified contamination ratios in honey and sugar syrup. Previous studies have demonstrated the efficacy of machine learning techniques, such as RF, SVM, artificial neural networks (ANNs), and XGBoost in identifying honey adulteration [9, 10]. Recent studies have applied AI techniques -such as colorimetric sensor arrays and two-dimensional correlation spectroscopy with deep learning to detect honey adulteration, achieving high accuracy and classification performance [8, 11]. In addition, deep learning techniques such as two-dimensional convolutional neural networks (CNNs) have demonstrated high accuracy in detecting sugar syrup adulteration [12]. Ensuring honey authenticity is crucial, as it directly influences the economic sustainability of local honey producers and protects consumers from fraudulent products [6, 7].

3 Methodology

3.1 Dataset

Table 1 presents the composition of the publicly available New Zealand honey adulteration dataset, which includes samples from seven brands (C1, C4, C5, C7, C8, C9, C10) and eleven botanical origins (Clover, MultiFloral, Manuka with UMF ratings 5, 10, 15, 20, ManukaBlend, BorageField, Kamahi, Rewarewa) [7]. To simulate real-world adulteration, pure honey was mixed with sugar syrup in varying concentrations (0, 5, 25, 50%). Each 7g sample was imaged using the SOC-710 hyperspectral imager (400–1000 nm, 5 nm intervals, 520 $\times$ 696 resolution) under controlled halogen lighting, with six repetitions for robustness. The dataset contains 8,675 samples, including 1,050 pure and 7,625 adulterated samples, with class sizes ranging from 450 to 1,350. Key honey types like Manuka and Clover are well-represented due to their global significance. Spectral features span wavelengths from 399.40 to 1063.79 nm, capturing brand, acquisition details, concentration levels, and botanical origins. Hyperspectral imaging reveals non-linear spectral changes due to adulteration, where machine learning techniques enhance detection accuracy.

Table 1 Pure and adulterated honey samples in dataset [7]

Brand	Class	Adulteration concentration					
		0%	5%	10%	25%	50%	Sum
C1	Clover	150	150	300	300	300	1200
C10	MultiFloral	150	150			150	450
	ManukaUMF5		150	150	150	150	600
	ManukaUMF15		150	150	150	150	600
	ManukaUMF20		150	150	150	150	600
C4	ManukaUMF10		150	150	150	125	575
C5	ManukaBlend		150		150	150	450
C7	BorageField	150	150	150	150	150	750
	Kamahi	150	150	150	150	150	750
	Rewarewa	150	150	150			450
C8	ManukaBlend	150	150	150	150	150	750
C9	Manuka	150	300	300	300	300	1350

3.2 Model Selection

Three machine learning models were chosen for their ability to handle non-linear data patterns: **Random Forest** improves predictive accuracy through ensemble averaging, making it effective for honey adulteration detection; **Logistic Regression** serves as a robust binary classifier for determining if a honey sample is pure or adulterated; and **Support Vector Machine** excels with high-dimensional data, effectively distinguishing separable classes such as Manuka, Clover, and Kamahi.

3.3 Methodology

Figure 1 illustrates the methodology of the current study. It consists of four separate experimental setups. Each of these setups is discussed as follows:

- **Experimental Setup 1: Baseline Model**
  To classify pure and adulterated honey, the dataset in Table 1 contains 8675 samples (1050 pure, 7625 adulterated) with 133 features. After pre-processing-binary conversion of the *Concentration class* attribute, label encoding, and one-hot encoding of the *Brand* attribute-131 features remain. The dataset is split into 80% training and 20% testing sets, both scaled using MinMax scaling. Five-fold cross-validation is applied to the training set, and binary classification is performed using SVM, LR, and RF. Performance metrics (mean accuracy, testing accuracy, precision, recall, and F1-score) are then obtained, with the baseline model shown in Fig. 1a.

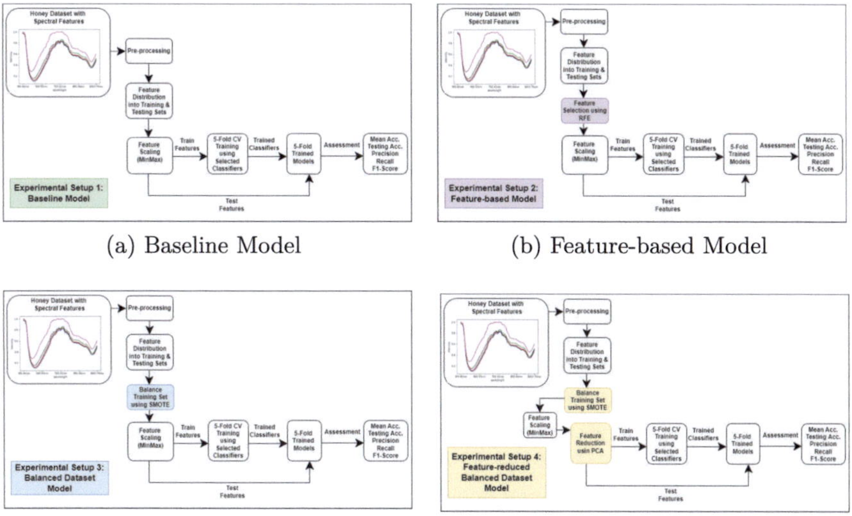

(a) Baseline Model (b) Feature-based Model

(c) Balanced Dataset Model (d) Feature-red. Balanced Dataset Model

Fig. 1 Proposed methodology

- **Experimental Setup 2: Feature-based Model**
  RFE selects important features by recursively training models and ranking features without modifying data. RFE is performed before feature scaling in experimental setup 2, where 75 most important features are selected out of 131 using LR. After obtaining the most important features, fivefold cross-validation is applied to the training set, and binary classification is performed using SVM, LR, and Rf. All the trained models are tested using a testing set containing 75 features. Performance metrics (mean accuracy, testing accuracy, precision, recall, and F1-score) are obtained, with the feature-based model shown in Fig. 1b.
- **Experimental Setup 3: Balanced Dataset Model**
  The honey adulteration dataset is imbalanced, with fewer pure honey samples, causing biased predictions toward adulterated honey. To address this issue, SMOTE is used to balance the dataset, which works by generating synthetic samples for the minority class (pure honey) rather than simply duplicating the existing data points. In experimental setup 3, SMOTE is performed on the training set to balance the dataset where each class has 6102 samples with a total of 12,204 samples. After balancing, both sets are feature scaled, followed by fivefold cross-validation and binary classification. Performance metrics (mean accuracy, testing accuracy, precision, recall, and F1-score) are then obtained, as shown in Fig. 1c.
- **Experimental Setup 4: Feature-reduced Balanced Dataset Model**
  The hyperspectral imaging dataset contains many features, so in experimental setup 4, PCA is applied for dimensionality reduction after SMOTE. PCA modifies the data to extract key features, producing principal components. By retaining 95%

variance, it reduces 131 features to 6 components. Then, fivefold cross-validation and binary classification are performed, and performance metrics (mean accuracy, testing accuracy, precision, recall, and F1-score) are obtained, as shown in Fig. 1d.

4 Results and Discussion

4.1 Baseline Model

Table 2 shows that SVM achieved the highest mean accuracy (0.969) and testing accuracy (0.976), followed by RF (0.961, 0.958) and LR (0.948, 0.958). Although all models performed well in detecting adulterated honey, LR and RF showed lower precision in classifying pure honey, as seen in Table 3. This is likely due to dataset imbalance, with only 12.11% pure honey samples compared to 87.89% adulterated samples.

Table 2 Baseline model accuracy

Technique	5-fold CV					Mean accuracy	Testing accuracy
	Fold 1	Fold 2	Fold 3	Fold 4	Fold 5		
SVM	0.965	0.971	0.973	0.963	0.977	0.969	0.976
LR	0.937	0.955	0.956	0.937	0.955	0.948	0.958
RF	0.973	0.963	0.979	0.914	0.976	0.961	0.958

Table 3 Baseline model performance metrics

Technique		Training			Testing		
		Precision	Recall	F1-Score	Precision	Recall	F1-Score
SVM	Pure	0.800	1.000	0.889	0.835	1.000	0.909
	Adult.	1.000	0.966	0.983	1.000	0.972	0.9860
LR	Pure	0.698	1.000	0.822	0.759	0.998	0.862
	Adult.	1.000	0.941	0.969	0.999	0.956	0.978
RF	Pure	0.744	1.000	0.853	0.746	0.995	0.853
	Adult.	1.000	0.952	0.975	0.999	0.953	0.976

Table 4 RFE model accuracy

Technique	5-fold CV					Mean accuracy	Testing accuracy
	Fold 1	Fold 2	Fold 3	Fold 4	Fold 5		
SVM	0.933	0.932	0.937	0.937	0.934	0.935	0.949
LR	0.976	0.977	0.984	0.989	0.979	0.981	0.986
RF	0.994	0.996	0.995	0.997	0.996	0.996	0.997

Table 5 RFE model performance metrics

Technique		Training			Testing		
		Precision	Recall	F1-Score	Precision	Recall	F1-Score
SVM	Pure	0.992	0.462	0.630	1.000	0.580	0.734
	Adult.	0.931	1.000	0.964	0.945	1.000	0.972
LR	Pure	0.988	0.854	0.916	0.985	0.901	0.941
	Adult.	0.980	1.000	0.989	0.986	1.000	0.992
RF	Pure	1.000	0.963	0.981	1.000	0.972	0.986
	Adult.	0.995	1.000	0.997	0.996	1.000	0.998

4.2 RFE Model

The performance of SVM, LR, and RF is summarized in Table 4. RF achieved the highest mean accuracy (0.996) and testing accuracy (0.997), demonstrating strong predictive power and class imbalance handling. LR was followed with a mean accuracy of 0.981 and a testing accuracy of 0.986, showing balanced precision and recall. SVM had the lowest mean accuracy (0.935) and testing accuracy (0.949), with a lower recall for pure honey, indicating challenges with class imbalance as shown in Table 5. Overall, RF performed best, while LR showed strong efficiency, and SVM required further tuning.

4.3 SMOTE Model

The SMOTE-balanced models performed well as shown in Table 6. RF achieved the highest testing accuracy (0.995) and F1-score (0.997) for adulterated honey. For pure honey, SVM recorded 0.883 precision and 1.0 recall (F1-score 0.938), while LR and RF had lower precision but high recall, with RF attaining the best F1-score (0.979) shown in Table 7. All models had perfect precision (1.0) for adulterated honey, with RF's recall at 0.995. Although RF showed some false positives, indicating potential for fine-tuning, overall, SMOTE effectively balanced the classes and improved performance.

Table 6 SMOTE model accuracy

Technique	5-fold CV					Mean accuracy	Testing accuracy
	Fold 1	Fold 2	Fold 3	Fold 4	Fold 5		
SVM	0.989	0.989	0.989	0.989	0.988	0.989	0.984
LR	0.981	0.985	0.986	0.989	0.988	0.986	0.982
RF	0.980	0.989	0.985	0.988	0.995	0.987	0.995

Table 7 SMOTE model performance metrics

Technique		Training			Testing		
		Precision	Recall	F1-Score	Precision	Recall	F1-Score
SVM	Pure	0.978	1.000	0.989	0.883	1.000	0.938
	Adult.	1.000	0.978	0.989	1.000	0.982	0.991
LR	Pure	0.972	1.000	0.986	0.872	1.000	0.932
	Adult.	1.000	0.972	0.986	1.000	0.980	0.990
RF	Pure	0.976	1.000	0.988	0.968	0.991	0.979
	Adult.	1.000	0.975	0.987	0.999	0.995	0.997

4.3.1 Discussion on SMOTE and Its Limitations

While SMOTE enhances performance and addresses class imbalance, it may also risk overfitting if synthetic samples don't accurately reflect the minority class-especially in small or complex datasets where noise or unrealistic patterns can emerge.

SMOTE generates synthetic samples by interpolating between minority class samples, which can reduce variability. In hyperspectral datasets, this may lead to synthetic samples that do not fully capture the real-world variability of honey adulteration. In this study, SMOTE was carefully implemented to balance the dataset while preserving feature relationships. Future work could explore advanced techniques like Borderline-SMOTE [13] or ADASYN [14], which focus on decision boundaries or density-based adjustments, and integrate SMOTE with feature engineering to enhance model robustness and reduce overfitting. Additionally, PCA was applied for dimensionality reduction to further improve performance and efficiency.

4.4 SMOTE with PCA Model

The SMOTE with PCA model showed a decrease in performance due to feature reduction as shown in Table 8. SVM maintained a better balance between classes, achieving the highest testing accuracy (0.959) and strong F1-scores as shown in Table 9. However, LR and RF exhibited significant drops in precision and F1-scores

Table 8 SMOTE with PCA model accuracy

Technique	5-fold CV					Mean accuracy	Testing accuracy
	Fold 1	Fold 2	Fold 3	Fold 4	Fold 5		
SVM	0.975	0.973	0.967	0.973	0.974	0.972	0.959
LR	0.850	0.853	0.856	0.852	0.855	0.853	0.805
RF	0.914	0.908	0.905	0.905	0.909	0.908	0.844

Table 9 SMOTE with PCA model performance metrics

Technique		Training			Testing		
		Precision	Recall	F1-Score	Precision	Recall	F1-Score
SVM	Pure	0.950	0.997	0.973	0.747	1.000	0.855
	Adult.	0.997	0.947	0.972	1.000	0.953	0.976
LR	Pure	0.807	0.929	0.864	0.381	0.958	0.545
	Adult.	0.917	0.777	0.841	0.993	0.783	0.876
RF	Pure	0.845	0.999	0.916	0.438	0.991	0.608
	Adult.	0.999	0.817	0.899	0.998	0.823	0.903

with RF particularly struggling with generalization. These results indicate that PCA removed essential information, making it less suitable for honey adulteration detection.

4.5 Comparison

Figures 2 and 3 compare mean and testing accuracies for four models (Baseline, RFE, SMOTE, and SMOTE with PCA) across SVM, LR, and RF. All techniques performed well with SMOTE, but RF with RFE achieved the highest mean accuracy (0.996) and testing accuracy (0.997). SVM maintained high testing accuracy across models, peaking at 0.984 with SMOTE. Overall, SMOTE and RFE yielded promising results, with SMOTE combined with PCA notably benefiting LR and RF. Table 10 shows the comparison of our RFE model with the ones reported in the literature [7] on the dataset. Our model shows a slightly lower training F1-Score (0.989) but a higher testing F1-Score (0.992) compared to Linear-SVM, indicating better generalization.

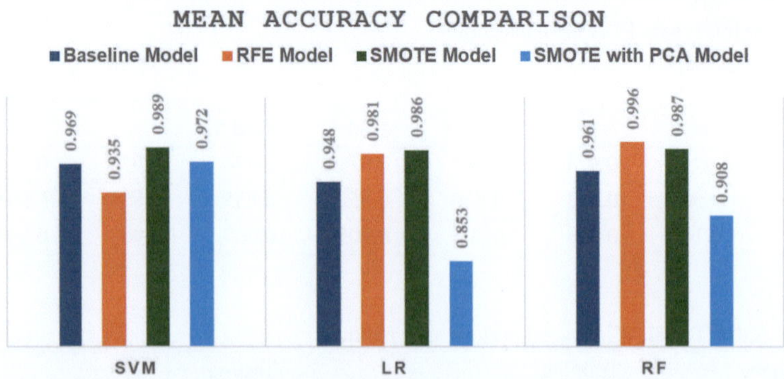

Fig. 2 Comparison of mean accuracy of all models

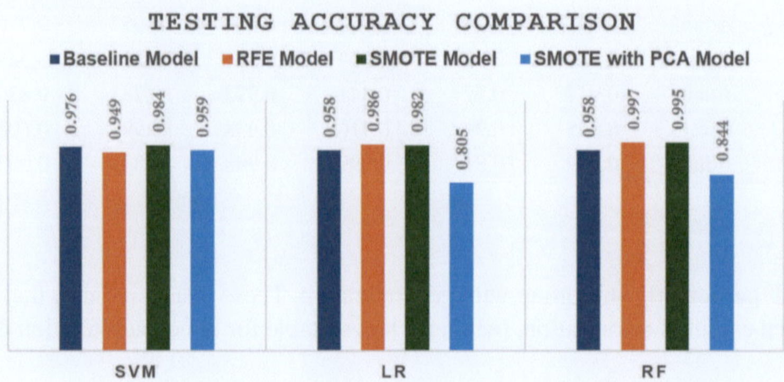

Fig. 3 Comparison of testing accuracy of all models

Table 10 Comparison of ML models with existing techniques from literature

Method	Year	Accuracy (%)	F1-Score	
			Training	Testing
Linear-SVM [7]	2023	0.999	1.000	0.940
RFE-RF (ours)	2024	0.996	0.989	0.992

5 Conclusion

Honey adulteration poses significant health risks. This study employed feature selection (RFE) and resampling (SMOTE) to enhance detection accuracy using SVM, LR, and RF models. The results showed that RFE improved classification by selecting key features, while SMOTE effectively addressed class imbalance. Among all models, RF with RFE achieved the highest accuracy, demonstrating the effectiveness of feature selection and balancing techniques in honey adulteration detection.

References

1. Fakhlaei R et al (2020) The toxic impact of honey adulteration: a review. Foods 9(11):1538
2. Guellis C et al (2020) Non-targeted method to detect honey adulteration: combination of electrochemical and spectrophotometric responses with principal component analysis. J Food Compos Anal 89:103466
3. Noviyanto A, Abdulla WH (2020) Honey botanical origin classification using hyperspectral imaging and machine learning. J Food Eng 265:109684
4. Shao Y, Shi Y, Xuan G, Li Q, Wang F, Shi C, Hu Z (2022) Hyperspectral imaging for non-destructive detection of honey adulteration. Vib Spectrosc 118:103340
5. Lindner P, Bermann E, Gamarnik B (1996) Characterization of citrus honey by deuterium NMR. J Agric Food Chem 44(1):139–140
6. Tomczyk M et al (2023) Tracking of thermal, physicochemical, and biological parameters of a long-term stored honey artificially adulterated with sugar syrups. Molecules 28(4):1736
7. Phillips T, Abdulla W (2023) A new honey adulteration detection approach using hyperspectral imaging and machine learning. Eur Food Res Technol 249(2):259–272
8. Masoomi S, Sharifi H, Hemmateenejad B (2024) A paper-based optical tongue for characterization of Iranian honey: identification of geographical/botanical origins and adulteration detection. Food Control 155:110052
9. Usman H, Amjad A, Khan MM, Salahuddin S (2024) Honey adulteration detection through hyperspectral imaging and machine learning. Int J Innov Sci Technol
10. Mehdizadeh SA, Abdolahzare Z, Karaji FK, Mouazen AM (2023) Design and manufacturing a microcontroller based measurement device for honey adulteration detection. J Food Compos Anal 116:105049
11. Wu X, Xu B, Luo H, Ma R, Du Z, Zhang X, Liu H, Zhang Y (2023) Adulteration quantification of cheap honey in high-quality manuka honey by two-dimensional correlation spectroscopy combined with deep learning. Food Control 154:110010
12. Brar DS et al (2024) Detection of sugar syrup adulteration in unifloral honey using deep learning framework: an effective quality analysis technique. Food Humanity 2:100190
13. Han H, Wang W-Y, Mao B-H (2005) Borderline-smote: a new over-sampling method in imbalanced data sets learning. In: International conference on intelligent computing. Springer, pp 878–887
14. He H, Bai Y, Garcia EA, Li S (2008) ADASYN: adaptive synthetic sampling approach for imbalanced learning. In: IEEE international joint conference on neural networks (IEEE world congress on computational intelligence). IEEE, pp 1322–1328

Revolutionizing Healthcare with IoT: Exploring Applications, Benefits, and Challenges

Meenakshi Jadhav, Ankush Kudale, Chandrani Singh, and Kirti Deng

Abstract Advancements in information and communication technology have led to the emergence of the Internet of Things (IoT), a revolutionary development that enables diverse devices to connect and communicate seamlessly with one another. By utilizing sensors and microprocessor chips, IoT facilitates the efficient exchange of data over the Internet. It was estimated that the number of Internet-connected devices would surpass 50 billion by 2020. In modern healthcare settings, IoT technology is increasingly employed for various applications, including remote patient monitoring, patient information management, telesurgery, and comprehensive healthcare administration. This integration of technology offers significant benefits, such as enhanced convenience and cost efficiency, for both medical professionals and patients. This paper investigates the integration of IoT in the healthcare industry, with a specific emphasis on Internet of Medical Things (IoMT) systems. It examines the benefits and drawbacks of these systems and provides a comprehensive assessment of the essential technologies required for their successful deployment and use. The aim is to enhance the effectiveness of IoMT in advancing healthcare services.

Keywords IoT · Radio frequency identification · Wireless sensor networks · Remote surgery · Medical devices · IoMT · MHA first section

M. Jadhav (✉) · A. Kudale · C. Singh · K. Deng
Sinhgad Institute of Management, Pune, India
e-mail: meena17jadhav@gmail.com

A. Kudale
e-mail: ankush.kudale@gmail.com

C. Singh
e-mail: singh.chandrani@gmail.com

C. E. R. Hewage et al. (eds.), *AI Applications in Cyber Security and Privacy of Communication Networks*, Lecture Notes in Networks and Systems 1453,
https://doi.org/10.1007/978-981-96-7400-8_6

1 Introduction

The Internet of Things (IoT) marks a significant transformation in information technology by merging the "Internet" with "Things," enabling a global network that connects billions of users through TCP/IP systems. IoT integrates physical entities—both living and inanimate—into IT infrastructure for data collection and transmission, focusing on tangible items equipped with sensors for tracking [1]. In healthcare, IoT enhances medical services by improving patient care, optimizing processes, and reducing costs through applications like continuous glucose monitoring and predictive maintenance. Hospitals, such as the Mayo Clinic and Humber River Hospital, leverage IoT and AI for tasks like tracking patient movements and automating logistics [2]. With the rise of chronic diseases and an aging population, there is a shift toward home-based care supported by IoT-enabled remote monitoring and telehealth solutions, allowing for continuous health monitoring and timely interventions [3]. As illustrated in below Fig. 1 and Table 1, this technology has the potential to reduce hospital monitoring costs and support the early detection of health issues. Since its emergence in the early twenty-first century, IoT has been recognized as a transformative force in ICT, revolutionizing various sectors such as healthcare, transportation, agriculture, education, and industry.

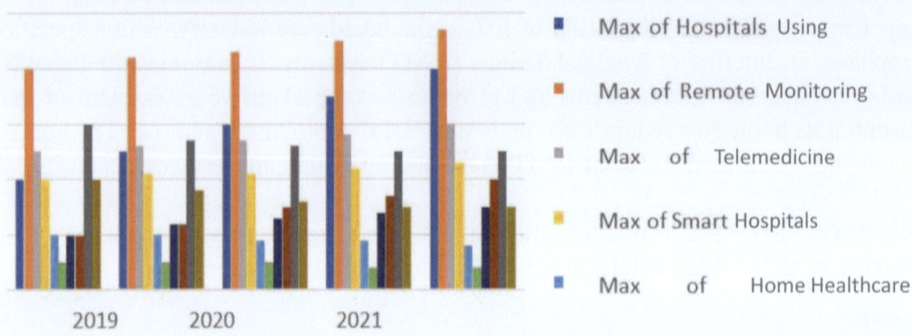

Fig. 1 Percentage data graph

Table 1 Last 5 years' data analysis for how IoT revolutionized the health care

Year	IoT devices deployed (millions)	Hospitals using IoT devices (%)	Remote monitoring usage (%)	Telehealth usage (%)	Smart hospitals usage (%)	Home healthcare usage (%)	Telesurgery usage (%)	Cost savings per hospital (millions)	Increase in workflow efficiency (%)	Security incidents (annual)	Data management issues (%)	Interoperability problems (%)
2019	15	20	40	25	20	10	5	$1.5	10	25	30	20
2020	18	25	42	26	21	10	5	$1.7	12	28	27	18
2021	22	30	43	27	21	9	5	$1.8	15	32	26	16
2022	27	35	45	28	22	9	4	$2.0	17	29	25	15
2023	33	40	47	30	23	8	4	$2.2	20	30	25	15

2 Problem Statement

2.1 Research Questions

- What key technologies underpin the Internet of Medical Things (IoMT)?
- In which primary fields is IoMT applied?
- What challenges impede the implementation of IoMT?
- What additional technologies can organizations utilize to address challenges associated with IoMT?
- What are the principal advantages of adopting IoMT?

2.2 Research Significance

Modern hospitals and advancements in technology have significantly increased the variety of data available on patients' health conditions. Medical centers utilize IoT technology to achieve strategic goals, such as maintaining agility, sustaining a competitive edge, reducing data redundancy, enhancing data consistency and sharing, making timely decisions, and enabling live diagnosis and monitoring [4].

3 Background

IoT connects smart devices to exchange data, improving efficiency and precision. IoMT integrates IoT with healthcare, enabling remote monitoring and home-based care, reducing costs while maintaining quality. It enhances telemedicine, remote education, and medical communication. Key technologies like RFID, WSNs, and cloud computing drive IoMT advancements [5].

4 Literature Review

The primary technologies involved in the enhancements to IoT radio frequency identification technology form the backbone of IoT. It employs active or passive radio frequency identification tags and readers to track objects or items. Each radio frequency identification device comprises two main components: an integrated circuit that processes and stores data, and an antenna that transmits and receives signals from the reader. This technology is widely adopted across various industries for its ability to deliver accurate and timely information about connected objects, reducing the need for large workforces and automating business processes [6].

Radio frequency identification tags significantly contribute to advancements in the pharmaceutical sector. They can be used with patients and medical equipment to monitor health conditions, including temperature, blood pressure, and heart rate. This allows for a reduction in traditional health monitoring costs and enables patients to remain comfortable at home instead of making frequent hospital visits. Additionally, radio frequency identification tags can be integrated into emergency ambulances to enable doctors to monitor and manage patient conditions remotely until they reach the hospital [7].

Wireless sensor networks (WSNs) consist of distributed sensors that track physical and environmental conditions such as vibration, light, and temperature. ECG, pressure, and other data can be transmitted wirelessly to a processing center without human involvement.

WSN is widely employed in several domains, such as military monitoring, environmental monitoring (e.g., forest fires), water quality and air pollution monitoring, and machine status monitoring.

4.1 Cloud Computing

Cloud computing is an advanced digital technology where multiple servers are integrated onto a single cloud platform, enabling resource sharing that can be accessed anytime and anywhere [8].

Combining IoT with cloud computing enhances efficiency and scalability by overcoming IoT's limitations in processing and storage capacity. This synergy is mutually beneficial: cloud computing provides unlimited processing and storage, while IoT enables dynamic interaction with real-world objects, regardless of location. Ultimately, cloud computing improves IoT services by increasing precision, efficiency, and competitiveness, driving research into cloud-integrated IoT models [9–11].

4.2 Internet of Medical Things (IoMT)

Research on IoMT includes applications such as home healthcare, hospital management, telemedicine, mobile health, alarm systems, and emergency health applications. Selecting the right IoMT applications is crucial for business model success. Establishing a robust infrastructure for these applications requires incorporating software engineering techniques and technologies in the business model design. However, few academics focus on SE technologies despite their ability to meet both functional and nonfunctional requirements [12, 13].

4.3 Hospital Health Services

The authors recommend using IoT technology to monitor and manage patients in Intensive Care Units (ICUs). This system allows for real-time observation of critically ill patients, aiding in determining appropriate medical settings and reducing costs. The proposed design includes

1. An "ICU bedside monitor" that tracks vital signs such as heart rate, oxygen saturation, blood pressure, breathing rate, and body temperature.
2. "Kinect sensors" that monitor patient mobility, facial expressions, and speech using a natural user interface to detect skeletal movements without being close to patients.
3. A "sensor board for environmental monitoring" that collects data on temperature, humidity, gases, movement, and vibration, transmitting this information to a server for access by doctors and nurses through a GUI [14].

4.4 At-Home Medical Care

The CogSense system enables real-time remote health monitoring via IoT, integrating data analysis, communication, and sensor processing. It uses cloud-based patient history, secure network protocols, and live supervision tools (cameras, ECGs, sensors). To reduce congestion, the Ihomesystem includes iMedBox (IoT hub), iMedPack (medication reminders), and Bio-Patch (vital sign monitoring) for efficient home healthcare.

4.5 Tele-Health

Spanakis developed the My Health Avatar (MHA) system, which employs IoMT technology for personalized digital health management. According to [15], individuals could carry a collection device throughout their lives to store all relevant information. MHA maintains a comprehensive database of medical and personal details, such as age, environmental factors, and genetic information. This system provides access to a wealth of data that can be utilized for medical research and knowledge acquisition. Additionally, MHA enables direct communication between doctors and patients, enhancing treatment and guidance. It also offers various features to support e-health monitoring, including.

Data collection involves using smartphones, social networks, and web information extraction to gather valuable data while filtering out unreliable information.

- Ontology helps manage the diversity in the collected data.
- Cloud-based design facilitates extensive resource sharing, including data and storage capabilities.

- A methodology for big data analytics and simulation has been developed to aid in decision-making.

H. J. Baek, G. S. Chung, K. K. Kim, and K. S. Park developed a sophisticated diagnostic chair that automatically tracks heart rate and blood pressure. This chair features electrodes embedded in the backrest and specialized textiles in the seat, enabling seamless registration of ECG data depending on the fabric's type. M-health Home healthcare is evolving with mobile technology integration. A. Ghose, C. Bhaumik, D. Das, and A. K. Agrawal advocate using mobile phones to collect data from medical device sensors, which is uploaded to IoT cloud servers for analysis. They also propose a mobile service for wheelchair users that employs wireless body sensor networks to monitor ECG, heart rate, blood pressure, and environmental conditions, enhancing reliability and interaction [16].

In contrast, C. Doukas and I. Maglogiannis address limited mobile storage by combining cloud computing with IoT to manage extensive datasets. Their system collects data from wearable medical sensors, uploads it to the cloud via a mobile sensor gateway, and utilizes a lightweight REST-based API for data exchange. Another approach combines IoT and cloud services for a protected, scalable electronic health solution using an IOIO microcontroller for ECG data capture, securely stored on a server, enhancing storage, reducing redundancy, and improving security with user authentication while lowering monitoring costs through an Android app.

However, some mHealth applications face security issues. J. Morak et al. propose a tele-monitoring system for elderly patients with chronic diseases, featuring a simple interface and utilizing Bluetooth and NFC to reduce reliance on wireless connections, although it has security weaknesses [17].

H. Chen, R. H. Chiang, and V. C. Storey introduce a novel protocol, YOAPY, to address challenges in mHealth, such as data transmission delays and privacy concerns, ensuring ongoing monitoring, robust connectivity, and secure management. Figure 2 illustrates the remote patient monitoring.

4.6 Key Challenges of IoMT

The main issues include managing the extensive flow of information between smart medical devices, preventing unauthorized access, protecting patient privacy, dealing with network congestion, and effectively managing big data analytics. These challenges represent considerable hurdles in the realm of IoT healthcare.

IoT enables smart device connectivity but faces data fragmentation, standardization, and compatibility challenges. UDA-IoT and semantic models enhance emergency medical data management [18]. Big data analytics aids healthcare decisions, ensuring security, scalability, and accuracy. The CARE system predicts risks and assists diagnosis, while IoT-driven emergency systems improve rapid response and data handling.

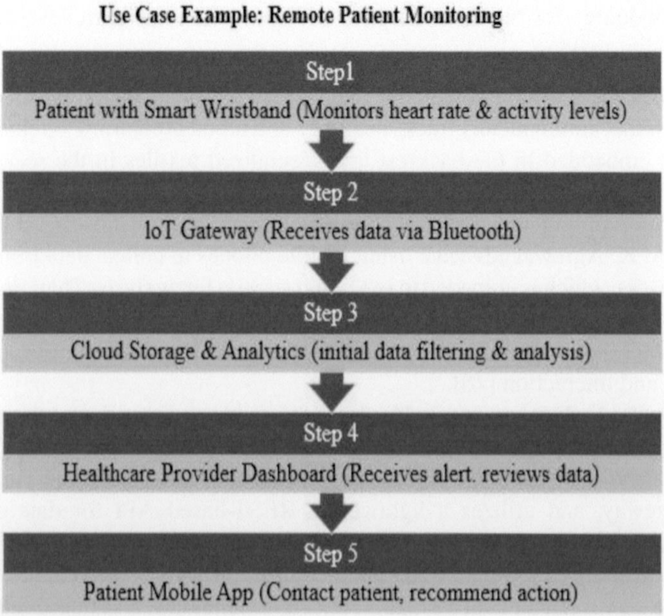

Fig. 2 Remote patient monitoring

Data transmission delays implementing the Internet of Medical Things (IoMT) enable real-time transfer of critical medical information, which is vital for remote surgeries that rely on sensor data for informed treatment decisions. The 5G network enhances IoMT with faster connectivity, improved management, and greater data capacity, boosting the quality and reliability of medical applications through shorter transmission intervals, as noted by West and Darrell. K. E. Skouby and P. Lynggaard propose an ISHSC infrastructure that integrates 5G, IoT, cloud technologies, and AI for intelligent networking and efficient scaling in smart homes [19].

4.7 *Privacy and Security*

M. J. Chang, J. K. Jung, M. W. Park, and T. M. Chung emphasize the critical need to prioritize security within the IoMT framework, given its dependence on telecommunications. The flexible and open nature of IoT systems brings about several security challenges, including unauthorized data access, remote control of smart devices, and misuse of personal information by external parties [20]. To address these concerns, it is recommended to adopt a user-managed, privacy-centric access control protocol, complemented by advanced privacy protection strategies.

4.8 Energy Consumption

Improving sensor efficiency is crucial for systems relying on live data collection from smart objects. This includes managing energy consumption to extend operational life and prevent communication disruptions. S. Chen et al. propose energy conservation methods such as using high-capacity batteries, implementing energy-efficient protocols, and maintaining access points. P. Kuila et al. explore ZigBee and WiMax radios in wireless sensor networks for energy savings. Additionally, selecting gateways with the lowest link cost relative to their distance from the Internet and remaining sensor energy is essential. Combining cloud and IoT technologies can further optimize energy usage and management.

4.9 Benefits of IoMT

Implementing IoT in healthcare reduces the need for physicians to travel from remote locations and alleviates the burden on hospitals by decreasing the number of patients requiring in-hospital care, thus freeing up resources for emergency rooms and post-surgery monitoring.

4.10 Improved Patient Outcomes

Access to a comprehensive knowledge base and live health data enables physicians to detect unusual diagnoses early, reducing patient mortality risk and supporting prompt, informed decisions. Many patients, especially the elderly and disabled, prefer receiving routine care at home for added comfort, while doctors and nurses can efficiently oversee patient care from their offices.

4.11 Advancing Medical Research

IoMT facilitates consultations between clinicians who are geographically distant by improving communication and collaboration among medical staff. Additionally, it enhances the management of information among specialists.

Table 2 Literature review table

Technology	Application	Benefits	References
RFID	Supply chain, logistics, healthcare	Real-time tracking, cost reduction, process automation	[11]
RFID in healthcare	Patient and equipment monitoring, emergency services	Reduced health monitoring cost, improved remote patient care	[11]
WSNs	Environmental, military, industrial monitoring	Automated data collection, real-time response	[11, 13],
WSNs in healthcare	Patient monitoring, ECG, and pressure data transmission	Minimized manual intervention, enhanced remote health management	[11, 14]
Cloud computing	IoT integration, healthcare	Scalability, precision, resource availability	[13, 14]
IoMT	Home healthcare, telemedicines, ICUs	Real-time monitoring, reduced cost, personalized care	[4, 15]
ICU monitoring	ICU patient monitoring	Enhanced real-time supervision, reduced hospital cost	[7]
mHealth	Remote patient monitoring	Scalable storage, real-time analytics, user friendly	[15, 16]
YOAPY protocol for mHealth	Secure and efficient data transmission	Enhanced privacy, minimized delays, robust connectivity	[22]

4.12 Healthcare Facilities with a Competitive Advantage

The rapid advancement of IoT and other ICT technologies motivates medical centers to adopt these innovations to maintain their competitive edge. Utilizing these technologies helps facilities leverage their benefits and distinguish themselves from their competitors. Table 2 illustrates the different IoT and ICT technologies, their applications, and benefits.

5 Real-Time Implementation of IoT-Enabled Smart Pill Box

Forgetfulness is a major barrier to medication adherence, crucial for maintaining health and preventing disease. Many struggles with prescribed regimens due to memory lapses or lack of caregiver supervision. An IoT solution, involving a sensor-enabled pill box connected via Wi-Fi, can address this issue by monitoring adherence and sending reminders to users and caregivers [21]. Similar in design to traditional pill boxes, the system includes a mobile app that allows users to schedule medication times and activate reminders. As mentioned in Figure 3, the sensors monitor

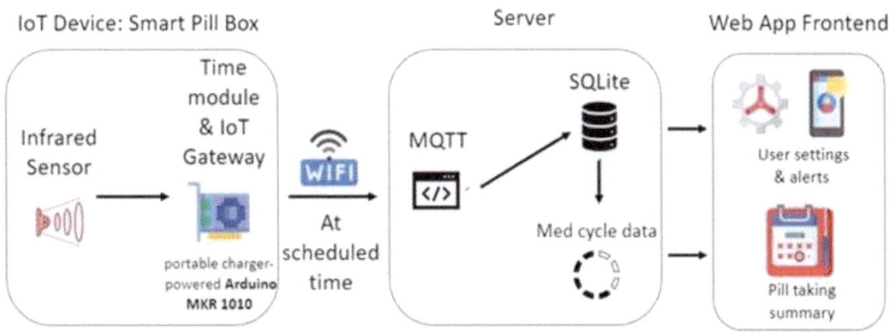

Fig. 3 IoT-enabled smart pill box

pill presence, sending reminders every 10 minutes if doses are missed, and notifying users to restock if pills remain in the box by day's end.

The smart pill box acts as a publisher, with mobile phone users and caregivers as subscribers. Its sensor monitors pill presence at regular intervals and sends data to the IoT gateway, which transmits the information to a backend system for recording in a server-based database. Users receive notifications if medication is detected late or missing, can pre-schedule alerts, and configure accounts for notifications. The backend also conducts basic data analysis to track and calculate medication cycles [21].

6 Conclusion and Future Enhancement

To effectively implement IoT technology in fields like healthcare, it's crucial to create innovative business models that outline the company's vision, goals, and SWOT analysis, addressing both functional and nonfunctional requirements. Nonfunctional aspects—such as security, reliability, accessibility, and usability—along with quality of service and power efficiency are vital for comprehensive system evaluation. Ignoring these can lead to suboptimal systems that fail to meet user needs, particularly in healthcare where security and privacy are paramount.

Many healthcare applications neglect these standards, impacting patient welfare and usability, especially for older or disabled users. Developing an IoMT system requires careful consideration of security and storage capacity due to the vast data collected; integrating cloud computing is recommended for addressing storage issues. Effective IoMT implementation can enhance patient outcomes by providing real-time insights for healthcare providers, enabling better informed decisions and improving access to care, especially in rural areas.

References

1. Relevant Software (2023) IoT in healthcare, relevant software blog, https://relevant.software/blog/iot-in-healthcare, Last accessed 2024/12/01
2. S&P Global Market Intelligence (2023) Medical IoT technology in US hospitals helps to reduce costs and improve care. S&P Global Blog. https://www.spglobal.com/marketintelligence/en/news-insights/blog/medical-iot-technology-in-us-hospitals-helps-to-reduce-costsand-improve-care, Last accessed 2024/12/01
3. Al-Majeed SS, Al-Mejibli IS, Karam J (2015) Home telehealth by internet of things (IoT). In: Proc. 28th canadian conference on electrical and computer engineering, pp 609–613
4. Yang L, Ge Y, Li W, Rao W, Shen W (2014) A home mobile healthcare system for wheelchair users. In: Proc. IEEE 18th international conference on computer supported cooperative work in design (CSCWD), pp 609–614
5. Gubbi J, Buyya R, Marusic S, Palaniswami M (2013) Internet of Things (IoT): A vision, architectural elements, and future directions. Futur Gener Comput Syst 29(7):1645–1660
6. Chiuchisan I, Dimian M (2015) Internet of Things for e Health: An approach to medical applications. In: Proc. international workshop computational intelligence for multimedia understanding, pp 1–5
7. Chiuchisan I, Costin HN, Geman O (2014) Adopting the internet of things technologies in health care systems. In: Proc. international conference and exposition on electrical and power engineering (EPE), pp 532–535
8. Kudale A, Kumar B (2014) Review—a study on authentication and access control for cloud computing. Int J Res Comput Sci & Manag 1(2), pp 13–15
9. Kudale A et al (2019) IoT application for smart agriculture. in: national conference on recent trends and advances in the field of information technology, pp 21–26. Karpagam Academy of Higher Education
10. Kudale A, Hemalatha M (2019) Emerging trends in cloud computing. Karpagam J Comput Sci 14(1):1–3
11. Farooq MU, Waseem M, Mazhar S, Khairi A, Kamal T (2014) A review on internet of things (IoT). Int J Comput Appl 13(1):241–257
12. Tendulkar B, Singh C, Kudale A (2023) Analysis of prediction performance of machine learning techniques on diabetes data set. In: Joshi A, Mahmud M, Ragel RG (eds). Inf Commun Technol Compet Strat (ICTCS 2022). ICTCS 2022. Lecture Notes in Networks and Systems, 623. Springer, Singapore. https://doi.org/10.1007/978-981-19-9638-2_12
13. Rao BP, Saluia P, Sharma N, Mittal A, Sharma SV (2012) Cloud computing for Internet of Things & sensing based applications. In: Proc. IEEE 6th international conference in sensing technology (ICST), pp 374–380
14. Botta A, De-Donato W, Persico V, Pescapé A (2015) Integration of cloud computing and internet of things: a survey. Futur Gener Comput Syst 56:684700
15. Spanakis EG, Kafetzopoulos D, Yang P, Marias K, Deng Z, Tsiknakis M, Dong F (2014) MyHealthAvatar: personalized and empowerment health services through Internet of Things technologies. In: Proc. IEEE 4th international conference on wireless mobile communication and healthcare (Mobihealth), pp 331–334
16. Baek HJ, Chung GS, Kim KK, Park KS (2012) A smart health monitoring chair for nonintrusive measurement of biological signals. IEEE Trans Inf Technol Biomed 16(1):150–158
17. Morak J, Kumpusch H, Hayn D, Modre-Osprian R, Schreier G (2012) Design and evaluation of a telemonitoring concept based on NFC-enabled mobile phones and sensor devices. IEEE Trans Inf Technol Biomed, 16(1), pp 17–23
18. Chen S, Xu H, Liu D, Hu B, Wang H (2014) A vision of IoT: Applications, challenges, and opportunities with china perspective. IEEE Internet Things J 1(4):349–359
19. Skouby KE, Lynggaard P (2014) Smart home and smart city solutions enabled by 5G, IoT, AAI and CoT services. In: Proc. IEEE international conference on contemporary computing and informatics, pp 874–878

20. Chang MJ, Jung JK, Park MW, Chung TM (2015) Strategy to reinforce security in telemedicine services. In: Proc. IEEE 17th international conference on advanced communication technology, pp 170–175
21. The Smart Pill Box—Taking medications the smart way | by yuchen yao | Medium
22. Antonio J, Miguel A, Izquierdo Z, Antonio F, Skarmeta F (2013) Interconnection framework for mHealth and remote monitoring based on the Internet of Things. IEEE J Sel Areas Commun 31(9):47–65

Improved Vulnerability Detection in Applications Using Machine Learning

Sunil Mane, Nehal S. Randive, Milind Godase, and Bhagwat Kaulwar

Abstract In recent years, the surge in complexity of software systems has underscored the critical importance of detecting and mitigating security vulnerabilities at the source code level. Function-level vulnerabilities, characterized by flaws within specific functions or methods in software code, present significant risks to system security. Conventional approaches to vulnerability detection often rely on manual inspection or static analysis techniques, which are labor intensive and prone to human error. This research project presents an open-source framework based on supervised machine learning techniques for detecting vulnerabilities in source. The framework is motivated by current developments in deep learning methodologies. A distinctive feature of this framework is its utilization of both contextual and non-contextual embedding methods for generating code representations. This duality enables the framework to discern polysemy in code contexts and enhance vulnerability detection accuracy. The project includes a meticulously curated dataset comprising vulnerable functions sourced from nine prominent open-source software projects, facilitating rigorous evaluation and benchmarking. The framework's usage is facilitated by a straightforward set of instructions. By leveraging the provided Python scripts, users can effortlessly train and test various neural network models with customizable settings, seamlessly integrating with different embedding methods. Furthermore, the framework supports the extraction of high-level representations from trained models, facilitating downstream tasks such as vulnerability classification using alternative classifiers like random forests. This research project not only advances the state of the art in function-level vulnerability detection but also provides a practical and accessible tool for software developers and security practitioners.

S. Mane · N. S. Randive
Department of Computer Science and Engineering, COEP Technological University, Pune, India
e-mail: sunilbmane.comp@coeptech.ac.in

M. Godase · B. Kaulwar (✉)
Sinhgad Institute of Management, Pune, India
e-mail: bhagwat_kaulwar@yahoo.com

M. Godase
e-mail: milindgodase@sinhgad.edu

C. E. R. Hewage et al. (eds.), *AI Applications in Cyber Security and Privacy of Communication Networks*, Lecture Notes in Networks and Systems 1453,
https://doi.org/10.1007/978-981-96-7400-8_7

Keywords Software security vulnerability · Vulnerability detection · Neural network · Bi-LSTM · Machine learning

1 Introduction

Pervasive integration of software systems in nearly each part of modern society has brought forth unprecedented convenience and connectivity. However, this ubiquity also exposes these systems to a myriad of security threats, ranging from data breaches to malicious attacks. Among these threats, vulnerabilities at the function level of software code represent a particularly insidious risk, potentially enabling attackers to exploit weaknesses within individual functions or methods to compromise system integrity and confidentiality.

Traditional approaches to detecting function-level vulnerabilities often rely on manual code inspection or static analysis tools, which are inherently labor intensive, error prone, and limited in scalability. As software systems become increasingly intricate and expansive, there arises a pressing demand for automated and scalable methodologies to adeptly identify and mitigate vulnerabilities within them. In recent years, machine learning, and particularly deep learning, has emerged as a powerful paradigm for automating various aspects of software engineering, including vulnerability detection. Deep learning models offer a promising avenue for enhancing the accuracy and efficiency of function-level vulnerability detection by leveraging their capacity to discern intricate patterns and representations from raw data.

1.1 Objective

This study seeks to design and assess a cutting-edge framework for identifying function-level vulnerabilities using supervised machine learning, focusing specifically on advanced deep learning approaches. This framework seeks to provide developers and security practitioners with a robust and accessible tool for identifying and mitigating function-level vulnerabilities in software code, which is built on the foundational base set by recent work in machine learning.

To achieve this objective, we draw inspiration from two seminar papers in this field: "Deep Learning-Based Vulnerable Function Detection: A Benchmark" and "A Context-aware Neural Embedding for Function-level Vulnerability Detection". Key components of our framework include a diverse set of neural network models, encompassing both mainstream architectures and novel approaches tailored specifically for vulnerability detection. In conjunction with the development of our framework, we have meticulously curated a dataset comprising vulnerable functions sourced from a variety of open-source software projects. This dataset works as a benchmark for assessing the performance of our framework and facilitating rigorous experimentation and analysis.

In the subsequent sections of this paper, we provide detailed insights into the design and implementation of our framework, including the methodology for dataset collection, model architecture, training procedures, and evaluation metrics. Furthermore, we present experimental results demonstrating the effectiveness and scalability of our approach, along with discussions on practical implications and future directions for research in function-level vulnerability detection.

2 Literature Review

2.1 *Variety of Machine Learning Algorithms to Detect Software Vulnerabilities*

Detection of software vulnerabilities is crucial for ensuring the security of systems, particularly during the development phase, to mitigate potential risks and financial losses. While traditional detection methods have been widely employed, there is a growing need to incorporate machine learning, including supervised and deep learning techniques, to address emerging vulnerabilities effectively [1]. However, existing studies often overlook this trend and fail to deliberate the evolving research interests in vulnerability detection such as the study demonstrated by Khan et al. [2], where the security assessments of open-source systems underscore the importance of holistic vulnerability detection frameworks. For bridging this gap, we present a comprehensive taxonomy of research interests in software vulnerability detection, highlighting methods, detection mechanisms, and dataset challenges. The analysis reveals a significant focus on refining detection methods, with less attention given to code and dataset issues, indicating areas for future exploration. Recent work by Ladisa et al. [3] further emphasizes the need for robust taxonomies to address evolving attack vectors in open-source software supply chains. Moreover, the study extends the taxonomy to include various machine learning approaches, such as supervised and deep learning, which demonstrate promising capabilities in identifying vulnerabilities like buffer overflow and SQL injection. The paper concludes by outlining potential avenues for future research, including the development of robust datasets and the exploration of multi-vulnerability detection techniques.

2.2 *Toward Large-Scale Vulnerability Discovery Using Machine Learning*

With the sustained growth of software complexity, the necessity of identifying security vulnerabilities in OSs has become increasingly apparent [4]. Modern OSs are distributed with thousands of binary executables, yet methodologies and tools for OS-scale program testing within a time budget remain elusive. This paper considers an approach utilizing lightweight static and dynamic features to predict if a test case likely contains a software vulnerability using machine learning techniques is presented. An extensive experiment was conducted to detect easily exploitable memory corruptions using 1039 Debian programs obtained from its bug tracker, collecting 138,308 unique execution traces and statically exploring 76,083 different subsequences of function calls. Reasonable accuracy was achieved in predicting which programs contained dangerous memory corruptions. Additionally, the authors developed and implemented "Vdiscover", a tool that utilizes state-of-the-art machine learning techniques to predict vulnerabilities in test cases. This tool will be released as open source to encourage research in vulnerability discovery at a large scale, together with "Vdiscovery", a public dataset that collects raw analyzed data.

2.3 Automated Software Vulnerability Discovery with Machine Learning

Thousands of security vulnerabilities are discovered in product software each time, either reported intimately to the Common Vulnerabilities and Exposures database or discovered internally in personal law [5]. Vulnerabilities frequently manifest subtly, escaping discovery by law pundits or inventors themselves. Using the wealth of open-source law available for analysis, there exists an occasion to learn the patterns of bugs that can lead to security vulnerabilities directly from data. In this paper, a data-driven approach to vulnerability discovery using machine literacy, specifically applied to C and C programs, is presented. A large dataset comprising hundreds of thousands of open-source functions labeled with the labors of a static analyzer is collected. Styles applied directly to source law are compared with those applied to vestiges uprooted from the figure process, with source-grounded models performing better. Deep neural network models are compared with more traditional models similar as arbitrary timbers, with the stylish performance achieved by combining features learned by deep models with tree-grounded models. The loftiest performing model achieves an area under the perfection–recall wind of 0.49 and an area under the ROC wind of 0.87.

2.4 A Survey of Machine Learning for Big Code and Naturalness

Probabilistic models of source law have the prospect to backing new tools in nearly every area of software engineering and program analysis [6]. This paper reviews being effort in the field, presenting a classification of probabilistic machine learning source law models and their operations. Utmost of the exploration contained in this review was led within the once many times, signifying growing attention in this area among the machine literacy, programmers, and software engineers' groups. Probabilistic models of source law nurture the instigative occasion of learning from being law, probabilistically logic about new source law vestiges, and transferring knowledge between inventors and systems.

2.5 End-To-End Vaticination of Buffer Overruns from Raw Source Law via Neural Memory Networks

This work proposes a memory network-grounded model for prognosticating buffer overruns in programming language analysis [7]. It's the first attempt to apply a deep literacy grounded approach to a problem in program analysis taking both syntactic and semantic knowledge. Performance results show that memory networks are superior to other models in working buffer overrun tasks across all difficulty situations. The model successfully learns the notion of figures and their quantitative comparisons from textual data in an end-to-end setting. Supply chain attacks, as analyzed by Ohm et al. [8], highlight the urgency of securing dependencies in software ecosystems. There's room for enhancement in colorful aspects, similar as expanding the model to cover different program analysis tasks and applying advanced variants of memory networks to handle colorful conditions in source law.

3 Methodology

3.1 Data Preprocessing

The dataset used in this exploration consists of vulnerable and non-vulnerable functions collected from nine open-source systems written in the C programming language. The source law functions are warehoused in a Zip train holding the systems used in the partner experimentations. The CVE ID is used to name each vulnerable function, while non-vulnerable functions are termed with filename format like <functionname>.c, where the function name will be the name of corresponding non-vulnerable function. Also, the Vulnerabilitiesinfo.xlsx train provides detailed information on the collected function-position vulnerabilities. These vulnerabilities are sourced from the National Vulnerability Database (NVD) till end of 2018.

3.2 Implementation Method

The implementation involves several Python scripts for analyzing the C scripts. Parallel processing techniques, such as those explored by Swapnil et al. [9], could

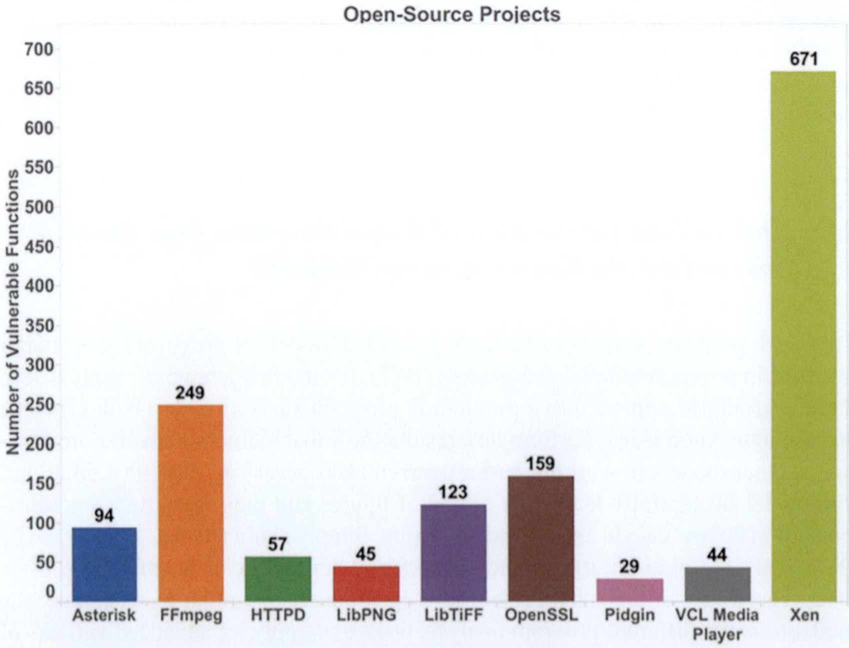

Fig. 1 For each of the sources the number of labeled examples of vulnerable functions in the datasets [10–12]

enhance scalability for large-scale code analysis, thereby enhancing the process of C language scripts. Although maximum utilization of parallel processing methods serves their own advantages, they also add complexities to certain extent, hence we stick to the current proposed implementation. These scripts are organized into the following components:

- main.py: This script handles training and testing of a stated network model. Users have a flexibility to apply different embedding methods and also switch between training and testing mode by providing different options or parameters.
- Obtain representations.py: This script is used for gaining elevated representations from a trained network model. Users can obtain representations from any layer of the model for various downstream tasks. The distribution of vulnerable examples across datasets is visualized in Fig. 1.

3.3 Workflow

The machine learning methodology involves training and testing neural network models using TensorFlow and Keras. The process is as follows:

Step (a) Train the Neural Network Model: A YAML configuration file is used to hold the parameters related to experiment/model settings. This gives flexibility to users to suitably adjust the settings by just altering the configuration file. The neural network can be trained once the configuration file is ready. Exploitable vulnerabilities within software systems pose significant security risks. Over time, there have been advancements in machine learning (ML) techniques aimed at automating and expediting vulnerability detection.

Many ML approaches focus on identifying vulnerable units of source code, such as lines or functions. However, we assert that a code segment's vulnerability is contingent upon its semantic context, including aspects like control flow and data flow. Effective detection relies heavily on contextual understanding. As shown in Fig. 2, the dataset exhibits a balanced distribution of vulnerability classes.

This research proposes a supervised approach leveraging pre-trained context-sensitive embedding (ELMo) to capture subtle contextual nuances. A bidirectional long short-term memory (Bi-LSTM) layer further refines these representations, allowing the model to identify complex code relationships across extended sequences.

This framework converts source code functions into compact embedding, facilitating effective feature extraction for machine learning classification. Our methodology exhibits improved detection accuracy over existing techniques. Furthermore,

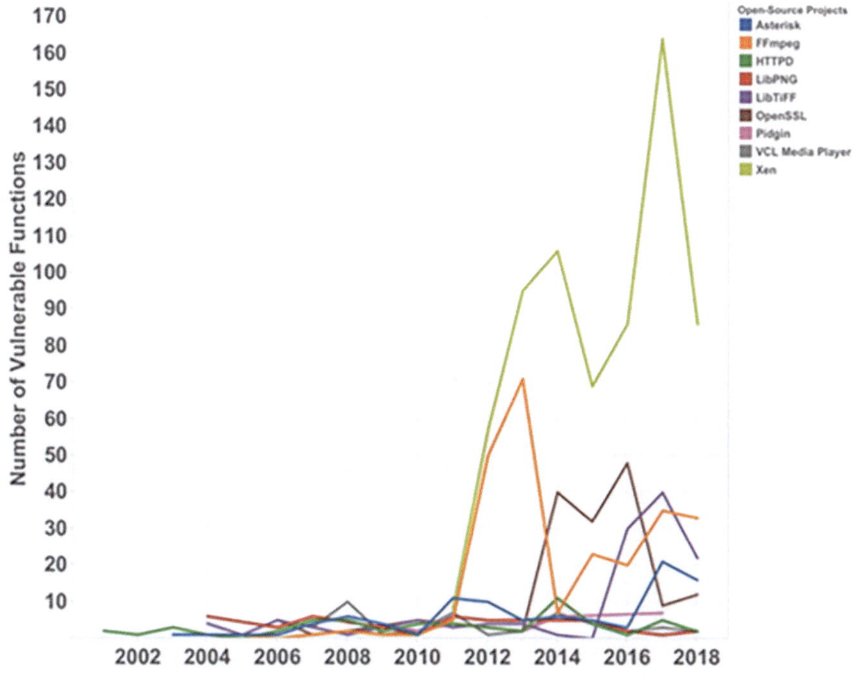

Fig. 2 The distribution of labeled vulnerable examples in the taken dataset

integrating our framework's feature representations enabled random forest and support vector machine models to achieve state-of-the-art performance, surpassing four baseline systems in our empirical evaluation.

3.3.1 Proposed Framework

The following section outlines the workflow of our proposed framework, detailing how it identifies and extracts potentially vulnerable code contexts within function boundaries.

Our proposed framework is a deep neural network tailored for function-level embedding extraction. It leverages labeled vulnerable data from software projects to identify vulnerable functions. The work flow consists of three stages:

Stage 1: Network Training—We train the network using labeled source code functions (vulnerable/non-vulnerable). This fine-tunes the built-in ELMo module on software code data.

Stage 2: Embedding Extraction—Labeled and unlabeled functions are fed into the trained network. We extract the output of one hidden layer (specifically, the third-last layer) as learned embedding. These embedding serve as distributed representations of input functions, reflecting code semantics and syntax indicative of vulnerable patterns.

Stage 3: Classification—Extracted function embedding is used as feature vectors for machine learning classifiers, enabling further classification.

These three stages are depicted in Fig. 3.

This scenario involves a three-stage process: firstly, training the neural network with labeled data to optimize its parameters. Secondly, feeding both labeled and unlabeled data through the trained network to extract outputs from a specific hidden layer, yielding function-level embedding, and finally leveraging the extracted embedding as feature sets for subsequent machine learning tasks, such as training a conventional classifier for enhanced classification capabilities.

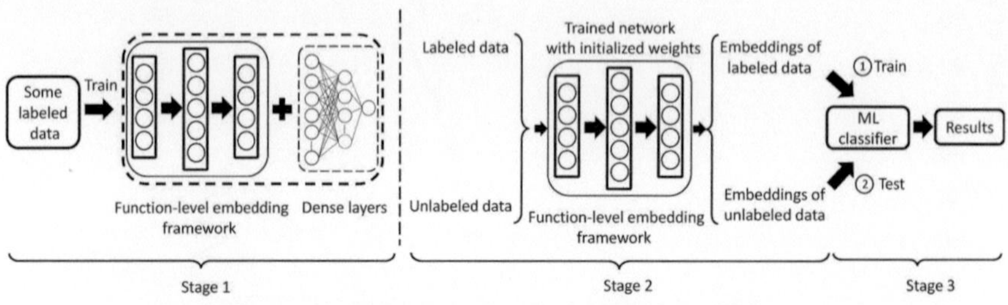

Fig. 3 The proposed framework is tailored for scenarios where labeled data for a software project are available

3.3.2 Code Context Analysis

To identify vulnerable programming patterns leading to security vulnerabilities, analyzing code context is essential. A common vulnerability arises when data flows from a legitimate source to a susceptible sink without proper validation (e.g., buffer size checks).

Effective detection requires a two-step approach: first, identify potentially vulnerable contexts, comprising at least a source and sink, and second verify the absence of adequate validation between the source and sink, ensuring data integrity before transmission.

3.3.3 Contextual Semantic Learning

Identifying potentially vulnerable contexts relies on learning precise code semantics that capture contextual meaning. Consider the example of adding an if statement before the sensitive sink print to validate the size of testStr. This modification remedies the vulnerability by preventing the sink's execution if testStr exceeds the array size.

The vulnerability of print depends on its preceding code context, specifically the presence of validation statements. Therefore, the same token print should be represented differently in distinct contexts (with or without parameter validation). Recognizing these differences is crucial for detecting vulnerable patterns.

Step 2: Testing: After training, users can test the network model using a separate test set via a specific command. Alternatively, users can employ specific test set by setting parameter in the config.yaml file, namely, using_separate_test_set to True.

Step 3: Obtaining High-Level Representations: Upon training completion, users can extract elevated representations from any layer of the skilled network model. These representations serve as features for various tasks, such as clustering or classification.

4 Results

4.1 Performance Evaluation

Figure 4 shows the accuracy of our proposed framework on the test dataset. As illustrated, our framework achieves a high accuracy rate of 95%, indicating its effectiveness in detecting vulnerabilities.

Figure 5 displays the precision and recall values of our framework across different vulnerability classes. The precision–recall curve demonstrates the trade-off between

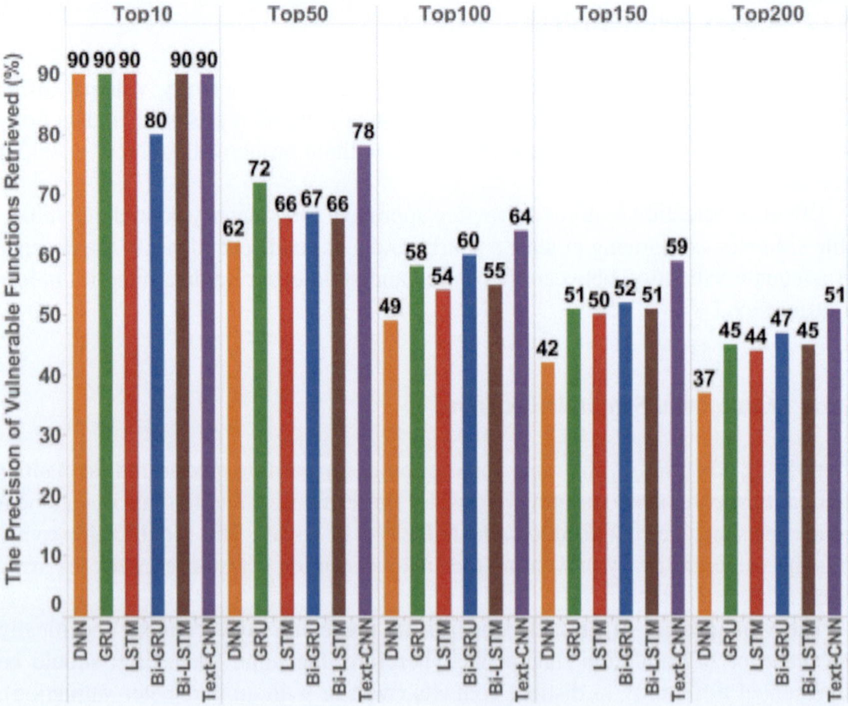

Fig. 4 Accuracy of the proposed framework on the test dataset

precision and recall, with our framework achieving a balance between the two metrics.

Overall, the results indicate that our framework performs well in detecting vulnerabilities across various scenarios, achieving high accuracy and maintaining a balance between precision and recall.

5 Future Scope

In the future, it is essential to explore advanced techniques for generating contextual embedding from language models. This includes investigating newer pre-trained models or developing custom embedding methods tailored specifically for source code analysis. Additionally, there is a need to integrate additional data sources, such as code comments, API usage patterns, or developer-authored documentation, to enrich the contextual information available for vulnerability detection.

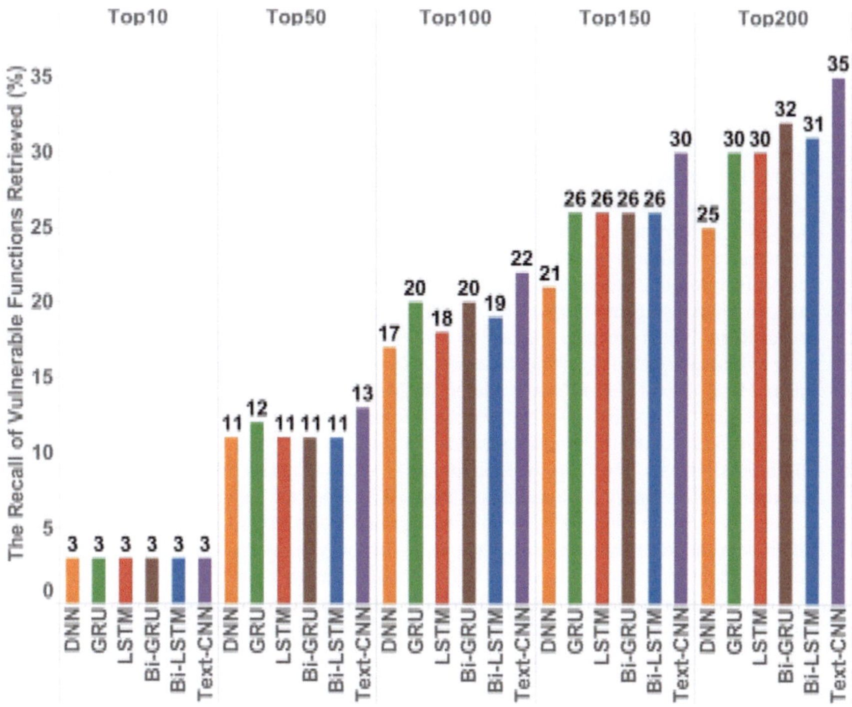

Fig. 5 Precision–recall curve of the proposed framework across different vulnerability classes

Furthermore, extending the research to analyze vulnerabilities in code written in languages other than C could provide valuable insights. Investigating the effectiveness of the proposed framework for detecting vulnerabilities in languages like Java, Python, or JavaScript would be beneficial.

Moreover, exploring the feasibility of deploying the vulnerability detection framework in real-time development environments is crucial. Additionally, insights from adversarial domains like social media malicious profile detection [13] could also inform vulnerability detection strategies. Techniques for integrating the framework into continuous integration/continuous deployment (CI/CD) pipelines for automated vulnerability scanning should be investigated. Moreover, hybrid approaches combining clustering and gradient boosting could further optimize model performance, as demonstrated by Verma et al. [14]. Opinion mining techniques could be explored to analyze developer comments and documentation, providing additional contextual cues for vulnerability detection [15].

Developing techniques for visualizing and understanding the model's decision-making process would enable developers to trust and interpret the detected vulnerabilities more effectively.

References

1. Gustavo Grieco, Guillermo Luis Grinblat, Lucas Uzal, Sanjay Rawat, Josselin Feist, Laurent Mounier (2016) Toward large-scale vulnerability discovery using machine learning. In: Proceedings of the Sixth ACM conference on data and application security and privacy, CODASPY '16, Assoc Comput Mach, page 85–96, New York, NY, USA
2. Idris Khan, Faraz Javed, Yasir Alenezi, Mamdouh (2019). Security assessment of four open source software systems. Indones J Electr Eng Comput Sci. 16. 860–881. https://doi.org/10.11591/ijeecs.v16.i2.pp860-881
3. Ladisa, Piergiorgio Plate, Henrik Martinez, Matias Barais, Olivier (2022) Taxonomy of attacks on Open-Source software supply Chains. https://doi.org/10.48550/arXiv.2204.04008
4. Komal Rege, Nikita Goenka, Pooja Bhutada, Sunil Mane (2013) Bluetooth communication using hybrid encryption algorithm based on AES and RSA. Int J Comput Appl
5. Hazim Hanif, Mohd Hairul Nizam Md Nasir, Mohd Faizal Ab Razak, Ahmad Firdaus, Nor Badrul Anuar (2021) The rise of software vulnerability: Taxonomy of software vulnerabilities detection and machine learning approaches. J Netw Comput Appl, 179:103009
6. Miltiadis Allamanis, Earl T Barr, Premkumar Devanbu, Charles Sutton (2018) A survey of machine learning for big code and naturalness. ACM Comput. Surv., 51(4)
7. Min-Je Choi, Sehun Jeong, Hakjoo Oh, and Jaegul Choo (2017) End-to-end predic- tion of buffer overruns from raw source code via neural memory networks. CoRR, abs/1703.02458
8. Ohm M, Plate H, Sykosch A, Meier M (2020) Backstabber's Knife collection: a review of open-source software supply chain attacks. In: Maurice C, Bilge L, Stringhini G, Neves N (eds). Detection of intrusions and malware, and vulnerability assessment. DIMVA 2020. Lect Notes Comput Sci, vol 12223. Springer, Cham. https://doi.org/10.1007/978-3-030-52683-2_2
9. Swapnil P Dravyakar, Sunil B Mane, Pradeep K Sinha (2014) Private content based multimedia information retrieval using map-reduce. Int J Sci, Eng Comput Technol
10. Shital A Hande, Sunil B Mane (2015) An analysis on data Accountability and Security in cloud. In: 2015 international conference on industrial instrumentation and control (ICIC). CollE Eng Pune, India. 28–30
11. Jacob A Harer, Louis Y Kim, Rebecca L Russell, Onur Ozdemir, Leonard R Kosta, Akshay Rangamani, Lei H Hamilton, Gabriel I Centeno, Jonathan R Key, Paul M Ellingwood, Marc W McConley, Jeffrey M Opper, Peter Chin, Tomo Lazovich (2018) Automated software vulnerability detection with machine learning. CoRR, abs/1803.04497
12. Aniket Bhadane, Sunil B Mane (2017) State of research on phishing and recent trends of attacks, I-Manager's J Comput Sci
13. Shruti Shinde, Sunil B Mane (2021) Malicious profile detection on social media: A survey paper. In: 2021 9th international conference on reliability, infocom technologies and optimization (Trends and Future Directions) (ICRITO)
14. Parag Verma, Shayan Anwar, Shadab Khan, Sunil B Mane (2018) Network intrusion detection using clustering and gradient boosting. In: 2018 9th International conference on computing, communication and networking technologies (ICCCNT)
15. Sunil B Mane, Kruti Assar, Priyanka Sawant, Monika Shinde (2017) Product rating using opinion mining. Int J Comput Eng Res Trends

CyberHER: Addressing Gender Inequality Through Digital Skills and Cultural Change in Cybersecurity

Shreya Macherla, Mohammad Haseeb Zafar, Sumaira Johar, and Fiona Carroll

Abstract This paper explores the ongoing issue of gender inequality, particularly in women's access to education, jobs, and digital skills in fields like cybersecurity. Findings from the CyberHER project, funded by the British Council and led by Cardiff Metropolitan University and the Institute of Management Sciences in Pakistan, highlight challenges such as societal norms, limited digital literacy, and online harassment that affect women's participation in technology. In many South Asian and Arab regions, policies to improve gender equality often lack strong local support and are mainly driven by outside pressure, making them less effective. Survey data and case studies from these areas show that true progress requires more than just policy changes; it needs a shift in cultural attitudes and local involvement. Projects like CyberHER aim to bridge the gender gap in cybersecurity by offering training, awareness programs, and collaboration opportunities, showing how community-led efforts can empower women in digital spaces. The study emphasizes that sustainable gender equality needs both policy backing and changes in social values to make lasting improvements.

Keywords Gender inequality · Cybersecurity · Digital skills · Societal norms · Empowerment

1 Introduction

Gender disparities in access to education, health, and resources persist globally, often beginning in early life, as boys tend to receive more family investment. This preference restricts opportunities for girls, particularly in education, leading to lower literacy rates and limiting socio-economic growth. Studies show that when girls

S. Macherla · M. H. Zafar (✉) · F. Carroll
Cardiff School of Technologies, Cardiff Metropolitan University, Cardiff, UK
e-mail: mhzafar@cardiffmet.ac.uk

S. Johar
Institute of Management Sciences, Peshawar, Pakistan

C. E. R. Hewage et al. (eds.), *AI Applications in Cyber Security and Privacy of Communication Networks*, Lecture Notes in Networks and Systems 1453,
https://doi.org/10.1007/978-981-96-7400-8_8

receive equal educational access, it promotes better health outcomes, economic advancement, and stronger community welfare [1].

This research, funded by the British Council, investigates gender equality initiatives across Brazil, Egypt, Indonesia, Jordan, Kenya, Myanmar, Nigeria, Pakistan, South Africa, Ukraine, and Vietnam. Economic and social welfare goals, initially outlined by Cardiff Metropolitan University (CMet) [2] in collaboration with the Institute of Management Sciences (IMSciences) [3], underpin these initiatives. Notably, the "CyberHER: Bridging the Gender Gap in Cybersecurity through Inclusive Pathways" project addresses gender inequality in cybersecurity, aiming to empower women in a field often dominated by men. In Pakistan, cultural norms, limited access to technology education, and few female role models exacerbate this gender imbalance.

In some Muslim-majority countries, such as Bangladesh and Pakistan, while Islamic teachings advocate for women's dignity and rights, traditional practices sometimes contradict these values, particularly in regions where tribal customs override religious teachings. Understanding these restrictions requires examining societal and political influences that shape gender roles. Additionally, some women inadvertently reinforce these limitations through specific social or political actions [4].

Expanding beyond Islamic contexts, this study compares educational access, employment, political participation, and digital literacy across developed and developing countries. Low female literacy is a key factor that limits women's political, economic, and social engagement and contributes to domestic violence.

In the digital space, this study explores cybersecurity, data privacy, and online safety for women, highlighting issues of online harassment and the potential of digital platforms for female activism. Social media is examined as both a site of conflict and an avenue for positive change in promoting women's rights.

The paper is structured as follows: Sect. 2 reviews literature; Sect. 3 outlines methodology; Sect. 4 discusses results and analysis; and Sect. 5 concludes with future directions.

2 Literature Review

Studies have consistently explored gender equality in developing countries, highlighting challenges across multiple domains. In education, the gender gap remains significant, especially at higher levels, as early marriage and cultural norms hinder girls' access to schooling. Global goals like the Sustainable Development Goals (SDGs) and the Dakar "Education for All" targets stress the importance of equal education [5]. However, true equality requires that both genders receive the same opportunities to succeed [6, 7]. In many cases, boys are favored due to their perceived role as future breadwinners. A 2019 survey in Pakistan showed 71% believed there was gender-based discrimination in education, favoring boys over girls [8]. Despite Pakistan signing the Convention on the Elimination of All Forms of Discrimination

Against Women (CEDAW) in 1996, social barriers continue to limit girls' educational access [9]. Arab nations also show similar patterns where, although progress has been made in primary enrollment, cultural norms and discriminatory laws still limit girls' opportunities [6, 10]. In countries like Afghanistan, political issues and gender-biased laws hinder girls' education, as seen with the recent ban on women's university attendance under the Taliban [11].

In the workplace, gender inequality persists, with women facing pay disparities, limited advancement opportunities, and challenges in male-dominated fields. Cultural expectations in developing countries further restrict women's career growth, impacting their economic status and mental health [10]. Even in progressive nations like Sweden, gender pay gaps remain, indicating that deep-rooted biases hinder true economic equality [12].

The digital gender divide, defined as disparities in technology access, further limits women. In developing countries, women are less likely to own smartphones or use the Internet, restricting their participation in the digital economy [13]. This gap also exposes women to online harassment and cybercrimes, particularly in India, where increased smartphone usage has been linked to rising cyberviolence against women [14]. Although companies like Google and Facebook provide gender-specific data to better understand these divides [15], addressing cultural norms is essential for meaningful change [13].

Cultural norms present additional challenges in achieving gender equality, particularly with Gender-Based Violence (GBV). In Pakistan, deeply ingrained patriarchal values contribute to GBV, including domestic violence and honor killings [9]. Furthermore, the rise of digital spaces has introduced online violence, such as cyberstalking and harassment, with countries like India showing a growing concern for women's safety in digital contexts [14].

International frameworks like CEDAW and the SDGs are vital in promoting gender equality, but cultural and enforcement barriers often limit their impact [6]. The COVID-19 pandemic has worsened gender disparities, pushing many women out of the workforce and increasing risks of GBV, demonstrating the need for a societal shift to achieve true equality by 2030 [11].

3 Methodology

This study employs a mixed-methods approach, combining quantitative and qualitative analyses to explore gender equality across various domains. For example, the *Patterns of Gender Equality at Workplaces* study uses cluster analysis and logistic regression to examine workplace equality through gender ratios, education levels, and salaries, refined with socio-demographic adjustments to understand how equality impacts mental health [9]. A systematic review of gender equality interventions conducted using the SPIDER (Sample, Phenomena of Interest, Design, Evaluation, Research Type) tool integrates both quantitative and qualitative insights, targeting change across micro-, meso-, and macro-levels in education, politics, and employment [7].

Addressing digital gender inequality, *Monitoring Digital Gender Inequality Using Facebook and Google Advertising Audience Estimates* uses data from Google and Facebook alongside indicators such as the Human Development Index. Regression models highlight gender gaps in Internet use, showing how online and offline factors intersect to create digital divides [10]. Qualitative findings reveal societal expectations that negatively impact women, suggesting areas for improvement on digital platforms.

Key project activities included a hybrid event, *Cyber Security Awareness for Women*, held at IMSciences to enhance accessibility and raise cybersecurity awareness. The event featured influential speakers and successfully engaged a diverse audience on the importance of gender inclusivity in cybersecurity. Additionally, a team from IMSciences visited Cardiff Metropolitan University for a networking session, facilitating a rich exchange of knowledge and promoting global collaboration in cybersecurity.

An online cybersecurity awareness short course was also delivered to students from IMSciences and Women University Mardan, with 50 participants. Surveys assessed the course's impact on students' interest and knowledge, and the course concluded with a poster session that received positive feedback. These activities supported the study's quantitative data collection and qualitative insights, providing a holistic view of gender equality and digital empowerment.

Figure 1 illustrates Pakistan's workforce and education gender gaps, with men forming 78.1% of the workforce and 28.7% male versus 22.1% female participation in secondary education. A global comparison highlights countries like Yemen and Iraq with lower gender equality [16–18]. Together, these methods present a holistic perspective on the journey toward gender equality.

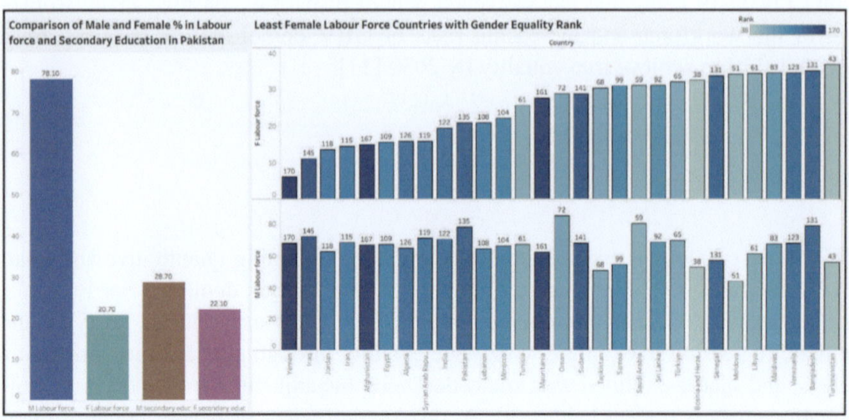

Fig. 1 Workforce and education gender gaps in Pakistan, with global comparison

4 Results and Analysis

According to data gathered from various open platforms and digital sites, there is a strong association with data on the Internet that predominantly highlights gender gaps. In many developing countries, the percentage of women using digital platforms lags behind that of men across various age groups. Studies have also documented how cultural norms impact women's digital use [19, 20].

This research applied multiple analysis approaches: exploratory data analysis from various online datasets; survey thematic analysis, which examines the results of the CyberHER survey conducted with IMSciences students prior to the short course engagement to assess their knowledge and interest in technology and cybersecurity; and K-means clustering analysis using the digital gender gap dataset. This multi-faceted approach allowed for a comprehensive comparison between qualitative and quantitative research methods.

4.1 Exploratory Data Analysis

Figure 2 illustrates the global gender disparity in Internet access, with darker shades representing countries with higher female-to-male Internet usage ratios, such as the UK, USA, and Australia. In contrast, countries like Pakistan and Yemen show lower ratios, emphasizing the digital gender gap and limited inclusivity for women in specific regions [16, 19].

Figure 3 visualizes female representation in national parliaments globally, using a color gradient to distinguish countries with high female representation, such as New Zealand, Norway, and Rwanda, from regions with low representation, such as

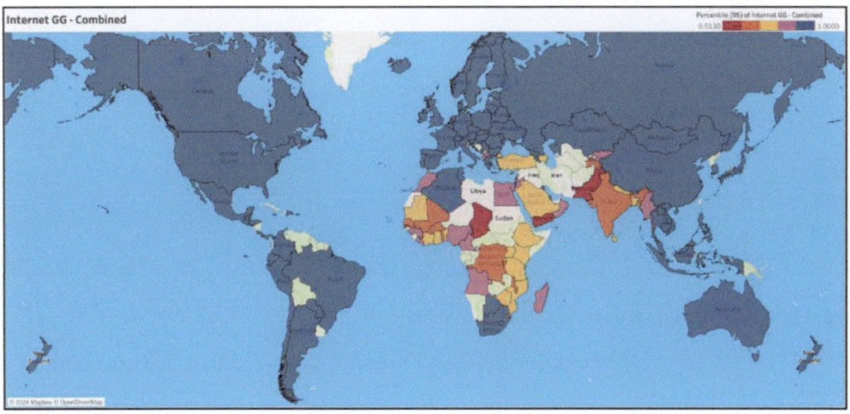

Fig. 2 Global gender disparity in Internet access

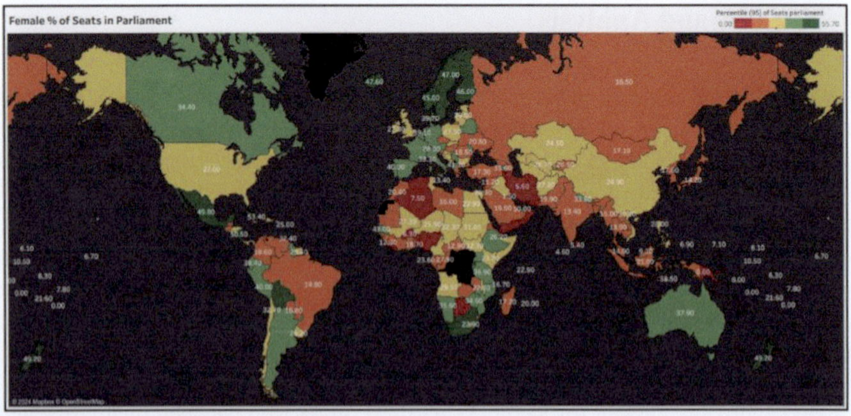

Fig. 3 Female representation in national parliaments

Yemen and Oman. This figure highlights the political gender gap, showing variations in gender inclusivity in governance worldwide [16, 17].

Figure 4 shows the link between human development and gender equality, with high-ranking regions like Scandinavia and Western Europe scoring well in both aspects. In contrast, areas like Sub-Saharan Africa and South Asia, including Afghanistan and Pakistan, display low ranks, indicating that economic development does not always guarantee progress in gender equality [16, 17].

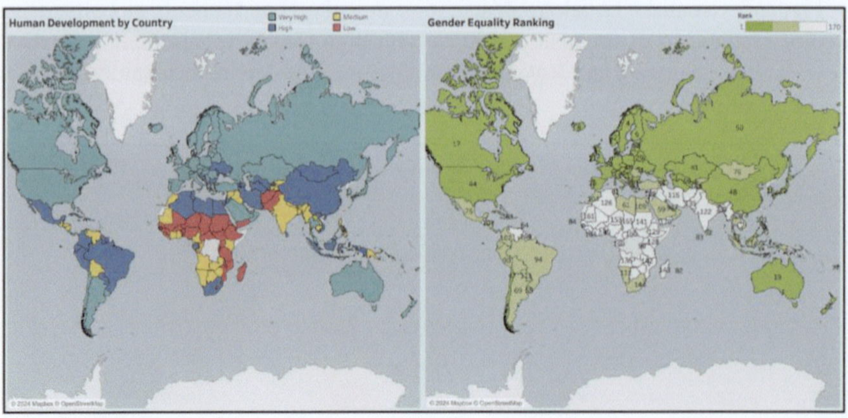

Fig. 4 Human development versus gender equality

4.2 Thematic Analysis of CyberHer GEP Survey

The CyberHer GEP Survey provides important insights into the challenges and support systems that affect women's involvement in cybersecurity. Conducted with students from IMSciences and Women University Mardan, the survey shows how societal expectations, limited access to education, workplace issues, and the need for mentorship and financial support shape women's career paths. The findings reveal a mix of barriers and possible opportunities that influence women's participation and success in cybersecurity, highlighting the need for specific, supportive actions. Below are five key findings from the survey.

Barriers in Education and Training: A major finding was the limited access to educational resources for women in underserved areas. "Insufficient access to education and training" was the most cited issue (Fig. 5), followed by "Stereotypes and social norms". This illustrates how structural and cultural factors deter women from entering cybersecurity careers, indicating the need for accessible local training programs.

Workplace Bias and Gender Discrimination: Gender bias in the workplace emerged as a significant barrier, with respondents noting discrimination in hiring, promotion, and wages. Many women felt they had to work harder than their male counterparts for the same recognition, underpinning the need for inclusive workplace policies. "Limited resources" and "time constraints" were among the top challenges (Fig. 6), which often discourage women from pursuing long-term careers in cybersecurity.

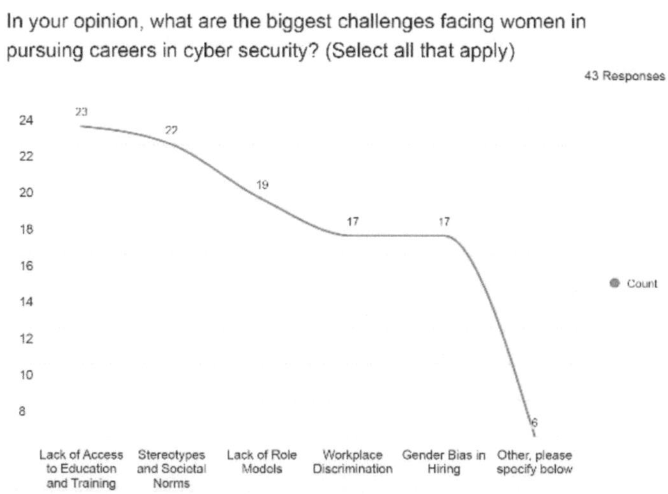

Fig. 5 Top barriers in cybersecurity: education access leads

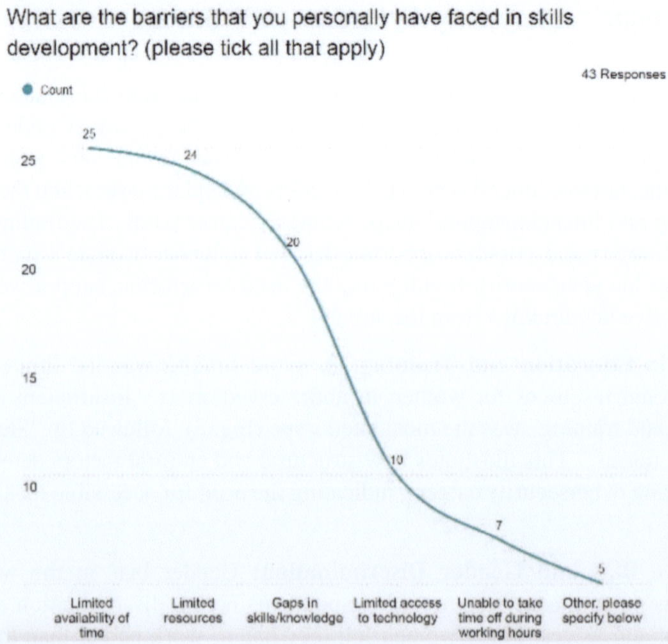

Fig. 6 Main learning obstacles: time and resources

Perception of Cybersecurity as Male-Dominated: The survey highlighted the strong perception of cybersecurity as a male-dominated field, with "male dominance" noted by the majority of respondents (Fig. 7). This perception, coupled with a lack of female role models, discourages young women from entering the field, pointing to a need for visible female leadership to break down these stereotypes.

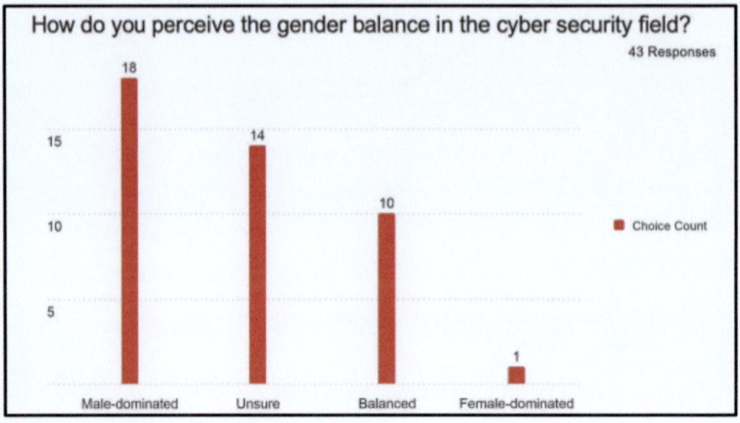

Fig. 7 Cybersecurity seen as male dominated

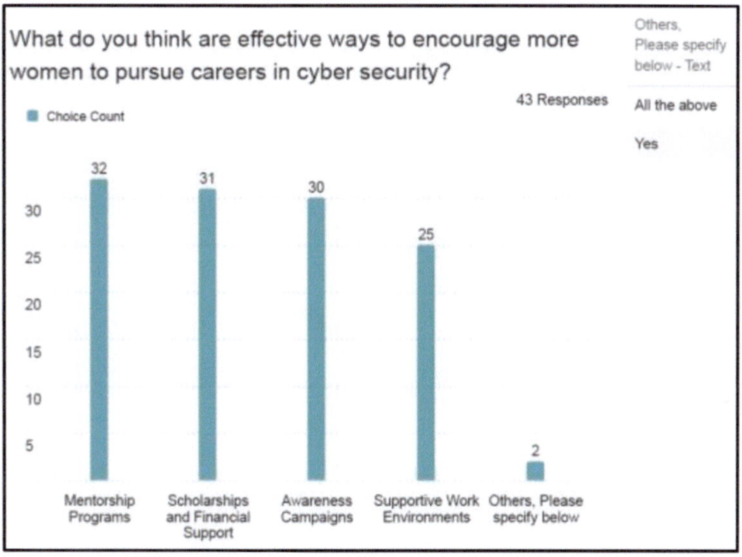

Fig. 8 Key supports: mentoring, financial aid prioritized

Supportive Interventions: Mentorship, networking, and financial support were identified as critical supports for encouraging women in cybersecurity. "Mentoring programs" and "scholarships" received high votes from respondents (Fig. 8), indicating that structured mentorship and financial assistance could significantly help women overcome career obstacles in cybersecurity.

Impact of CyberHer and Similar Programs: The CyberHer program was widely appreciated, with 53% of respondents "strongly agreeing" on its positive impact (Fig. 9). However, participants recommended more practical, hands-on training to build technical skills. This feedback suggests that while CyberHer addresses key gaps, additional technical workshops could further enhance its impact.

These findings highlight both the challenges faced by women in cybersecurity and the effective support structures that can enable their success, providing actionable insights for future initiatives aimed at closing the gender gap in this field.

4.3 Clustering Analysis

The dataset considered for this analysis is based on the digital gender gap across countries [19]. K-means clustering analysis was performed on the dataset, using the elbow method and silhouette scores as performance metrics. The elbow method indicates that the optimal number of clusters is K = 4, as inertia drops sharply up to this point and then levels off. Although Fig. 10 shows results with K = 10 clusters, the

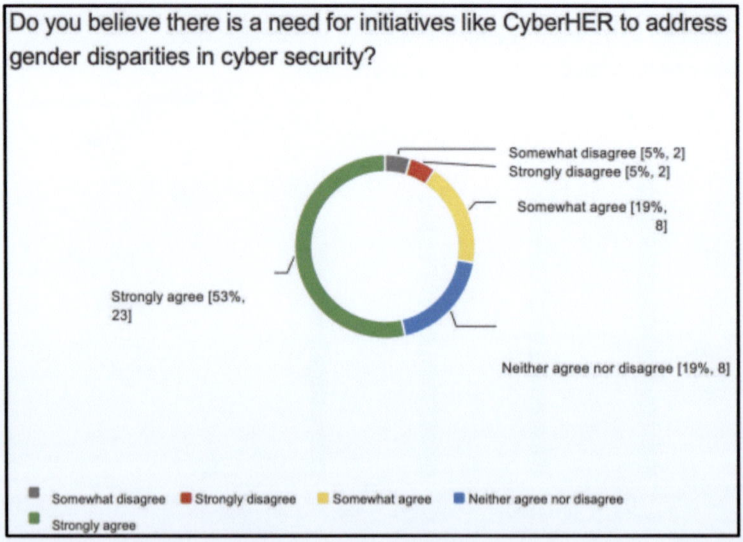

Fig. 9 Strong agreement on CyberHER's positive impact

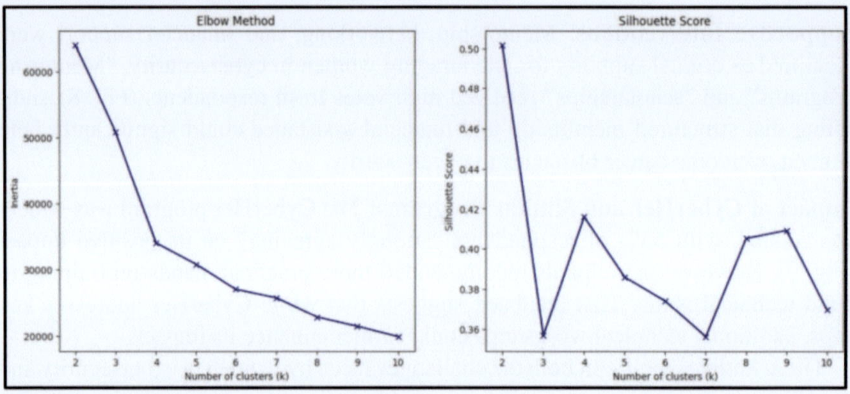

Fig. 10 Optimal digital gender gap clustering at K = 4

final decision was to use K = 4, as adding more clusters beyond this point provides diminishing returns in clustering effectiveness. The silhouette score, which measures how well separated the clusters are, peaks at K = 2 but begins to fluctuate at higher values of K. Although K = 2 shows the highest silhouette score, it may be overly simplistic for effective clustering. A balance between the number of clusters and clustering quality is achieved at K = 3 or K = 4, with K = 4 being the most suitable choice, offering a good balance between inertia reduction and cluster cohesion and separation.

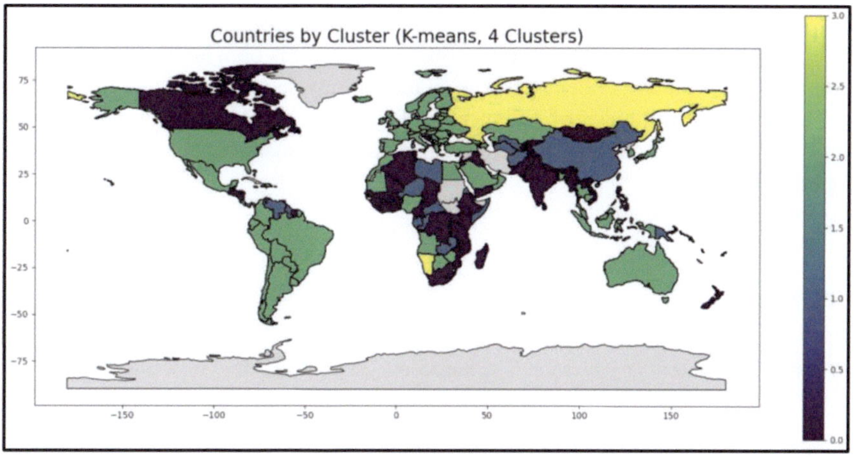

Fig. 11 Global digital gender gap clusters by socio-economic traits

The world map in Fig. 11 displays countries divided into four clusters using K-means, likely based on shared socio-economic, political, or developmental traits. North America and Western Europe mostly fall into the same cluster (green), indicating similar characteristics. In contrast, Asia and Africa show more diverse patterns, reflecting differences in development and economic conditions. Most African countries are part of the purple cluster, possibly pointing to common features like lower economic development. While the specific factors used for clustering are not detailed, elements such as GDP, population size, or political stability may have influenced the grouping. Some regions are uncolored, likely due to missing data.

5 Conclusion and Future Work

This study highlights the complex challenges of gender inequality in cybersecurity, particularly in developing regions where limited access to digital education, societal norms, and digital harassment create significant barriers for women. Through the CyberHER project, we observed that while policy initiatives aim to bridge these gaps, many fall short due to weak local commitment and limited cultural adaptation. Effective change requires more than legal frameworks; it demands deeper community involvement and shifts in cultural attitudes. Community-driven efforts, like those implemented through CyberHER, demonstrate the potential of targeted programs in providing practical skills and building supportive networks that encourage women to pursue careers in cybersecurity.

Future work will focus on expanding CyberHER's model to other underrepresented areas, adapting its framework for different cultural contexts. Future initiatives should explore integrating technical workshops, mentorship, and policy collaboration to strengthen local engagement. Further research will investigate the effectiveness of digital skill training and community support programs on a larger scale to assess long-term impacts on gender inclusion in technology sectors. Additionally, studies will evaluate the influence of enhanced policies that enable collaborative partnerships with local educational institutions, aiming for sustainable gender equity in digital fields.

References

1. International Labor Organization: The impact of care responsibilities on women's labor force participation. ILO (2024)
2. Cardiff Metropolitan University: Cardiff Metropolitan University. https://www.cardiffmet.ac.uk/Pages/default.aspx (Accessed October 2024)
3. Institute of Management Sciences: IMSciences. https://imsciences.edu.pk/ (Accessed October 2024).
4. Othman N (2006) Muslim women and the challenge of Islamic fundamentalism/extremism: An overview of Southeast Asian Muslim women's struggle for human rights and gender equality. Women's Stud Int Forum 29(4):339–353
5. Subrahmanian R (2005) Gender equality in education: Definitions and measurements. Int J Educ Dev 25(4):395–407
6. Sika N (2011) The Millennium Development Goals: Prospects for gender equality in the Arab world. J Int Women's Stud 12(3):27–42
7. Guthridge M, Kotsadam A, Nese P, John M, Gleditsch NP (2022) Promoting gender equality: A systematic review of interventions. Soc Justice Res 35(4):318–343
8. Rabia M, Tamime RA, Fatehkia M, Kashyap R, Weber I (2020) Monitoring digital gender inequality using Facebook and Google advertising audience estimates. Demogr Res 43(27):779–816
9. UN Women: National report on the status of women in Pakistan. UN Women (2023)
10. UNESCO: Gender imbalances remain in digital skills and STEM careers. https://en.unesco.org/news/gender-imbalances-remain-digital-skills-and-stem-careers (Accessed October 2024)
11. Nature: Gender equality: the route to a better world. Nature 621(7980), 8 (2023)
12. Elwér S, Harryson L, Bolin M, Hammarström A (2013) Patterns of gender equality at workplaces and psychological distress. PLoS ONE 8(1):e53246
13. Kashyap R, Fatehkia M, Tamime RA, Weber I (2019) Monitoring digital gender inequality using Facebook and Google advertising audience estimates. Data2X
14. Narayani A (2024) Women's safety in digital space. Indian J Public Adm 70(3):546–561
15. World Bank Group: Gender Data Portal. https://genderdata.worldbank.org/ (Accessed October 2024)
16. Shreya M (2024) Gender inequality visualizations. Tableau Public. https://public.tableau.com/app/profile/mac0173/vizzes (Accessed October 2024)
17. United nations development program: gender inequality index. http://hdr.undp.org/en/content/gender-inequality-index-gii (Accessed October 2024)
18. International labour organization: World employment and social outlook: May 2024 update. ILO (2024)
19. University of Oxford: Using digital traces to measure digital gender inequality in real-time. https://www.sociology.ox.ac.uk/article/using-big-data-to-measure-digital-inequalities (Accessed October 2024)
20. UN Regional Collaborative Platforms: Out-of-school children of primary school age, female. https://data.unicef.org/topic/education/primary-education/ (Accessed October 2024)

Understanding the Impact of Risk Perception on Cybersecurity Training Effectiveness in Small and Medium Enterprises (SMEs)

Jonathan Ayodele, Amir Javed, and Anete Poriete

Abstract Historically, SMEs have been perceived by cybercriminals as more vulnerable targets, primarily due to their often limited resources in cybersecurity. This perception places them at an elevated risk, underscoring the urgency to comprehend and address their risk perceptions. Economically, SMEs are pivotal, but despite their substantial contribution to the global economy, they often find themselves exposed to a myriad of cybersecurity threats. This research delves into the cybersecurity risk perceptions among SMEs, aiming to bridge the observed gap between the perceived escalating digital threats and its influence on their preparedness levels. Through a comprehensive survey, the study provides insights into SMEs' engagement with cybersecurity training programs, the effectiveness of these initiatives, and the overarching factors influencing their risk perceptions. It also gages the level of cybersecurity risk awareness and assesses the effectiveness of existing training programs. The insights derived from this research have the potential to inform policy and regulatory frameworks. Such informed perspectives can lead to the creation of policies and regulatory guidelines, specifically tailored to buttress cybersecurity within this critical sector.

Keywords Cybersecurity · Risk perception · Cybersecurity training · Cybersecurity preparedness · Cyber-resilience · Human factors in cybersecurity

---

J. Ayodele (✉) · A. Javed
Cardiff University, Cardiff, UK
e-mail: ayodelejona@gmail.com

A. Poriete
CyberSmart Ltd, London, UK

C. E. R. Hewage et al. (eds.), *AI Applications in Cyber Security and Privacy of Communication Networks*, Lecture Notes in Networks and Systems 1453,
https://doi.org/10.1007/978-981-96-7400-8_9

1 Introduction

Recently the significance of cybersecurity has exponentially risen, especially within the realm of business where cyber-threats have not only intensified but diversified. Small and medium enterprises (SMEs), with their limited resources, often bear the brunt of cyber-attacks. Recent statistics show that approximately 43% of cyber-attacks target SMEs [1], with 4 in 10 businesses experiencing a cyber-attack in the last 12 months [2] and a startling 60% of small businesses that fall victim to such attacks close down within 6 months of the breach [3]. Similarly, in the UK Government 2023 cybersecurity breaches survey, it was found that 32% of businesses have experienced a cyber-attack in the past year [4], with SMEs reporting higher instances. Such statistics underscore the vulnerability of SMEs in the face of cyber-threats and the immense importance of cybersecurity awareness and training. Small and medium enterprises (SMEs) play a crucial role in the global economy, but they often face cybersecurity challenges due to limited resources and awareness.

The intrinsic connection between risk perception and cybersecurity behavior has been a pivotal aspect of research in large corporations. However, SMEs, despite being at the frontline of cyber-threats, remain under-researched. Understanding this dynamic is crucial not just for the survival of individual SMEs but also for the overall economy given that SMEs account for a significant proportion of employment and economic activity. As businesses increasingly rely on digital platforms and technologies to drive their operations, they also become susceptible to a myriad of cyber-threats [5]. These threats, ranging from data breaches to ransomware attacks [6], have the potential to disrupt business operations, compromise sensitive data, and tarnish reputations [7]. The global landscape of cyber-threats is ever-evolving [8], with cybercriminals employing sophisticated techniques and strategies to target vulnerabilities in digital systems [9]. Furthermore, the human element is also critical. Employee and stakeholder awareness, risk perception, and adherence to protocols significantly impact SME's cybersecurity resilience [10].

Preliminary findings suggest that while there is a general awareness of cyber-threats among SMEs [11], there may be gaps in their understanding of the intricacies of these threats and the best practices to mitigate them.

This research aims to investigate how risk perception influences cybersecurity practices in SMEs, especially in scenarios where individuals perceive cybersecurity as a low risk or no risk at all. The findings will contribute to enhancing cybersecurity preparedness and resilience in SMEs.

1.1 Objectives

The primary objectives of this study are as follows:

1. To assess the risk perception of individuals and decision-makers in SMEs concerning cybersecurity threats.

2. To examine the relationship between risk perception and the willingness to invest in cybersecurity awareness training and security measures in SMEs.
3. To identify the barriers and challenges faced by SMEs in improving their risk perception and adopting effective cybersecurity practices.
4. To develop practical recommendations for enhancing risk perception, cybersecurity awareness, and the implementation of security measures in SMEs.

1.2 Research Questions

1. How does the risk perception of SMEs influence their receptiveness and responsiveness to cybersecurity awareness training?
2. Do cybersecurity awareness training programs shape the risk perception of SMEs?
3. What are the effects of low-risk perception on SMEs?
4. Does awareness training programs impact SMEs' risk perception and subsequent adoption of cybersecurity measures?

2 Related Work

Risk perception refers to the way individuals or groups view and understand the potential dangers or threats they might face. It is the cognitive evaluation of risk that is influenced by individual and social factors, including personal beliefs, experiences, emotions, and social trust [12]. Studies such as Grima et al. [13] categorize risk perception as a multidimensional construct that encompasses both intuitive and affective reactions to hazards. It is often influenced by biases, such as judgments based on experiences and values [14]. The significance of understanding risk perception is further amplified when considering its impact on behavior. As noted by Cano and Salzberger [15], risk perception is instrumental in determining risk-addressing behaviors. It acts as a catalyst, influencing the degree to which individuals are motivated to undertake actions to mitigate potential threats.

2.1 Cybersecurity Risk Perception

Cybersecurity risk perception refers to how individuals and organizations perceive the risks associated with cybersecurity threats [16]. It involves understanding and evaluating the potential harm and likelihood of cyber-attacks and their impact on information systems, data, and operations.

Research and studies have been extensively conducted to delve into the risk perceptions of cybersecurity. These studies have identified dimensions of cyber-risk perception, such as perceived benefit, optimistic bias, and news sentiment, which influence how individuals perceive and respond to cyber-risks [17]. In addition, risk perception in cybersecurity can be swayed by factors such as perceived control and understanding of the risks [18]. This means that individuals who feel they have a grasp on cybersecurity measures may perceive lower risks. Additionally, research has shown that factors like government actions and adaptation and learning effects can impact societal assessment of cyber-risk sources [19]. In a study conducted by Maisikeli [20] that compared UAE cyber-activity and perception data with data from nine developed nations, it was found that the cyber-risk behavior is comparatively the same as in the other nine developed nations.

However, there can be a disparity between perception and actuality in terms of cyber-risk. This discrepancy can be ascribed to natural optimism bias and a deficiency in comprehending cyber-risks [21]. Such biases can lead to individuals underestimating the potential impact of cyber-threats. Similarly, companies that have been spared from a cyber-incident might struggle to fully grasp the risks. As a result, they might undervalue cybersecurity, prioritizing other concerns such as profitability or market expansion, leading to skewed risk perceptions [22]. Cyber-risk perception is a multifaceted research domain, shaped by determinative factors and cognitive processes, this suggests that understanding risk perception requires a holistic approach, considering both psychological and technical aspects. According to a study conducted by [16] on "The Microfoundations of State Cybersecurity: Cyber Risk Perceptions and the Mass Public", it was discovered that exposure to a personally relevant data breach increases risk perception and willingness to engage in safer online practices. Van Schaik et al. [23] analyzed USA and UK students' risk perceptions and precautionary behavior related to cybersecurity, the study revealed a noticeable similarity in the pattern of mean scores across hazards for both nations. The influence of hazard on risk perception was found to be small to moderate and significant for UK students. In contrast, for US students, the effect was minimal and on the verge of being significant.

2.2 Gaps and Limitations in Existing Literature

Current literature offers substantial insights into risk perception and cybersecurity training in SMEs. However, more research is needed to understand the nuanced interactions between these two factors. In addition, the majority of research has focused on the technical aspects of cybersecurity training, leaving human factors relatively underexplored. Future research should also consider the influence of cultural and organizational factors on risk perception and training effectiveness.

3 Methodology

The research employed a mixed approach, leveraging a well-structured questionnaire to gather data. As articulated by Rahman [24], quantitative research is inherently deductive, allowing for specific hypotheses to be tested against collected data, aligning with the positivist paradigm. This approach was selected due to its capacity to amass data from a broad spectrum of respondents, thereby offering a holistic view of risk perception's role in shaping cybersecurity practices in SMEs and as highlighted by Tuffrey et al., [25] questionnaire facilitates standardized data collection, ensuring consistency and reliability in the gathered responses. Adopting a deductive research approach, the study began with a clear set of objectives and hypotheses drawing from the extensive literature review conducted in the previous chapter. However, the choice of a questionnaire also comes with its challenges. There's the potential risk of respondents treating the exercise as a test, aiming to provide the "right" answers rather than their genuine perceptions. This challenge is further amplified in a domain-like cybersecurity, where there might be a perceived "correct" stance on issues. Moreover, the self-administered nature of questionnaires might lead to incomplete responses or potential biases. Yet, the benefits of scalability, consistency, and the ability to reach a wider audience make it a compelling choice for this research. The primary data for this research was collected using a structured questionnaire which was favored over other methods like interview as it is generally deductive and allows for empirical verification of specific hypotheses on a particular dataset and ensures objectivity by minimizing the influence of the researcher's biases and subjective preferences [26]. Also, due to time constraints, it would be impossible to conduct interviews. The questionnaire was designed to directly address the study's research questions. To minimize potential biases, participants were assured of anonymity and questions were phrased neutrally to avoid leading responses. The data gathered from the survey was analyzed using a combination of descriptive, inferential, and thematic analysis. The study surveyed 70 respondents across various sectors. The inclusion and exclusion criteria for participants were clearly defined. According to the European Union, a small enterprise is defined as a company with fewer than 250 employees and has a turnover of less than or equal to €50 million [27]. Only those who were representatives or decision-makers within SMEs and/or had some level of engagement in the SME ecosystem were considered. Micro-enterprises (fewer than 10 employees) and large organizations (more than 250 employees) were excluded to maintain focus on SMEs. Conversely, large corporations that did not align with the SME definition were excluded from the study. A random sampling technique was employed to administer the questionnaire, ensuring an unbiased representation of the target population. This method was deemed appropriate as it provided each SME representative an equal opportunity to be part of the study, thereby enhancing the generalizability of the findings.

Based on the research questions, the following hypotheses were derived and tested:

Null Hypothesis (H0): An individual's risk perception within SMEs has no influence on their susceptibility to awareness training and the subsequent implementation of security measures.

Alternative Hypothesis (H1): An individual's risk perception within SMEs has a significant influence on their susceptibility to awareness training and the subsequent implementation of security measures.

Null Hypothesis (H0): Cybersecurity awareness training programs do not shape the risk perception of SMEs.

Alternative Hypothesis (H1): Cybersecurity awareness training programs significantly shape the risk perception of SMEs.

3.1 Ethical Considerations

Due to the research involving human respondents, adherence to ethical guidelines was priority. Prior to the commencement of data collection, an application for ethical approval was submitted and subsequently granted by the School Research Ethics Committee (SREC) with reference: COMSC/Ethics/2023/105. Furthermore, to bolster the research's ethical foundation, I successfully completed the Research Integrity Training and obtained a certificate underscoring my commitment to uphold the highest ethical standards in the study [28].

4 Results

Out of 70 respondents, the demographic data revealed diverse sectoral representation, respondents came from industries including IT and technology, retail, education, hospitality, health and social care, manufacturing, and others. Key findings highlighted notable differences in cybersecurity risk awareness and engagement with training programs. Notably, IT, cybersecurity, and technology had significant representation, with employee counts varying between 1 and 80. Sector-specific analysis indicated that SMEs in the IT sector had higher risk awareness, while those in retail and hospitality were more likely to underestimate threats due to limited prior exposure to incidents. Future analyses could explore the influence of organizational maturity, such as years in operation and technological adoption, on these trends.

The reliability of the questionnaire was analyzed using Cronbach's alpha and it is shown in Table 1. The findings which are discussed in further detail clearly demonstrated that the questionnaire's Cronbach's alpha value was a reasonable 0.88. A substantial degree of internal consistency and overall instrument reliability are shown by this numerical value, which serves as a vital indicator.

Table 1 Reliability analysis

Cronbach's alpha	No. of items per section
0.961	4
0.872	5
0.932	4
0.923	2
0.789	4
0.88	19

The study evaluated the impact of cybersecurity awareness training and risk perception on cybersecurity practices among SMEs. A reliability analysis of the questionnaire yielded a Cronbach's alpha of 0.88, indicating high internal consistency. Descriptive statistics showed a relatively high awareness of cybersecurity threats among respondents (mean: 4.67, SD: 0.93) but low participation in training programs, attributed to budget constraints and a perception that SMEs are less likely targets for cyber-attacks.

Regression analysis explored the effect of risk perception on training effectiveness. The coefficient for risk perception (7.098e-05, $p = 0.999$) was not statistically significant, suggesting that changes in perceived risk have minimal impact on the effectiveness of cybersecurity training. Employee engagement had a coefficient of -8.618e-02 ($p = 0.108$), implying a weak negative relationship with training effectiveness, though not statistically significant. SME characteristics showed no significant impact on training effectiveness either, with a coefficient of -3.576e-02 ($p = 0.628$), further underscoring that risk perception alone does not drive training effectiveness.

Pearson correlation analysis revealed statistically significant relationships among the key variables. Risk perception (RP) showed a strong positive correlation with cybersecurity training effectiveness (CTE) at $r = 0.748$, significant at the 0.01 level, indicating that higher risk perception tends to align with increased training effectiveness. Similarly, RP correlated positively with SME characteristics ($r = 0.728$) and employee engagement (EE) ($r = 0.612$), both at the 0.01 level, suggesting a linked increase across these factors. CTE and EE also demonstrated a positive correlation ($r = 0.621$), significant at the 0.01 level, indicating that as training becomes more effective, employee engagement in cybersecurity practices tends to rise.

SMEs with higher risk perception reported more frequent updates to their cybersecurity measures. This proactive approach aligns with previous findings suggesting that perceived threat levels drive cybersecurity actions. Industries perceiving greater cyber-risks, such as IT and finance, demonstrated higher engagement with advanced security measures compared to sectors like hospitality and social housing. The findings highlight the interconnectedness of risk perception, employee engagement, and SME characteristics within cybersecurity training contexts, suggesting that targeted, industry-specific strategies may help enhance training effectiveness among SMEs with diverse cybersecurity needs.

5 Discussion

This study highlights the nuanced relationship between risk perception and cybersecurity training effectiveness in SMEs, emphasizing that although risk perception is high, actual engagement with training is limited. The findings align with prior research showing that SMEs often underestimate their vulnerability to cyber-threats, primarily due to financial constraints [29] and a misconception that only larger companies are targeted. The study observed that SMEs with a higher risk perception were more proactive in updating security measures, consistent with theories suggesting perceived risk motivates protective behaviors.

Interestingly, while awareness training has some positive impact, the regression analysis suggests limited effectiveness due to sporadic or generic training approaches. This supports literature advocating for industry-specific, regularly updated, and interactive training methods, such as gamification, to improve retention and practical application. In addition, while regression analysis indicated weak relationships between risk perception and training effectiveness (e.g., coefficient for risk perception = 7.098e-05, $p = 0.999$), these findings suggest that perceived risks alone may not drive behavior change and effective engagement with training programs. This underscores the importance of complementary factors, such as employee engagement and leadership buy-in, in improving training outcomes [18]. For policymakers, this indicates an opportunity to design targeted incentives, such as subsidized training programs [30], to bridge the gap between awareness and action [31]. Future research should explore sector-specific training methodologies to further enhance effectiveness and scalability.

6 Limitations

One of the primary constraints was the sample size and representation. While the initial aim was to engage with 100 SMEs, the study was able to gather data from only 70 respondents. This reduced sample might not fully capture the diverse range of risk perceptions and training effectiveness across the broader SME sector, potentially affecting the generalizability of the findings. The geographical scope of the research was another limitation. Conducted primarily within a specific region, the findings might be particularly indicative of the cybersecurity perceptions and practices of SMEs within that locale. This geographical concentration suggests that the results might not be universally applicable to SMEs operating in different cultural or regulatory environments.

7 Recommendations and Future Work

To enhance cybersecurity in SMEs, regular training, risk-based strategies, and awareness initiatives are essential to address evolving threats and dispel misconceptions about low-risk exposure. Tailored approaches should prioritize advanced detection for IT, cost-effective solutions for retail, compliance for healthcare, and operational security for manufacturing. Leveraging free or low-cost resources, such as cybersecurity toolkits and training programs offered by governmental or non-profit organizations, can help mitigate budgetary limitations. Leadership attitudes significantly influence security culture [32]. with targeted training for decision-makers on the importance of cybersecurity. Such initiatives can significantly elevate risk perception and drive sustained improvements in cybersecurity practices.

Future research should expand the participant pool, integrate regional and cultural factors, and adopt qualitative methods for richer insights. Adopting conditional logic and inclusive survey designs can improve data quality and capture industry-specific challenges.

References

1. Verizon: Verizon data breach investigations report. https://www.verizon.com/business/resources/reports/2019/2019-data-breach-investigations-report.pdf, Last accessed 2025/01/28
2. Rae A, Patel A (2019) Defining a new composite cybersecurity rating scheme for SMEs in the U.K. In: Lect Notes Comput Sci, 11879, pp 362–380. Springer, Heidelberg. https://doi.org/10.1007/978-3-030-34339-2_20.
3. Morgan S (2021) Cyberwarfare in the C-suite: Cybercrime facts and statistics. Cybersecur Ventur
4. GOV.UK: Cyber security breaches survey 2023. https://www.gov.uk/government/statistics/cyber-security-breaches-survey-2023/cyber-security-breaches-survey-2023, Last accessed 2023/10/06
5. Mohamed H, Ali H (2021) Finding solutions to cybersecurity challenges in the digital economy. In: Fostering Innovation and Competitiveness With FinTech, RegTech, and SupTech, pp. 80–95. IGI Global. https://doi.org/10.4018/978-1-7998-4390-0.ch005.
6. Al Obaidan F, Saeed S (2021) Digital transformation and cybersecurity challenges. In: Adv Cybersecur Digit Transform, pp 203–226. IGI Global. https://doi.org/10.4018/978-1-7998-6975-7.ch011
7. Sandhu K (2021) Advancing cybersecurity for digital transformation. In: Adv Cybersecur Digit Transform, pp 1–17. IGI Global. https://doi.org/10.4018/978-1-7998-6975-7.ch001
8. Shinde SS, Ansurkar G (2023) Upcoming threats in cybersecurity. Int J Sci Res Sci Technol. 10:806–816. https://doi.org/10.32628/IJSRST523102121
9. Shanker AK, Usha G (2017) Cyber threat landscape in cyber space. In: proceedings of the international conference on electronics, communication and aerospace technology (ICECA), pp. 375–380. https://doi.org/10.1109/ICECA.2017.8203709
10. Kettani H, Wainwright P (2019) On the top threats to cyber systems. In: proceedings of the 2nd international conference on information and computer technologies (ICICT), pp 175–179. https://doi.org/10.1109/INFOCT.2019.8711324
11. Shojaifar A, Fricker SA (2020) SMEs' confidentiality concerns for security information sharing. IFIP Adv Inf Commun Technol. 593:289–299. https://doi.org/10.1007/978-3-030-57404-8_22

12. Wilson RS, Zwickle A, Walpole H (2019) Developing a broadly applicable measure of risk perception. Risk Anal 39:777–791. https://doi.org/10.1111/RISA.13207
13. Grima S, Hamarat B, Özen E, Girlando A, Dalli-Gonzi R (2021) The relationship between risk perception and risk definition and risk-addressing behaviour during the early COVID-19 stages. J Risk Fin Manag. 14:272. https://doi.org/10.3390/JRFM14060272
14. López-Vázquez E, Marván ML (2018) Introduction to risk psychology. In: Risk Psychology, pp 1–12. https://doi.org/10.1007/978-3-319-73799-7_1
15. Cano S, Salzberger T (2017) Measuring risk perception. In: Consum Percept Prod Risks Benefits, pp 191–200. https://doi.org/10.1007/978-3-319-50530-5_10
16. Kostyuk N, Wayne C (2021) The microfoundations of state cybersecurity: Cyber risk perceptions and the mass public. J Glob Secur Stud 6. https://doi.org/10.1093/JOGSS/OGZ077.
17. Larsen MH, Lund MS (2021) Cyber risk perception in the maritime domain: A systematic literature review. IEEE Access 9:144895–144905. https://doi.org/10.1109/ACCESS.2021.3122433
18. Bada M. Sasse AM, Nurse JRC (2019) Cybersecurity awareness campaigns: Why do they fail to change behaviour?
19. Perols RR, Murthy US (2021) The impact of cybersecurity risk management examinations and cybersecurity incidents on investor perceptions and decisions. AUDITING: J Pract & Theory 40, 73–89. https://doi.org/10.2308/AJPT-18-010
20. Maisikeli S (2020) UAE cybersecurity perception and risk assessments compared to other developed nations. In: Proceedings of the 3rd international conference on information and computer technologies (ICICT), pp. 432–439. https://doi.org/10.1109/ICICT50521.2020.00075
21. Stedmon N (2020) The impact of cybersecurity threats on the 2020 US elections
22. Munyoka W, Maharaj MS (2019) Privacy, security, trust, risk and optimism bias in e-government use: The case of two Southern African development community countries. S Afr J Inf Manag. 21. https://doi.org/10.4102/SAJIM.V21I1.983
23. van Schaik P, Jeske D, Onibokun J, Coventry L, Jansen J, Kusev P (2017) Risk perceptions of cybersecurity and precautionary behaviour. Comput Hum Behav 75:547–559. https://doi.org/10.1016/J.CHB.2017.05.038
24. Rahman MS (2016) The advantages and disadvantages of using qualitative and quantitative approaches and methods in language "testing and assessment" research: A literature review. J Educ Learn 6:102. https://doi.org/10.5539/JEL.V6N1P102
25. Tuffrey C, Bateman BJ, Colver AC (2013) The questionnaire of young people's participation (QYPP): A new measure of participation frequency for disabled young people. Child Care Health Dev 39:500–511. https://doi.org/10.1111/CCH.12060
26. Queirós A, Faria D, Almeida F (2017) Strengths and limitations of qualitative and quantitative research methods. Eur J Educ Stud. https://doi.org/10.46827/EJES.V0I0.1017
27. European Union: SME definition. https://single-market-economy.ec.europa.eu/smes/sme-fundamentals/sme-definition_en, Last accessed 2025/01/28
28. Wenner DM (2015) The social value of knowledge and international clinical research. Dev World Bioeth 15:76–84. https://doi.org/10.1111/DEWB.12037
29. Chidukwani A, Zander S, Koutsakis P (2022) A survey on the cybersecurity of small-to-medium businesses: Challenges, research focus and recommendations. IEEE Access 10:85701–85719. https://doi.org/10.1109/ACCESS.2022.3197899
30. Rombaldo C, Becker I, Johnson S, Rombaldo Junior C (2023) Unaware, unfunded and uneducated: A systematic review of SME cybersecurity. arXiv:2309.17186
31. Song J, Park MJ (2024) A system dynamics approach for cost-benefit simulation in designing policies to enhance the cybersecurity resilience of small and medium-sized enterprises. Inf Dev. https://doi.org/10.1177/02666669241252996
32. Ifinedo P (2012) Understanding information systems security policy compliance: An integration of the theory of planned behavior and the protection motivation theory. Comput Secur 31:83–95. https://doi.org/10.1016/J.COSE.2011.10.007

Advancements in Quantum-Resistant Cryptography: Safeguarding Future Digital Communications

Ramesh D. Jadhav, Chandrani Singh, Rupali Taware, and Sarvesh Bendre

Abstract The emergence of quantum computing presents a critical threat to classical cryptographic systems, endangering the security of current digital communication frameworks. This study goals to identify and evaluate quantum-resistant cryptographic algorithms, focusing on lattice-based, hash-based, and multivariate polynomial cryptography, which are designed to withstand quantum-based attacks. Using a comparative research methodology, we assess the effectiveness, efficiency, and scalability of these algorithms across various applications, including financial transactions, cloud storage, and government communications. The outcomes highlight the potential of these algorithms to secure sensitive data against quantum threats while addressing integration challenges in large-scale infrastructures. The findings emphasize that adopting quantum-resistant cryptographic measures is essential to protect global digital communications, maintaining cybersecurity resilience in the face of advancing quantum technologies. This study contributes to the field by offering insights into practical post-quantum solutions, underscoring the necessity for proactive defenses to sustain digital trust as quantum computing evolves.

Keywords Quantum-resistant cryptography · Post-quantum algorithms · Digital communication security · Quantum computing threats · Lattice-based cryptography · Cybersecurity resilience

R. D. Jadhav (✉) · C. Singh · R. Taware · S. Bendre
Sinhgad Institute of Management, Pune, India
e-mail: rameshdjadhav@gmail.com

C. Singh
e-mail: directormca_siom@sinhgad.edu

R. Taware
e-mail: rupalitaware.siom@sinhgad.edu

S. Bendre
e-mail: to6tr@sinhgad.edu

C. E. R. Hewage et al. (eds.), *AI Applications in Cyber Security and Privacy of Communication Networks*, Lecture Notes in Networks and Systems 1453,
https://doi.org/10.1007/978-981-96-7400-8_10

1 Introduction of the Study

As quantum computing technology rapidly advances, traditional cryptographic methods are at risk, and as they rely on quantum computers, they have the potential to solve various mathematical problems efficiently. Conventional encryption techniques like RSA and ECC are particularly vulnerable, as quantum algorithms, such as Shor's algorithm, can decrypt these systems exponentially faster than classical computers. This looming threat underscores the urgent need for quantum-resistant cryptography, also known as post-quantum cryptography, Bernstein et al. [1], which seeks to advance new cryptographic methods capable of withstanding quantum attacks [1].

Quantum-resistant cryptography explores algorithms designed specifically to counteract the computational power of quantum systems. Leading solutions contain lattice-based, code-based, hash-based, and multivariate polynomial cryptographic schemes, each presenting unique advantages in terms of security, performance, and resilience against quantum decryption [2]. For instance, lattice-based cryptography, Regev [3] favored for its scalability and robustness, relies on the hardness of lattice-based problems, which are difficult for even quantum computers to solve. Similarly, hash-based techniques use complex hashing structures that remain infeasible for quantum decryption, and code-based cryptography leverages error-correcting codes to secure data. These cryptographic approaches [3] McEliece [4] are designed to maintain strong security in the quantum era and are being widely studied for applications in banking, government communications, and other critical infrastructure sectors.

Adopting quantum-resistant algorithms, however, presents substantial challenges. Beyond the technical difficulties of designing secure and efficient algorithms, there are issues of scalability, compatibility with current digital systems, and ensuring seamless integration into existing infrastructures. Furthermore, transitioning global networks to quantum-safe protocols requires extensive collaboration among governments, businesses, and academic institutions to establish standards, support development, and facilitate smooth deployment [4].

This research aims to investigate recent advancements, address the practical challenges of implementation, and analyze the future potential of quantum-resistant cryptography. By examining the most promising algorithms and their real-world applications, this study underscores the necessity of proactive measures to secure digital communications before quantum technology reaches maturity. Quantum-resistant cryptography is not only a technical solution but a strategic imperative to ensure continued trust and security in global digital ecosystems as we enter the quantum computing era.

2 Objective of the Study

- To analyze and compare the effectiveness of leading quantum-proof cryptographic algorithms.
- To evaluate the challenges and feasibility of deploying quantum-resistant cryptographic solutions.
- To explore the potential applications and future impacts of quantum-resistant cryptography across critical sectors.

3 Hypothesis of this Study

- Quantum-resistant cryptographic algorithms can effectively protect digital systems from quantum computing threats.
- These algorithms will provide security, efficiency, and scalability comparable to traditional methods.
- Adopting quantum-resistant strategies in advance strengthens the enduring security of digital infrastructure systems.

4 Scope of the Study

- Examine prominent quantum-resistant cryptographic methods, including lattice-based, hash-based, and multivariate polynomial approaches, with an emphasis on security, efficiency, and real-world applicability.
- Analyze the performance, scalability, and integration challenges of quantum-resistant solutions within existing digital systems across critical sectors, including finance, government, and cloud storage.
- Provide actionable insights for cybersecurity professionals and policymakers to effectively prepare for and adapt to the implications of quantum computing on digital security.

5 Problem Statement of the Study

Quantum computing threatens traditional cryptographic systems like RSA and ECC. This study highlights the need to develop and integrate quantum-resistant algorithms to ensure long-term security and privacy in digital communications.

6 Literature Review and Gap Analysis

The prompt advancements in quantum computing have sparked significant interest in the field of quantum-resistant cryptography, also known as post-quantum cryptography. Traditional cryptographic algorithms, such as RSA, DSA, and ECC, once deemed secure, are now increasingly vulnerable to quantum attacks due to the efficiency of quantum algorithms like Shor's algorithm, which can factor large integers and solve discrete logarithms in polynomial time [2]. This vulnerability has prompted researchers to explore alternative cryptographic schemes that can withstand the computational power of quantum systems [2].

Lattice-based cryptography has emerged as a leading contender in the realm of quantum-resistant solutions, thanks to its scalability and robust resistance against quantum attacks. Regev [3] established that lattice-based problems, particularly the LWE (Learning with Errors) and SVP (Shortest Vector Problem), remain computationally difficult for quantum computers, rendering them suitable for encryption purposes [3]. Further advancements by Gentry et al. in [5] have demonstrated the versatility of lattice-based techniques, which support both standard encryption and fully homomorphic encryption, illustrating their potential in post-quantum cryptography [5].

Hash-based cryptography also presents viable solutions, leveraging hash functions to create security mechanisms that do not depend on number-theoretic assumptions vulnerable to quantum attacks. Early methods, such as Lamport signatures [6] and Merkle trees [7], have proven resilient to quantum threats. Recent optimizations have further improved the efficiency of these techniques, making them suitable for digital signatures and authentication in contemporary applications [1, 6, 7].

Code-based cryptography, originating from the McEliece cryptosystem [4], relies on error-correcting codes and has consistently demonstrated resistance to quantum decryption attempts. The foundational problem of decoding a general linear code remains complex for quantum algorithms, ensuring the continued relevance of this approach [4, 9].

Multivariate polynomial cryptography operates on the premise that solving systems of multivariate polynomial equations is difficult, providing resilience against quantum attacks due to their non-linear complexity [8]. While this method is less prevalent than lattice- and hash-based approaches, it represents a promising area of active research, especially for specific applications where efficiency gains may be realized [8].

Despite progress in quantum-resistant cryptography, several gaps persist, including the need for comprehensive studies on practical implementation and performance benchmarks in real-world systems. Additionally, there is a lack of exploration into how these algorithms integrate with existing infrastructures. Standardized frameworks for assessing and certifying these solutions are essential, along with adaptations for specific applications to meet diverse security requirements.

7 Research Methodology

The research design for the study on quantum-resistant cryptography will employ a mixed-methods approach. This design combines quantitative analysis of algorithm performance and scalability with qualitative insights from expert interviews and case studies. By integrating both data types, the study purposes to provide a complete understanding of practical implementations, challenges, and the effectiveness of various quantum-resistant algorithms in real-world applications.

To address the problem of quantum vulnerabilities in traditional cryptographic systems, a post-quantum cryptographic framework must be developed that includes the following key components.

7.1 Algorithm Selection and Evaluation

- Lattice-Based Cryptography: Algorithms based on the hardness of lattice problems (such as Learning with Errors) offer promising resistance against quantum attacks. Lattice cryptography can be used for secure key exchanges and digital signatures.
- Code-Based Cryptography: Algorithms that rely on error-correcting codes, like the McEliece cryptosystem, are believed to be resistant to quantum attacks and can serve as alternatives for secure encryption.
- Hash-Based Cryptography: Methods like hash-based signatures (e.g., Merkle signature schemes) are quantum resistant and provide high-security assurance for digital signatures.
- Multivariate Quadratic Equations: Algorithms that rely on the difficulty of solving large systems of multivariate quadratic equations, which is considered quantum resistant, can be applied to both encryption and authentication. As shown in Fig. 1, post-quantum cryptography [9].

7.2 Mathematical Security Proofs

- Develop mathematical proofs that rigorously analyze the security of the selected algorithms against both quantum and classical attacks. This will help in establishing a theoretical foundation for each algorithm's resilience.
- Use reductions to demonstrate that breaking these cryptographic schemes would require solving specific mathematical problems known to be hard for both quantum and classical computers.

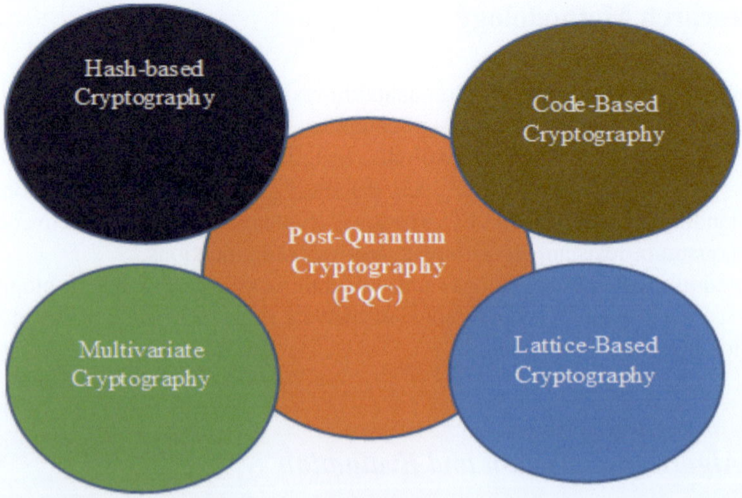

Fig. 1 Post-quantum cryptography

7.3 *Hybrid Cryptographic Approach*

- Implement a hybrid system that combines traditional and post-quantum cryptography for a transitional phase. For example, using both RSA and a lattice-based encryption algorithm allows a gradual migration while testing the security and efficiency of quantum-resistant algorithms.

7.4 *Performance Optimization*

- Many quantum-resistant algorithms are computationally intensive. Designing optimized algorithms and leveraging efficient mathematical structures (like ring or module lattices) can improve performance.
- Evaluate trade-offs between security and computational efficiency to make these algorithms practical for real-world applications, especially in resource-constrained devices.

7.5 *Implementation and Integration*

- Develop a framework for integrating quantum-resistant algorithms into current communication protocols (such as TLS, VPNs, and IoT security layers).

- Provide a standardized approach to facilitate the deployment of these algorithms within existing systems and networks, ensuring minimal disruption to current security infrastructure.

7.6 Compliance and Standardization

- Collaborate with standardization bodies like NIST, which is already working on a list of recommended post-quantum algorithms, to ensure compatibility and regulatory compliance.
- Regularly update the framework to align with evolving standards and emerging research on quantum computing threats.

7.7 Mathematical Solution Design

A robust mathematical solution would involve

- Constructing cryptographic primitives from chosen quantum-resistant mathematical problems.
- Applying cryptographic reductions that show breaking these primitives would be equivalent to solving hard mathematical problems like lattice reductions, which are infeasible for quantum computers.
- Formalizing proofs of security against both quantum and classical adversaries to provide assurances that the cryptographic schemes will remain secure in a post-quantum era.

By combining these strategies, a structured approach to post-quantum cryptography can be established, ultimately preserving digital security as quantum computing advances.

7.8 Mathematical Security Proofs

Mathematical security proofs in post-quantum cryptography often rely on demonstrating that breaking the cryptographic scheme is as hard as solving a known difficult mathematical problem. Here are some mathematical structures, equations, and formulations used in security proofs for lattice-based and other quantum-resistant cryptographic systems.

7.8.1 Lattice-Based Cryptography (LWE—Learning with Errors)

The LWE (Learning with Errors) problem is fundamental to lattice-based cryptography. It is defined as follows:

For a secret vector s ∈ Z^n q given a series of linear equations with errors:

$$\mathbf{A} \cdot \mathbf{s} + \mathbf{e} = \mathbf{b}\,(\mathbf{mod}\,\mathbf{q}) \tag{1}$$

where

- A is an m × n matrix with entries in Zq,
- e is a noise vector with small integer entries (typically from a Gaussian distribution),
- b is a vector in $Z^q{}_m$ that represents the "noisy" output of the system.

Security Proof: Breaking an LWE-based encryption system (i.e., recovering s given A and b) is at least as hard as solving the SVP (Shortest Vector Problem) in lattices, which is considered hard even for quantum computers. This reduction provides a basis for the security of LWE-based systems.

7.8.2 Ring-LWE Problem

Ring-LWE is a variant of LWE that operates in a polynomial ring, making it more computationally efficient:

$$\mathbf{a(x)} \cdot \mathbf{s(x)} + \mathbf{e(x)} = \mathbf{b(x)}\,(\mathbf{mod}\,\mathbf{f}\,(\mathbf{x})) \tag{2}$$

where

- a(x), s(x), e(x), and b(x) are polynomials in the ring Zq[x] / (f (x));
- f(x) is often chosen as a cyclotomic polynomial to enhance algebraic structure and efficiency;
- e(x) is a noise polynomial with small coefficients.

Security Proof: Solving Ring-LWE is reducible to finding short vectors in structured lattices (Ideal-SVP), which is conjectured to remain hard for quantum computers. This makes it secure under similar hardness assumptions as LWE, but more efficient in practice.

7.8.3 Hash-Based Digital Signatures (Merkle Trees)

Hash-based digital signature schemes, like Merkle trees, rely on the one-wayness of hash functions, which are resistant to both classical and quantum attacks (quantum computers would need $2^{n/2}$ steps to break a 2^n secure hash).

For example, in a Merkle tree:

$$\mathbf{H} = \mathbf{hash}\,(\mathbf{H_L} \| \mathbf{H_R}) \tag{3}$$

where

- H is the root hash;
- H_L and HR are hashes of the left and right child nodes, respectively;
- || denotes concatenation.

Security Proof: The security of Merkle tree-based schemes comes from the hash function's pre-image resistance. Quantum algorithms like Grover's can only achieve a quadratic speedup, so a hash with 2^{256} security is reduced to 2^{128} under quantum attacks, which is still secure.

7.8.4 Multivariate Quadratic Equations (MQ)

In MQ-based cryptography, security trusts on the hardness of solving systems of multivariate quadratic equations, well defined as follows:

Given a set of m quadratic polynomials:

$$\mathbf{p_i(x_1, x_2 \ldots, x_n) = 0\,for\,i = 1 \ldots, m} \tag{4}$$

where p_i are polynomials over a finite field F_q.

Security Proof: Solving MQ systems (i.e., finding (x_1, x_2……., xn) that satisfy all equations) is known to be NP-hard, even for quantum algorithms. The hardness of MQ problems forms the security basis for encryption and signature schemes based on multivariate equations.

7.8.5 Code-Based Cryptography (McEliece and Niederreiter Schemes)

Code-based cryptographic systems are built upon the difficulty of decoding a random linear code. Particularly for error-correcting codes like the Goppa code:

$$\mathbf{Y = c + e} \tag{5}$$

where

- y is the received vector,
- c is the code word, and
- e is the error vector with a limited number of non-zero entries.

Security Proof: The problem of decoding a random linear code (i.e., finding c given y and knowledge of the code parameters) is hard for both classical and quantum computers. Thus, the McEliece cryptosystem is considered quantum resistant.

Each of these cryptographic schemes has associated mathematical problems whose security proofs reduce to known hard problems:

- **LWE and Ring-LWE**: Shortest Vector Problem (SVP) and ideal-SVP in lattices.
- **Hash-Based Signatures**: Pre-image resistance of hash functions.
- **Multivariate Quadratic (MQ) Schemes**: NP-hardness of solving multivariate systems.
- **Code-Based Cryptography**: Decoding random linear codes is NP-hard.

These formulas and problem setups form the basis for security proofs in post-quantum cryptography, with mathematical reductions linking them to well-studied computational hardness assumptions.

7.8.6 Future Digital Communication Frameworks

This diagram represents a future digital communication framework focusing on hybrid quantum security as shown in Fig. 2.

Figure 2 illustrates a hybrid quantum security framework which fortifies future digital communications by integrating advanced cryptographic technologies. Quantum Random Number Generators (QRNGs) leverage quantum mechanics to produce high-entropy random numbers, ensuring unpredictability. Quantum Key Distribution (QKD) provides ultra-secure key exchange, forming the backbone of Quantum-Resistant Cryptography (QRC) to counter-quantum computing threats. Post-Quantum Cryptography (PQC) employs complex mathematical problems resistant to both classical and quantum attacks, enhancing encryption security. Quantum Key Vault (QKV) securely manages and distributes cryptographic keys, reinforcing data integrity. Together, these technologies establish a root of trust, ensuring authentication, encryption, and robust cybersecurity. This framework is essential for safeguarding critical infrastructure, financial systems, and next-generation digital

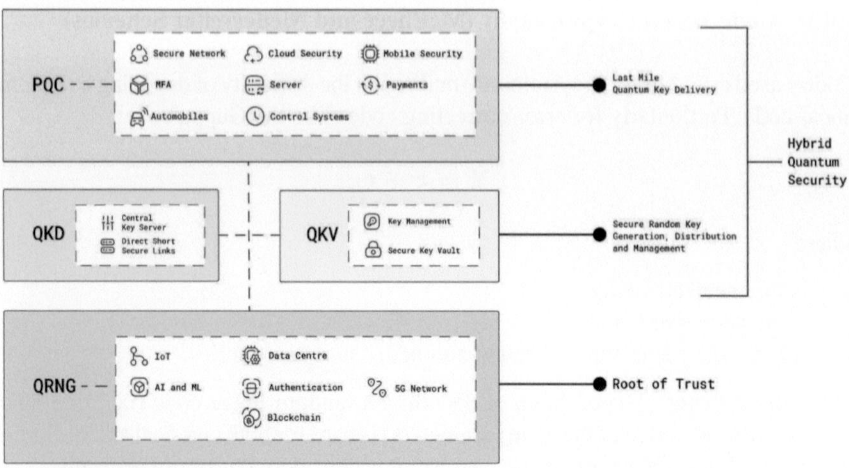

Fig. 2 Future digital communication frameworks

networks. By integrating quantum and classical cryptographic methods, it guarantees resilience against evolving cyber-threats, securing the digital world of the future [10].

8 Further Research

Future digital communications require scalable post-quantum cryptographic algorithms for seamless integration across diverse systems. Advancing Quantum Key Distribution (QKD) over long distances and high-speed Quantum Random Number Generators (QRNGs) is crucial for real-world security. Developing hybrid security architectures that merge classical and quantum cryptography will enhance resilience against cyber-threats. Optimizing quantum-safe key management and distribution will be key to securing next-generation networks and critical infrastructure. This research will drive the future of quantum-resilient digital communications.

9 Conclusion

This study highlights the urgent need to transition toward a secure framework incorporating these post-quantum algorithms. By selecting, optimizing, and integrating quantum-resistant schemes, we can maintain privacy and data integrity even in the face of advancing quantum technologies. However, ongoing research, standardization, and performance tuning will be critical as post-quantum cryptography evolves, helping organizations secure their systems in a post-quantum world.

The findings underscore that each method has unique advantages, but further development is needed for seamless integration and performance optimization across diverse digital systems. This research emphasizes the importance of adopting and standardizing these quantum-resistant algorithms to secure sensitive data in the coming quantum era.

References

1. Bernstein DJ, Lange T, Peters C (2017) Post-quantum cryptography. Nature 549(7671):188–194
2. Shor PW (1994) Algorithms for quantum computation: Discrete logarithms and factoring. In: Proceedings of the 35th Annual ACM symposium on theory of computing, pp 124–134. ACM
3. Regev O (2005) On lattices, learning with errors, random linear codes, and cryptography. In: Proceedings of the 37th Annual ACM Symposium on Theory of Computing, pp 84–93. ACM
4. McEliece RJ (1978) A public-key cryptosystem based on algebraic coding theory. Deep Space Network Progress Report 42:114–116

5. Gentry C, Sahai A, Waters B (2009) Homomorphic encryption from learning with errors: Conceptually, a pretty simple idea. In: Proceedings of the 4th Theory of Cryptography Conference (TCC), pp 50–67. Springer
6. Lamport L (1979) Constructing digital signatures from a one-way function. IEEE Trans Inf Theory 22(4):464–465
7. Merkle RC (1989) A digital signature based on a conventional encryption function. In: Proceedings of the 2010 IEEE symposium on security and privacy, pp 122–134. IEEE
8. Patarin J (1996) A fast and simple way to construct multivariate quadratic public key cryptosystems. In: Proceedings of the 1st international workshop on post-quantum cryptography, pp 35–50. Springer
9. Monroe J (2020) The state of code-based cryptography in a post-quantum world. J Cryptol 33(2):553–575
10. QNu Labs (n.d.) Request a demo. QNu Labs. Retrieved November 11, 2024, from https://www.qnulabs.com/request-a-demo.

Understanding the Complex Interplay of Social Media Privacy: Understanding Oversharing and Recommending Future Research

Matthew Tomlinson, Fiona Carroll, Sandeep Sengar, and Vibhushinie Bentotahewa

Abstract Social media platforms have transformed the landscape of communication and interaction, ushering in an era of unprecedented interconnectedness. However, this digital revolution has raised significant privacy and data security concerns. The "privacy paradox" encapsulates the delicate balance individuals must strike between sharing personal information for social validation and safeguarding themselves from potential threats. This paper delves into the multifaceted domain of social media privacy, addressing user behaviour, platform architecture, and susceptibility to exploitation. Through an extensive literature review, the paper explores concepts such as social engineering, oversharing, and the concerns voiced by social media users. Additionally, it highlights ongoing research efforts aimed at mitigating privacy risks. An exploratory analysis further reveals the vulnerabilities associated with seemingly innocuous images shared on social media, emphasising caution when disseminating personal content online. Building upon these insights, the paper presents a series of recommendations designed to address the pervasive issue of oversharing on social media, encompassing user education, platform design enhancements, and technological interventions. Ultimately, the paper advocates for a comprehensive approach to fortifying privacy in the digital age, calling for collaborative efforts from users, platform providers, and researchers. Furthermore, it outlines potential avenues for future research, recognising the evolving nature of social media and the continuous need for innovation in privacy protection.

M. Tomlinson (✉) · F. Carroll · S. Sengar · V. Bentotahewa
Cardiff School of Technologies, Cardiff Metropolitan University, Cardiff, UK
e-mail: MDTomlinson@cardiffmet.ac.uk

F. Carroll
e-mail: FCarroll@cardiffmet.ac.uk

S. Sengar
e-mail: SSSengar@cardiffmet.ac.uk

V. Bentotahewa
e-mail: ViBentotahewa@cardiffmet.ac.uk

C. E. R. Hewage et al. (eds.), *AI Applications in Cyber Security and Privacy of Communication Networks*, Lecture Notes in Networks and Systems 1453,
https://doi.org/10.1007/978-981-96-7400-8_11

Keywords Social media · Privacy paradox · Oversharing · Sharenting

1 Introduction

The advent of social media platforms has significantly shaped the contemporary era of digital interconnectedness, which has fundamentally transformed the dynamics of communication, experience sharing, and relationship building. However, this unprecedented interconnectivity has raised concerns about privacy and data security. Users often navigate a delicate balance between sharing personal information for social validation and safeguarding themselves from potential risks, leading to what is known as the privacy paradox. This paper comprehensively examines social media privacy, addressing the complex interplay between individual behaviours, platform architecture, and susceptibility to exploitation. The paper begins with a thorough review of existing literature, exploring concepts such as social engineering, oversharing, the privacy paradox, and the concerns expressed by social media users. This review also emphasises current research efforts aimed at mitigating privacy risks. Subsequently, an exploratory analysis is presented, delving into the depth of insights that can be derived from seemingly innocuous images shared on social media. This analysis highlights the often-underestimated vulnerabilities associated with visual data, stressing the importance of a cautious approach to distributing personal content online. Drawing on the insights from the literature review and exploratory analysis, this paper outlines recommendations to address the prevalent issue of oversharing on social media. These recommendations encompass a range of strategies, including user education, platform design refinements, and technological interventions. The paper underscores the critical need for a comprehensive approach to fortify privacy in the digital age. It advocates for a collaborative effort involving vigilant and responsible users, platform providers prioritising user privacy, and researchers to cultivate a more secure and privacy-conscious social media environment. Lastly, the paper explores the future direction of the researcher's research, recognising the ongoing evolution of social media, which necessitates continuous exploration and innovation in privacy protection.

2 Literature Review

2.1 Social Engineering

Ariyo and Zheng [1] discuss how the effectiveness of a social engineering attack can be directly linked to the amount of information that threat actors can obtain from their social networking sites. They explained that the FBI Internet Crime Complaint Centre received 791,790 and 847,376 complaints in 2020 and 2021, respectively, resulting in potential losses exceeding 4.2 billion and 6.9 billion. They discussed that during the COVID-19 pandemic, it was found that social media users were revealing more sensitive information during the pandemic which was successfully being used to launch engineering attacks. Information considered once private and sensitive, such as health information, was becoming a more accepted topic to discuss online early in 2020 and 2021. The FBI further reported an increase of 167% in data breaches since pre-pandemic levels. Future work should examine the correlation between social media users who actively overshare online and fall foul of social engineering attacks. With the rise of social media over the past 20 years, further enabling threat actors to create ever more convincing social engineering attacks.

2.2 *Privacy Paradox*

The privacy paradox refers to individuals expressing significant privacy concerns when using Internet-connected devices and social media platforms but still failing to safeguard their online privacy. They regularly share personal information online, which contradicts their stated privacy concerns [2, 3]. High privacy concerns are characterised by a perceived loss of control and lack of protection of personal information [4]. When focusing on social media networks and the privacy paradox, Salleh et al. [5] discovered, in a survey of 482 undergraduate students, that sharing personal information through these sites was not perceived as risky behaviour. Instead, it was more important for these individuals to share this information as it enhanced the appearance of their profiles, making it easier for friends and family members to find and confirm the authenticity of their profiles on social networking sites.

Knowles and Conchie [6] discuss how trust and motivation come hand in hand concerning the privacy paradox. They argue that individuals' motivation to use specific platforms and share information inadvertently increases their trust in a particular service. Despite users knowing that a specific platform may not be trusted explicitly, if that individual is currently a member, they're less likely to remove themselves from that platform.

Despite extensive research on this topic, a clear understanding of why the paradox exists has yet to be reached. A recent study by Hirschprung [3] involving 427 participants concluded that technological literacy did not affect whether an individual would encounter the paradox. The research shows that social media users are aware of the information they post online and express concerns specifically about specific types of information, mainly medical history, telephone numbers, and location information. Many suggest that this has an impact on what they post online. However, these concerns do not match the results found in the privacy paradox.

2.3 Oversharing

Brammer et al. [7] examined the behaviours and perceptions associated with oversharing on social media sites. They found that previous research had focused on the benefits and frequency of social media use, but oversharing is still vastly understudied. Their study looked at 410 university students within the United States and their interpersonal communication behaviour through various social media platforms. It looked to identify factors influencing their motivation to communicate through these platforms. They broadly define oversharing as individuals sharing too much about themselves and their lives on social media. Interestingly, the participants within the study defined oversharing as posting too often online. Within the research, it was further identified that 93.6% of respondents had actively seen someone oversharing on social media platforms, with 27.8% knowingly oversharing on these platforms and 33.4% of respondents indicating that divulging medical information on social media platforms would be deemed as oversharing.

Sharenting is a term used in the context of sharing information online about a child, which could be images, videos, or news from their parents on social media sites. It has been in use since circa 2012 [8]. Fox et al. [9] explored the implications of sharenting by first-time fathers. They found that while fathers perceive their and their children's information as sensitive, they are willing to share it with marketers when prompted online. Men appear less concerned about information perceived privacy than women, backing up previous literature on this topic and engaging in privacy-focused behaviour. They also found that first-time fathers engage in sharenting to express and showcase themselves as an involved father. However, they found that the mother was the primary sharer, with the father often collaborating with their wives to share on the topic, with the wife taking more photos and updating family and friends on social media than the male counterpart. They also found that these first-time fathers exhibited actions in line with the privacy paradox, where they expressed concerns about sharing personal information but still engaged in sharenting. Risk–benefit evaluations, situational cues, and a desire for self-identity influenced sharing of personal information.

Research has shown that the instant gratification social media users feel after posting updates to their profiles can lead to disregarding security measures. The approval and positive reactions they receive often overshadow their worries about privacy. This tendency to prioritise seeking validation and recognition over online security highlights the significant impact of social dynamics on behaviour in the digital age [10].

Social media users may take time to construct their posts and think about the information they're saying, not just for their privacy concerns but also because they want to be seen in a good and positive light. They often pair these posts with images, where the adage suggests a picture paints 1000 words. Throughout a social media user's lifetime, they will post somewhere in the region from 9828 images [11]. While social media users may take time to frame the perfect picture, very often, the individual taking this image is more fixated on ensuring that the foreground on the subject of this image is ideal rather than thinking about the kind of information that can be gathered from these images.

2.4 *Concerns of Social Media Users*

A study conducted by Van Schaik et al. [12] examined how social media users perceive threats. The study found that out of 201 Facebook users surveyed, cyberbullying was perceived as the main threat to their privacy and online security, followed by the sharing of telephone numbers. The study also revealed that browsing the Internet was considered the least risky online activity out of 18 categories related to privacy and security. Results also showed that 72% of users had taken precautions when sharing their telephone numbers online. In comparison, 17% indicated that they had not taken any precautions, and 11% were unsure whether they had taken precautions. Furthermore, 43% and 37% of users took precautions when sharing a new post or reposting an existing one, respectively, while 29% and 37% stated that they took no precautions. Additionally, 28% and 26% were unsure about what precautions they were taking.

Indeed, one topic that is regularly overlooked in terms of privacy on social media sites is the potential for high-resolution images and videos to be used for biometric recognition without the consent of the owner of those images and videos. Barni et al. [13] found that specific iris recognition algorithms could perform biometric recognition using pictures taken from social media sites. This could be used for persuasive identification recognition and tracking individuals without consent. They proposed using Generative Adversarial Networks (GAN) to automatically generate novel images of a high visual realism in which all biometric information associated with that individual in the Irish region can be removed. The method replaces the individual's iris with a synthetic biometric characteristic randomly generated by the GAN, which preserves the original iris's visual aspect and colour pigmentation. Previous work within this field had only specifically looked at the biometric identification of facial traits and used simple mitigation strategies such as blurring or pixelation. Only a few studies have concentrated explicitly on iris deidentification. They hope that from their research, they can optimise an algorithm that will improve users' privacy while allowing them freedom of expression without manipulating an image beyond what they created.

2.5 *Existing Frameworks and Assessment Tools*

Pensa and Di Blasi [14] were the first researchers that this paper could find that had looked at developing a theoretical framework for privacy self-assessment. Within this framework, they took factors from social media users such as their age, gender, hometown, and the number of friends they have on their social media network.

Finally, they compute a risk score by combining all this information to provide the social media user with a risk factor based on the probability that their social media usage pertains to a likely threat to their privacy.

Khan et al. [15] explored the impact of age, gender, frequency of Internet access and cyber-protection behaviour on the level of self-disclosure. The research formulated an algorithm that could sufficiently predict the level of self-disclosure by an individual based on these categories. They also discovered that the cyber-security awareness programmes alone were not inherently enough due to the volatile landscape of security threats. As the threat to an individual evolves, so must the lessons that we give. An awareness programme created today will not be as effective as an awareness programme in 3 years. By targeting specific groups of individuals based on their self-disclosure score, resources and funding can best target individuals where it is required.

Wu and Tamilselvan [16] explored the creation of a personal privacy risk assessment framework. The framework aims to analyse the privacy risk of a person when some of the patch boots are disclosed on the framework-specific categories as personally identifiable information into three groups: passive public personally identifiable information, protected and private. The framework used considers two factors: the amount of PII information publicly available on the Internet and the sensitivity of those attributes, and the assessment looks at the willingness of people to give out their personal identity information as a direct correlation with the perception they have of the PII's privacy the privacy score is then calculated based on the willingness measure which looks at how willingly an individual will share that information and the resolution power of the person identifying information. During their research, they found 100% willingness to share their name, but individuals considered their address a much more private matter. Only 10% were willing to share that information freely online. Results also showed that 17% of individuals were willing to share their date of birth online compared to 84% of individuals who did not wish to share their personal phone numbers online. The research continues by developing a risk matrix that measures the number of PII available online against an individual. The higher the number of PII available, the greater the risk score for their privacy.

3 Exploring the Design Process

To truly understand the volume of information that can be extracted from an image, the researcher conducted an exploratory analysis of one image uploaded online through the photo-sharing platform Flickr. The image used was randomly picked using Google's open image dataset.

Upon examining Fig. 1, we observe an unassuming image of a decorated house prepared for a seasonal occasion. Although such images are commonly shared on social media, they have the potential to divulge specific information.

Performing a simple reverse image search for this image will yield various related listings on Google. In Fig. 2, you can identify a specific property that resembles that

Fig. 1 Front of property

of Fig. 1, and by using property number 445, it is possible to identify this as a potential match. The provided link will take you to a property sales website (www.trulia.com) with a detailed description of the property's layout, past sales information, financial details, and interior/exterior photos. The information provided on the site confirms that the property's exact location has been confirmed.

While viewing the property listing, the exterior images revealed that the property has rear access through a minor exterior access road. Park and Lee [17] found that 39.6% of criminals prefer using a rear door to a property, compared with only 14.4% preferring using the front door. Using Google Street View, we can visualise the front and rear access to the property, allowing criminals to get a feel for the area and potential areas of cover.

Potential exit routes from the location can be explored using a map view. The property listing also includes various details, such as an estimated monthly running cost of approximately $4347, providing insight into the individual's financial situation. Additionally, it offers information about local schools, which a potential threat actor could use to ascertain the general vicinity of the individual child's school, potentially for malicious purposes such as abduction or blackmail.

The information mentioned raises significant privacy concerns for social media users because it includes details that users wish to keep private [16]. It has been demonstrated that social media users do not have to explicitly share this information, as their images can inadvertently reveal more information than expected.

Fig. 2 Results from a Google reverse image search

4 Recommendations

We have reached a significant turning point. An entire generation has documented its whole life through social media, from the moment it was born to its first day of school to its first job. Every aspect of Generation Z and Alpha has had its life captured and shared online. This presents a significant issue for several reasons, not only due to the extensive personal information that can be obtained about this individual but also the account settings they may have on various platforms. Consider common security questions for password resets, such as "What was the name of your first school?" "What is your favourite colour?" "What is your mother's maiden name?". All this information has been recorded since the day this individual was born. This is not, however, a problem specific to Gen Z and A, but an issue that has the potential to impact all and cyber-security professionals must do more to protect individuals.

Despite the growing awareness of potential threats associated with sharing personal information on social media, users continue to do so. Merely conducting security awareness sessions is not enough to address this issue. In the rapidly evolving cyber-landscape, professionals face the challenge of timely identification and mitigation of various threats associated with social media usage. To address this problem, the researcher proposes the creation of a social media vulnerability database and risk matrix. This database will catalogue vulnerabilities arising from the content that users post on social media. Subsequently, the proposed database will utilise a risk matrix also in development to assess potential threats and their impact, providing an overall risk score for users' content.

5 Conclusion

In conclusion, the literature review reveals the alarming impact of social engineering on social media users. The study by Ariyo and Zheng [1] highlights the correlation between oversharing on social networking sites and the effectiveness of social engineering attacks, with increased data breaches reported during the COVID-19 pandemic. Furthermore, the concerns of social media users are underscored by the privacy paradox, where individuals express significant privacy concerns but continue to share personal information online, influenced by trust and motivation. This is further compounded by the lack of awareness of potential security threats and the implications of oversharing, especially for Generation Z individuals. Mitigation research, as explored by Pensa and Di Blasi [14] and Khan et al. [15], offers potential solutions through frameworks for privacy self-assessment and self-disclosure prediction. Moreover, the design analysis of image extraction from social media platforms highlights the inadvertent disclosure of sensitive information, presenting a significant threat to personal privacy and security. These findings emphasise the urgent need for heightened awareness, improved privacy policies, and targeted security awareness training to mitigate the risks faced by social media users. Addressing these privacy concerns and implementing effective mitigation strategies become increasingly imperative to safeguard personal information and online security as the global population of social media users rises.

6 Future Work

6.1 Studies

The research will use three studies to strengthen the hypothesis that social media users are unaware of the information criminals can extract from their uploaded images. The first study will involve a group setting where participants will be asked to express their feelings towards images to determine whether they can identify potential risks. The second study will use eye tracking to record users' gaze patterns when viewing images, helping to identify if users can recognise these vulnerabilities and contributing to the research on user awareness. Lastly, a third study will use eye tracking to evaluate the effectiveness of the warning system developed during this research.

The research will use Google's Open Image Dataset V7 with images licensed under Creative Commons 2.0. Eye-tracking studies will be conducted using GazePoint's GP3 hardware and the corresponding GazePoint analytic software.

6.2 Computer Vision

The development of a social media vulnerability database and a risk matrix alone will not be sufficient to address the problems posed by the high volume of information posted by users online. As the research progresses, the development of an object detection tool will start. The tool will be developed using YOLOv8, which is an existing computer vision model and will be trained on Google's Open Image Dataset. The tool will be trained using images identified as containing potential vulnerabilities highlighted in the SMVD. Social media users can pass their posts through the tool, which will analyse the image to identify potential vulnerabilities and highlight these to the end user. The initial iterations of the tool will identify ten potential vulnerabilities as an initial proof of concept.

6.3 Human–Computer Interaction

The project must leverage the Social Media Vulnerability Detection (SMVD) tool, the risk matrix, and analysis from computer vision technology to develop a comprehensive warning system. This system should effectively convey warning messages to users, ensuring they are well informed and educated. The primary goal is to empower users to consciously decide whether to continue uploading their posts online based on the insights provided.

References

1. Ariyo O, Zheng J (2022) A study on security and privacy risks of self-disclosure on social networking sites during COVID-19 pandemic. In: 2022 IEEE international conference on Big Data (Big Data) [Preprint]. https://doi.org/10.1109/bigdata55660.2022.10021102
2. Chhabra S, Mukherjee J, Agarwal MN (2023) 'Understanding privacy paradox: An experimental approach to study privacy concerns and actual disclosure among app users'. J Organ Comput Electron Commer 33(3–4):97–116. https://doi.org/10.1080/10919392.2023.2251328
3. Hirschprung RS (2023) 'Is the privacy paradox a domain-specific phenomenon'. Computers 12(8):156. https://doi.org/10.3390/computers12080156
4. Stevic A et al (2022) 'Privacy concerns can stress you out: Investigating the reciprocal relationship between mobile social media privacy concerns and perceived stress'. Communications 47(3):327–349. https://doi.org/10.1515/commun2020-0037
5. Salleh N et al (2013) Examining information disclosure behavior on social network sites using protection motivation theory, trust and risk. J Internet Soc Netw Virtual Communities, pp 1–11. https://doi.org/10.5171/2012.281869
6. Knowles B, Conchie S (2023) 'Un-paradoxing privacy: Considering hopeful trust'. ACM Trans Comput-Hum Interact 30(6):1–24. https://doi.org/10.1145/3609329
7. Brammer SE, Punyanunt-Carter NM, Duffee RS (2022) 'Oversharing on social networking sites: A contemporary communication phenomenon'. Comput Hum Behav Rep 8:100236. https://doi.org/10.1016/j.chbr.2022.100236
8. Oxford English Dictionary, s.v. sharenting (n.). https://doi.org/10.1093/OED/7727760589

9. Fox AK, Hoy MG, Carter AE (2022) 'An exploration of first-time dads' sharenting with social media marketers: implications for children's online privacy'. J Mark Theory Pract 31(2):185–196. https://doi.org/10.1080/10696679.2021.2024441
10. Van Oosten JMF, Vandenbosch L, Peter J (2023) 'Predicting the use of visually oriented social media: The role of psychological well-being, body image concerns and sought appearance gratifications'. Comput Hum Behav 144:107730. https://doi.org/10.1016/j.chb.2023.107730
11. Study Finds (2022) Average lifetime online footprint of social media users in the United Kingdom as of July 2022. Statista. Statista inc. https://wwwstatista-com.ezproxy.cardiffmet.ac.uk/statistics/1402267/uk-online-footprintsocial-media-users/ (Accessed: September 09, 2024)
12. Van Schaik P et al (2018) 'Security and privacy in online social networking: Risk perceptions and precautionary behaviour'. Comput Hum Behav 78:283–297. https://doi.org/10.1016/j.chb.2017.10.007
13. Barni M et al (2021) 'Iris deidentification with high visual realism for privacy protection on websites and social networks'. IEEE Access 9:131995–132010. https://doi.org/10.1109/access.2021.3114588
14. Pensa RG, Di Blasi G (2017) 'A privacy self-assessment framework for online social networks'. Expert Syst Appl 86:18–31. https://doi.org/10.1016/j.eswa.2017.05.054
15. Khan NF et al (2021) 'Social media users and cybersecurity awareness: predicting self-disclosure using a hybrid artificial intelligence approach'. Kybernetes 52(1):401–421. https://doi.org/10.1108/k-05-2021-0377
16. Wu N, Tamilselvan R (2023) A personal privacy risk assessment framework based on disclosed PII. In: Proceedings—2023 7th International Conference on Cryptography, Security and Privacy, CSP 2023. Institute of Electrical and Electronics Engineers Inc., pp 86–91. https://doi.org/10.1109/CSP58884.2023.00021.
17. Park SY, Lee KH (2021) 'Burglars' choice of intrusion routes: A virtual reality experimental study'. J Environ Psychol 74:101582. https://doi.org/10.1016/j.jenvp.2021.10158

Enhancing Cyber-Security and User Experience in Digital Banking Platforms: A Comprehensive Approach

Stephen A. Ekeh, Sabeen Tahi, and Linda Mohaisen

Abstract The banking industry has experienced a rapid shift from traditional face-to-face services to online banking, with more than 50% of worldwide banking services currently being conducted through the Internet. Digitalization has had a profound impact on the way customers engage with banking services on a daily basis. This paper underscores the main objectives of the web-based platform, to provide a user-friendly interface, and contemporary cyber-security approaches to meet the shifting customer preferences, thereby safeguarding user data and privacy. It highlights the cyber-security enhancement needed to mitigate against any possible future cyber-threats, a comprehensive implementation of APIs and the use of CoPilot as mandatory steps in the design phase of the digital platform. This paper doesn't just serve as a blueprint for future developers of secured digital platforms, but also ensures compliant with all appropriate regulations. The scope of this papers covers browser and device compatibility, improve accessibility, and strong security and fraud protection system. It seamlessly integrates third-party services, adopts AI-driven features, and uses agile development to improve user experience.

Keywords Digital platforms · Security approaches · Vulnerability tests

S. A. Ekeh (✉) · S. Tahi
Department of Computer Science, Cardiff School of Technologies, Cardiff Metropolitan University, Cardiff, UK
e-mail: st20253071@outlook.cardiffmet.ac.uk

S. Tahi
e-mail: stahir@cardiffmet.ac.uk

L. Mohaisen
Department of Information Technology, Faculty of Computing and Information Technology, King Abdulaziz University, Jeddah, Saudi Arabia
e-mail: lmohaisen@kau.edu.sa

C. E. R. Hewage et al. (eds.), *AI Applications in Cyber Security and Privacy of Communication Networks*, Lecture Notes in Networks and Systems 1453,
https://doi.org/10.1007/978-981-96-7400-8_12

1 Introduction

The introduction of technology has brought about a substantial revolution in traditional banking processes. Traditional physical banking institutions have extended their operations into the digital sphere to meet the increasing need for convenient, easily available, and secure financial services. An analysis conducted by [1] reveals that digitalization significantly alters how people interact and engage in daily commercial activities such as banking services. The new shift has led to the developments of digital platforms such as financial applications promoting a profound impact on improving financial services globally. According to [2], the banking industry is experiencing a transformative change in its operations due to the increasing demand for digital banking services among millennials and Generation Z individuals.

This paper proposes a comprehensive approach to identifying the key security objectives, performing a detailed vulnerability assessment, implementing proper mitigation measures, consider the security of application programming interfaces, by safeguarding user data and privacy, and incorporate strong authentication and authorization techniques. Thus, by addressing these issues in advance, the developers will be able to design an enhanced secure system for digital financial platforms.

The methodology that is being suggested in this paper is well grounded and based on the vast literature review that critiques the current cyber-security threats faced by digital platforms and identifies the gaps. It includes security issues, risks, development processes, and factors that concern the user authentication, authorization, and privacy.

2 Literature Review

The literature evaluation of innovation models focuses on four critical key attributes that elaborate the concept of Internet Banking and its defining qualities. This section evaluates the diffusion of innovation, the technological adoption paradigm, and the trust factor. Segmentation is also analyzed as an innovation concept in Internet banking.

2.1 Advancement in Technology Innovations and Shift in Customer Preference

Financial institutions have recently embraced online banking as an essential factor to maintain competitiveness. Arcand et al. [3] assert that integrating mobile banking is necessary to tap into the growing population of Generation Z and millennials. According to [4], 45% of individuals who use banking services consider mobile banking one of the top three factors influencing their selection of financial institutions.

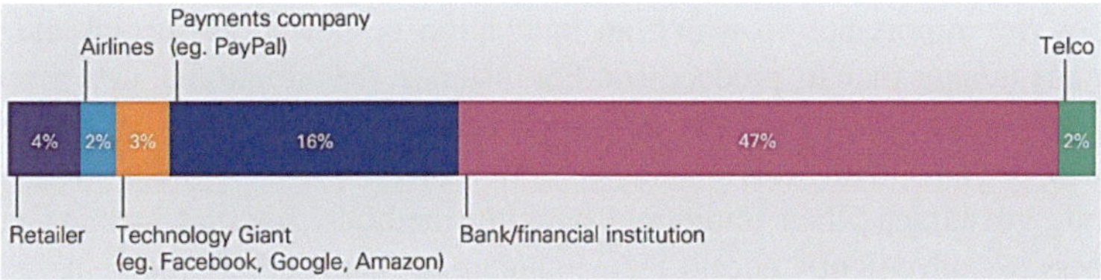

Fig. 1 The survey results on trust in banks, with 47% trusting their banks to keep data secure

Rapid technological advancements in web-based technologies, cloud computing, cyber-security, and data analytics are critical in strengthening the development of robust banking technological solutions. Dianat et al. [5] highlight that the emergence of blockchain technologies, cloud computing, and artificial intelligence has tremendously contributed to a rapid shift in the operational landscape of financial institutions. Technological developments and innovations have enhanced efficiency, bolstering security measures and delivering tailored consumer experiences. However, the cyber-security threats remain a high-level challenge. Figure 1 gives the outcome of a survey on those who trust the bank to keep their data.

2.2 *Diffusion of Innovation*

Ghelani et al. [6] define diffusion of innovation as the process by which new ideas, technologies, or practices spread and are adopted by individuals or groups within a social system. The model conceptualizes the adoption of Internet Banking as a social construct that gradually spreads throughout the community. Hendriyani and Raharja [7] argue that individuals display variation in their willingness to accept Internet Banking as an innovation. According to [8], the adoption rate of Internet Banking is impacted by five aspects within the theoretical framework. Some factors affecting web-based banking platforms include relative advantage, compatibility, complexity, trialability, and observability. This model is frequently observed in information system's research to elucidate the phenomenon of user acceptance and integration of novel technologies [9]. Recent studies have demonstrated that these attributes are critical in shaping the acceptance and proliferation of Internet-based technologies, with the relative advantage, compatibility, and complexity being the most prominent characteristics of Internet banking.

2.3 *Relative Advantage*

Relative advantage refers to the extent to which an innovation or technology is viewed as superior to existing alternatives. The concept of comparative advantage describes the importance to which an innovation is viewed as special and offers greater advantages than its predecessor. For instance, the adoption of new inventions is positively influenced by relative advantage [10]. In Internet Banking, comparative advantage pertains to an individual's perception that Internet Banking offers greater economic advantages than traditional banking methods. Internet banking enables customers to conveniently access their financial accounts at any moment and from any location. The product offers significant benefits to the consumer. Relative benefit has been recognized as an important factor influencing customers' inclination to use Internet Banking services. The transmission of the comparative advantage to consumers is of great importance for financial organizations such as commercial and retail banks.

3 Methodology

The project methodology employs a systematic approach to ensure the successful implementation of the web-based online banking platform. The following steps are necessary to implement a systematic approach:

Requirement Analysis: Evaluating, analysis, and documenting the specific needs and expectations of users, regulators, and stakeholders.

Design Phase: Creating a comprehensive plan that encompasses technical architecture, user interface, security measures, and regulatory compliance. This phase is very critical as the SDLC must be adhered to, as well as the implementations of embedded security protocols.

Development and Testing: Building the platform based on the design, conducting rigorous testing at each stage to identify and rectify any issues or bugs. Testing will ensure security and data protection.

Implementation: Deploying the platform in phases or all at once, depending on the readiness and requirements, ensuring a smooth transition.

Monitoring and Optimization: Constantly monitoring the platform's performance, gathering user feedback, and making necessary improvements or optimizations.

Security Measures: Implementing robust security protocols and continuously updating them to address evolving threats. The platform will employ data encryption, multi-factor authentication, and patching.

Compliance Adherence: Keeping abreast of regulatory changes and ensuring that the platform remains compliant with relevant laws and standards.

Throughout these stages, a collaborative and iterative approach will be adopted, involving stakeholders, developers, designers, and security experts. Regular reviews and feedback loops will be integral to refine and enhance the platform's features and functionalities. This methodology aims to ensure a holistic and efficient approach to developing and maintaining a robust web-based banking platform.

4 Design and Architecture

System Architecture Design—Define the system's architecture with scalability, performance, and security. This involves developing the banking platform, including coding and designing the platform's security measures. Sarker [11] asserts that data encryption and multi-factor authentication are important to the system's security at this stage. The platform system architecture design will consider integration compatibility with software applications available on the cloud. For instance, the plan will consider compatibility with Android and other operating systems on the Internet, such as Microsoft software applications.

4.1 Integration of APIs and Connections to External Services

Incorporating application programming interfaces (APIs) and external services enables apps to access a vast network of resources, providing users with a wider array of capabilities. Application Programming Interfaces (APIs) refer to a collection of protocols and technologies that facilitate the exchange of information and functionality between diverse systems and applications. This enables the seamless sharing of data and services among these entities. Application Programming Interfaces (APIs) can connect with various resources, including databases, cloud storage systems, weather services, and social networking platforms. Integrate APIs for uninterrupted connectivity with payment gateways, third-party services, and banking systems. Ensure API compatibility and security to facilitate data exchange.

4.2 Security Framework and Compliance Planning

A security framework encompasses a collection of principles, regulations, and standards that facilitate the safeguarding of an organization's assets against potential threats. The framework presented offers a systematic approach to assess risks, detect vulnerabilities, and establish protective measures to mitigate the possible consequences of security breaches. Compliance planning is the routine procedure to guarantee that an organization fulfills or surpasses the stipulated regulatory and industry standards. The process entails identifying, evaluating, and mitigating risks associated with regulatory compliance to prevent the imposition of penalties, fines, and harm to one's reputation. Integrating a security framework and compliance planning enables organizations to establish a strong and dependable security stance, safeguarding their assets and guaranteeing adherence to pertinent legislation and standards. Construct an all-encompassing security framework incorporating secure data storage, authentication, encryption, and authorization techniques.

4.3 Comprehensive Testing

Implementing thorough testing protocols in web-based banking systems is paramount to guarantee the system's security, functionality, and usability. The process encompasses the integration of many testing methodologies, including functional testing, performance testing, security testing, and user acceptance testing. Conduct various testing methodologies, such as user acceptability (UAT), unit, integration, and performance testing. When feasible, implement automated testing instruments to accelerate the testing process.

4.4 Static Analysis

It is possible to use static analysis tools that can detect potential vulnerabilities without source executing code it. This can be very useful when it comes to the identification of vulnerabilities at the initial stages of the software development process. Some methods include source code scanning and threat modeling.

Code review automation can be carried out by integrating tools like SonarQube into the build process, and we can identify vulnerabilities in the development phase. This is done by developing custom rule sets to make sure that the tools are well tuned to meet the needs of the application. So, the tool will be able to catch the vulnerabilities at the initial level.

4.5 Deployment Strategy and Rollout

Reference [12] highlights various deployment methodologies for implementing a web-based banking platform, encompassing the phased, big bang, and gradual approaches. The phased approach entails the systematic division of the site and the subsequent implementation of specific elements or parts at distinct intervals. Develop a phased deployment strategy to protect users and internal stakeholders from potential harm and guarantee a seamless transition. Vigilantly oversee the deployment process to resolve any issues that may arise promptly.

4.6 Continuous Improvement and Maintenance

Utilize analytics, user behavior data, and surveys to collect user feedback to identify areas requiring enhancement. Continuous improvement develops a strategy of periodic updates, feature enhancements, and problem fixes in response to user feedback and technological developments.

5 User Interface (UI) and User Experience (UX)

The dashboard and navigation in banking platforms provide an aesthetically pleasing and user-friendly dashboard that comprehensively summarizes accounts, transactions, and services. Efficient navigation facilitates smooth transitions between various sections. The user interface is the gateway to marketing a banking system. A pleasing interface enhances customer satisfaction and attracts better uptake of bank products.

5.1 Account Management

Capabilities to administer preferences and access account information, statements, and transaction histories. Customer convenience is met through providing account accessibility where they can navigate and obtain necessary information for their consumption.

5.2 Transaction Services

Interfaces for administering beneficiaries, transferring funds between accounts, paying bills, and establishing recurring transactions. Dianat et al. (2019) argue that transaction services through a web-based platform enable customer convenience, leading to increased productivity.

5.3 Security Features

Some features, such as multi-factor authentication, secure password protocols, and notifications for suspicious activity or login attempts, enhance safety. These security features protect customer data from malicious and unauthorized access. Efficient security features improve data encryption to maintain confidentiality.

5.4 Banking System Integration

There is optimal integration with fundamental banking systems to facilitate transaction processing, account information updating, and synchronization maintenance. System integration enhances operational efficiency and convenience for customers and bank employees.

5.5 Third-Party Integrations and APIs

APIs facilitate connectivity with credit bureaus, payment gateways, and other financial services to enable additional functionalities. This makes it easier for customers to make payments through various platforms and financial services.

5.6 Scalable Architecture

Building a resilient and expandable infrastructure that can accommodate growing user demands and technological progress is essential. Besides, a dynamic system is maintained to support future integration with diverse applications.

6 Security Measures Encryption Protocols

Implementing well-implemented protocols for end-to-end encryption that safeguard sensitive information throughout storage and transmission is critical for a web-based baking platform. It prevents any security breaches that may lead to manipulation of customer data.

6.1 Authentication and Authorization

A system that employs multi-layered authentication mechanisms to guarantee that access to accounts and transactions is restricted solely to authorized individuals. The authentication of user login details and customer accessibility is essential to protect organizational data.

6.2 Regular Security Audits

Consistent security audits include ongoing penetration testing, security assessments, and updates to mitigate emergent threats. These regular audits are necessary to ensure the effectiveness of penetration tests and security assessments.

6.3 Support Feedback Mechanisms and Continuous Improvement

Established mechanisms facilitate user input, enabling ongoing service enhancements and the overall user experience. Customer feedback is evaluated, and insights lead to developing new software applications or system improvements to meet customer satisfaction.

6.4 Enhancements and Updates

The system ensures consistent problem fixes, feature additions, and updates in response to user input, technological developments, and security standards. These updates are essential to identify any system vulnerabilities and safeguard customer information.

6.5 Customer Support

Channels for responsive customer support that assist users with inquiries, troubleshooting, and platform usage guidance. These customer support mechanisms enable smooth onboarding of new customers and enhance customer satisfaction.

6.6 Full-Time Access

Unrestricted access to banking services allows customers to manage their finances on schedule. The development of a web-based banking platform enhances account accessibility and convenience access through the Internet.

6.7 Cross-Device Functionality

Seamless access across multiple devices (PCs, tablets, smartphones) for a consistent user experience is a significant achievement. Customers can access their accounts through various devices, making it convenient and reliable. Convenience accessibility improves customer service experience, increasing productivity.

6.8 Adherence to Regulations

Strict compliance with industry regulations and standards like KYC, AML, GDPR, and PCI DSS, ensuring legal compliance and trustworthiness. Upon consideration of all the regulatory guidelines, the banking platform complies with all the regulations, including providing data safety and compliance.

6.9 Operational Efficiency

Reduction in operating costs, paperwork, and branch visits, optimizing resource utilization and enhancing efficiency.

7 Conclusion

Technology integration, especially web-based banking platforms, has revolutionized how customers access and interact with banking services. The growing demand for convenient, secure, and accessible financial solutions drives this change. The rise of web platforms in banking has played a key role in meeting changing customer needs. Technological advancements such as cloud computing, artificial intelligence, and blockchain have played a vital role in improving operational efficiency, enhancing security measures, and delivering a tailored guest experience.

Based on the security design review of the forthcoming financial application, there is a detailed analysis and recommendation to the design of the application with focus on the protection of customer's data and compliance with the set regulations. Key findings are given below:

Security Objectives Identification: The security objectives that have been identified are confidentiality, integrity, availability, and compliance with the UK General Data Protection Regulation (UKGDPR) and Payment Card Industry Data Security Standard (PCI DSS).

Vulnerability and Mitigation Strategies: All the vulnerabilities explained in each zone and the suggested mitigation for each vulnerability like input validation, smart pointers, and encryption.

Also, by embedding security testing into all the software development process, static and dynamic analysis, threat modeling, and continuous code inspection.

Data Protection and Privacy: Highlighted the need for the encryption of data stored, the reduction of data retained, the encryption of data in transit, user permission, and data retention policies.

Authentication and Authorization: The measures that had been proposed include the use of Multi-Factor Authentication (MFA), secure storage of passwords, role-based access control (RAC), tokenization, strong session management, audit trail, security training, third-party security assessment, and secure API.

8 Recommendations

The following are recommendations for the development of a comprehensive security strategy that achieves the identified security objectives for confidentiality, integrity, availability, and UKGDPR compliance.

In order to enhance the security of user account, the following should be implemented: integration of strong user authentication mechanisms such as the use of multi-factor authentication (MFA) and password management.

Define strong authentication mechanisms through tokenization and secure session management.

The following are secure session management practices that should be adhered to: logs and monitor authentication and authorization events to identify and deal with suspicious activities.

Ensure that everyone in the development team undergo through security training to ensure that they know the need of implementing secure authentication and authorization.

If the following recommendations are included in the development process, then financial institutions will be able to enhance the security of their financial application and protect customer's data effectively while also meeting the set of legal requirements. It is crucial to integrate these security principles at the very beginning of the project and then keep on monitoring the project to adapt it to the potential changes in the threat environment. Thus, it will be possible to create a secure application and at the same time enhance trust and confidence of its users and partners.

Acknowledgements This research was supported by the UK-Saudi Challenge Fund 2024, British Council, under the project titled "Data Fusion Dynamics: A Collaborative UK-Saudi Initiative in Cybersecurity and Artificial Intelligence."

References

1. Patel KJ, Patel HJ (2018) Adoption of Internet banking services in Gujarat: An extension of TAM with perceived security and social influence. Int J Bank Market 36(1):147–169
2. Al-Gharaibah OB (2020) Predictors of E-banking service adoption in Malaysia using an extended technology acceptance model. Int J Contemp Manag Inf Technol 1(1):23–29
3. Arcand M, PromTep S, Brun I, Rajaobelina L (2017) Mobile banking service quality and customer relationships. Int J Bank Mark 35(7):1068–1089

4. Chawla D, Joshi H (2017) Consumer perspectives about mobile banking adoption in India–a cluster analysis. Int J Bank Mark 35(4):616–636
5. Dianat I, Adeli P, Jafarabadi MA, Karimi MA (2019) User-centered web design, usability, and user satisfaction: The case of online banking websites in Iran. Appl Ergon 81:102892
6. Ghelani D, Hua TK, Koduru SKR (2022) A model-driven approach for online banking application using angularJS framework. Am J Inf Sci Technol 6(3):52–63
7. Hendriyani C, Raharja SUJ (2018) Analysis of building customer engagement through e-CRM in Indonesia's digital banking era. Int J Econ Policy Emerg Econ 11(5):479–486
8. Islam JU et al (2020) Impact of website attributes on customer engagement in banking: a solicitation of stimulus-organism-response theory. Int J Bank Mark 38(6):1279–1303
9. Patel KJ, Patel HJ (2018) Adoption of Internet banking services in Gujarat: An extension of TAM with perceived security and social influence. Int J Bank Mark 36(1):147–169
10. Rahi S, Abd M (2018) Ghani, "The role of UTAUT, DOI, perceived technology security, and game elements in Internet banking adoption." World J Sci Technol Sustain Dev 15(4):338–356
11. Sarker B, Sarker B, Podder P, Robel MRA (2020) Progression of internet banking system in Bangladesh and its challenges. Int J Comput Appl 177(29):11–15
12. MSMEs Bank in Implementing Crowdfunding. In Proceedings of the 2nd International Conference of Business, Accounting and Economics, ICBAE 2020, 5–6 August 2020, Purwokerto, Indonesia

Author Index

C. E. R. Hewage et al. (eds.), *AI Applications in Cyber Security and Privacy of Communication Networks*, Lecture Notes in Networks and Systems 1453,
https://doi.org/10.1007/978-981-96-7400-8

MIX
Papier aus verantwortungsvollen Quellen
Paper from responsible sources
FSC® C105338

If you have any concerns about our products,
you can contact us on
ProductSafety@springernature.com

In case Publisher is established outside the EU,
the EU authorized representative is:
Springer Nature Customer Service Center GmbH
Europaplatz 3, 69115 Heidelberg, Germany

Printed by Libri Plureos GmbH
in Hamburg, Germany

Sociologies of Formality and Informality

STUDIES IN SOCIAL SCIENCES, PHILOSOPHY AND HISTORY OF IDEAS

Edited by Andrzej Rychard

VOLUME 12

PETER LANG
EDITION

Adriana Mica / Jan Winczorek / Rafał Wiśniewski (eds.)

Sociologies of Formality and Informality

PETER LANG
EDITION

Bibliographic Information published by the Deutsche Nationalbibliothek
The Deutsche Nationalbibliothek lists this publication in the Deutsche Nationalbibliografie; detailed bibliographic data is available in the internet at http://dnb.d-nb.de.

Library of Congress Cataloging-in-Publication Data
Sociologies of formality and informality / Adriana Mica, Jan Winczorek, Rafał Wiśniewski, (eds.).
pages cm. — (Studies in social sciences, philosophy and history of ideas, ISSN 2196-0151 ; volume 12)
ISBN 978-3-631-65328-9
1. Social institutions. 2. Social systems. 3. Organizational sociology. 4. Sociology—Philosophy. I. Mica, Adriana. II. Winczorek, Jan. III. Wiśniewski, Rafał, 1977-
HM826.S63 2015
306--dc23

2015009648

This Publication was financially supported by the Polish Sociological Association.

ISSN 2196-0151
ISBN 978-3-631-65328-9 (Print)
E-ISBN 978-3-653-04521-5 (E-Book)
DOI 10.3726/978-3-653-04521-5

Peter Lang Edition is an Imprint of Peter Lang GmbH.

Peter Lang – Frankfurt am Main · Bern · Bruxelles · New York · Oxford · Warszawa · Wien

This publication has been peer reviewed.

www.peterlang.com

Acknowledgements

Some of the edited papers in this volume were presented at *Formality and Informality: From Decoupling to Entanglement. Workshop on the Formal and the Informal in Law, Institutions and Economy*, a scholarly conference that was organized by the Polish Sociological Association (Sociology of Law Section and Warsaw Department) and the University of Warsaw (Faculty of Law and Administration and Institute of Applied Social Sciences), with the support of the Ministry of Science and Higher Education of the Republic of Poland, and which took place on 13–14 May 2013 in Warsaw. This was the first occasion for discussing some of the ideas and concepts which were further developed in this volume. It was then followed by a period of intense writing and consulting between the editors and the selected and invited contributors to this project. We hope that the ensuing volume gives justice to all these efforts, original ideas and the scholarly potential of the theme undertaken by the edited work.

The publication of the book has been made possible thanks to the support from a number of institutions, including the Polish Sociological Association and the University of Warsaw, and the contribution to, and supervision of, the manuscript works by Barbara Damentka, Jarosław Kilias and Michael Landry. Last but not least, the editors would like to thank the participants in the workshop discussions and the contributing authors for taking part in this book project and for their input to the conceptualizing of various sociologies of formality and informality.

Contents

Adriana Mica, Jan Winczorek and Rafał Wiśniewski

Sociologies of Formality and Informality

Formality and informality are indelible elements of social life. They also happen to bear a special relationship to each other. On the one hand, the persistence of modern societies depends on a great variety of formal structures such as formal organizations and formal rules, which organize collective life and guide individual actions. On the other hand, today's societies are also perpetuated by a wealth of informal practices, including ones performed within and around formal institutions. It is indeed trivial to observe that every formal rule, organization or interaction is accompanied by an informal counterpart.

Conversely, every informal practice, institutionalized or occasional, takes place in a formal environment. Undeniably, this dialectics has many practical consequences. It also renders formality and informality as interesting objects of study for sociologists, traditionally inclined to lurk behind official facades. Has a social institution been successful because it had formal traits or just to the contrary, because it included informal elements? What are the informal undercurrents and preconditions of formal life? Is the informal side of an institution reasonably tamed by rationally crafted formalities or stifled by irrational bureaucracy?

This interest in the formal and the informal spans across many sociological disciplines. It has a firm place in the sociology of organizations, sociology of law, sociology of culture, development studies, sociology of work, and discourse analysis. Already this disciplinary multiplicity constitutes a sufficient reason to speak of sociologies of formality and informality rather than about a single sociology of these phenomena. As it often happens, representatives of different sociological trades are not necessarily in agreement as to what counts as formal or informal and what role they actually play in the phenomena studied.

For these reasons, the view of formality and informality and their linkage in sociology is complex and multifaceted. Anyone who intends to present the state-of-the-art in this field thus runs the risk of omitting some intricacies of theoretical baggage. One way to ensure that actual synthesis is provided is to start with the criteria that are used by particular sociological discourses in depicting the relationship between formality and informality. Stinchcombe (2001, 5–9), for example, advanced a typology of informality in the context of law and organizations comprising: "informally embedded formality", "formality being constructed" and "classical informality". At least one of these categories, if not two, could be hijacked for the purpose of creating a categorization of the general perspectives on the interlinkage between formality and informality. In turn, we could identify specific streams for conceptualizing formality and informality within these discourses – i.e. as revealed, for example, by the Böröcz (2000) and the Guha-Khasnobis, Kanbur and Ostrom (2006, 5) categorizations of literatures on informality. Böröcz (2000), for instance, identified two categorizations: "the school of »generic informality«" and the school of "sectoral informality". While coming from a public policy analysis viewpoint, Guha-Khasnobis, Kanbur and Ostrom (2006, 5) highlighted two framings that are nonetheless of sociological relevance: "the reach of official governance" and "the degree of structuring". In a similar vein, we could further differentiate among ways of depicting informality in terms of forms of constraint (new institutionalism – see North 2000), level of abstraction (sociology of law – see Stinchcombe 2001) and degree of freedom in interpretation of role requirements (sociology of culture, research of social cooperation – see Misztal 2000).

We propose a typology comprising: *the sociology of informally embedded formality, the sociology of formally embedded informality, the sociology of the interaction between formality and informality* and *the sociology of the emergence and transformation of formality and informality.* Learning from the analysis of a seminal author in economic sociology, we could benefit from looking at these notions as being meta-assumptions grounding the sociology of formality and informality – as in Portes (2010, 13), the four directions of investigating formality and informality presuppose distinct "«lenses» through which reality is grasped and explored". Still, more than in the case of economic sociology, in the sociology of formality and informality these meta-assumptions appear as superficially competing. We say competing because, when rendering the classification, it was almost as if we took Beckert's (2006) distinction between the "interpenetration" and "embeddedness" approaches to the relationship between the economy and society in economic sociology, and adjusted and extended it to depict views on the linkage between formality and informality in contemporary sociology. The following brief outline of the four frameworks for approaching this problem will probably give an initial idea of the extent to which these perspectives seem to be competing or not. Its purpose is also to show that the chosen contributors are authors whose work is illustrative of distinct types of theoretical framings and presents sites of inquiry as various as possible.

Part I concerns the *sociology of informally embedded formality* – that is, sociology relying on, or bringing in, informality-related explanatory mechanisms in the study of formality, formalization and formal organization. As is visible in the first contribution, in the sociology of law and the sociology of organizations, this stream of research is highly indebted to Stinchcombe (2001). In this chapter, Robert Dingwall, another established contributor to this stream, revisits the arguments made in the paper co-authored with Phil M. Strong, *The Interactional Study of Organizations* (1985), in the context of new developments in framing formality brought by new institutionalism, and inhabited institutionalism theorists in particular. The chapter promotes the research of the interactional construction of organizational formality. The notion of *charter*, which is proposed for framing the formal dimension of organizational life, aims to restore some balance in the study of formality. The idea is to study formality in a way that would not give in to the informality aspects to the extent that it would end up considering that "formality is all a fraud" – as Stinchcombe (2001, 1) observed that sociologists usually do – yet also not overlook the input of people towards the construction, negotiation, display and challenge of an organization's charter. In Chapter II, Grażyna Skąpska and Grzegorz Bryda interpret findings pointing to an obvious discrepancy between the opinions of lawyers and non-lawyers concerning the implementation of the rule of law in Poland. The discussion on the issue offers the occasion to touch on two related topics. First, the research looks at the reconsideration of the rule of law in the XXI century, subsequent to jolting social changes and the uncertainty facing regulation and implementation. Second, the authors discuss the need of an *empirical account of the rule of law* grounded in social experiences, in local memory and local knowledge. Although not framed in terms of "a charter", the chapter comes very close to the study of formality in the framing advocated by Dingwall and Strong. What evidently counts as an advantage of Skąpska and Bryda's paper, however, is that they interpret the relation and engagement with the rule of law charter by various social actors.

Part II presents contributions from the *sociology of formally embedded informality*. Depending on the case study, this sociological investigation results in a general recognition of the structural embeddedness of informality in the degree of regulation, costs of complying with the rules institutionalized by the state, or the ability and scope of regulation enforcement (see Fernández-Kelly and Garcia 1991; Sassen 1997; Portes and Haller 2005; Centeno and Portes 2006; Kus 2006; Portes 2010; Kanbur 2012). In the first paper in this section, Liela Groenewald redefines the notion of informal settlement in such a way that this would be more representative of the experiences of ordinary, poor people living in informal settlements in the global South and in particular in southern Africa. She insists on mainly three aspects: the interconnection between formality and informality; class structure and conflict of interest; and tenure insecurity and precariousness (primarily contributed by the state, the formal domain). Groenewald makes a point regarding the third characteristic, namely that the insecurity felt in relation to the formal domain pops up as the primary factor in the self-reflective conceptualization of the residents of informal settlements. She also indicates how this aspect "creates conceptual problems for a purely repressive state response". In the next contribution, Stef Adriaenssens, Dieter Verhaest and Jef Hendrickx also deal with a multidimensional definition of informality, and informal work in particular. In their case, however, the causal priors are all located in formality, in types of regulation. The authors advance a pilot study of multidimensionality – a binary depiction of informality in relation to labor regulation and taxation. The topic is quite relevant because the *multidimensionality of informality* has the potential to reveal both the pros and cons of defining informality as a violation and lack of protection by regulatory structure.

Part III is dedicated to the *sociology of the interaction between formality and informality* – that is, sociology which is less interested in clearly delimiting the formal and informal domains, and more in establishing types of relationships between formal and informal institutions, and in revealing their mutual conditioning, entanglement or decoupling (see Meyer and Rowan 1977; North 1990; Pejovich 1999; Lauth 2000; 2004; Misztal 2000; 2005; Nee and Ingram 2001; Helmke and Levitsky 2006; Pejovich and Colombatto 2008; Bromley and Powell 2012; Van Assche, Beunen and Duineveld 2014). In the first paper in this section, Barbara Misztal continues and revises the understanding of informality that she originally drafted in the book, *Informality: Social Theory and Contemporary Practice* (2000). In addition to Erving Goffman and Norbert Elias, she now also builds on Michel Foucault in the study of re-patterned configurations of formality and informality. The paper analyzes changes in the relationship of informality and formality in the contemporary setting, and the consequences of these developments in terms of the emergence of new types of informality (*formalized* and *instrumental informality*), and of the sustainment of cooperation and the exercise of social control. In the second contribution, Mikko Lagerspetz discusses the relationship between formal policies and informal practices now prevailing in the Estonian minority incorporation regime. The advanced case-study on the process of changing the Russian gymnasiums' language of tuition allows him to make some inferences about the mechanisms and possible consequences of decoupling in the political field. In the third work, Hans-Joachim Lauth, in a similar vein, examines the relationship between rule of law and informal legal systems in functioning and deficient types of *Rechtsstaat*. He specifies that the interaction between the systems differs in relation to the political regime types, and puts forward a categorization of competing legal systems on the basis of evidence from authoritarian regimes and young democracies – *hybrid legal system* and the *deficient rule of law*.

Lastly, Part IV deals with the *sociology of the emergence and transformation of formality and informality*. This stream studies how interaction processes effect the transformation of existing institutions or the emergence of new ones; and it also follows processes for the formalization of informal institutions, as well as sequences of informalization and the relaxation of formal rules (Knight 1992; Tsai 2006; Grzymala-Busse 2010; Carruthers 2012; Haldar and Stiglitz 2013, 113). In the first contribution to this part, Timothy Eccles depicts both the processes of the construction and deconstruction of formality in the context of deregulation of building control in the UK – that is, fragmentation of authority in the field. Eccles shows that transformation or lack of authority does not equal informalization, or a move away from formality. He introduces two concepts: *meta-formality* and *trans-informality*. The former pertains to the situation when various, competing, authorities act in a rational-legal manner, without a single dominant authority. The latter meanwhile is employed when informality moves towards formality, in the sense that the rational-legal approach is adopted into certain informal systems. The notions are important because they encourage thinking "outside" the formal-informal distinction/continuum. By pointing out that formality is no more unitary and homogenous than informality is, Eccles brings to our attention the interaction between formal and formal institutions and systems, in addition to that between formal and informal ones. In the second chapter, Aleksandra Herman deals with a phenomenon recalling Eccles's trans-informality – processes of the reconfiguration of power at the local level which entail the absorption of informal political forces in the formal domain, and the blurring of boundaries between the formal and the informal in the political field. She looks at how separate social institutions operate at the bottom level of self-governance and considers the political potential of informality in the local environment. We can risk a comparison between the two approaches: Eccles is interested in the manner in which processes of construction and deconstruction of formality within the field of regulation lead to new types of formality and informality, while Herman looks at how similar processes, within the field of local politics, lead to the blurring of boundaries. In the third chapter, Francisco Linares advances a computational simulation analysis of the effect of the network topology on the emergence of informal norms of resistance among peer workers. Although the contribution clearly gravitates towards the area of the sociology of the emergence and transformation of informality, before its conclusion it makes inferences about the role played by workers' formal organizations within firms as well. The findings confirm the author's intuitions about the potential for *computational simulation analysis in the sociology of formality and informality*.

What does this brief outline of the book tell us in relation to the sociology of formality and informality in general, and about these edited papers in particular? Regarding the former, it certainly shows us that there is a sustained and conceptually quite developed stream of research into the relationship between formality and informality. We underlined herein the aspects of (two-way) embeddedness, interaction, and emergence and transformation, but surely it is just a matter of time before other treatments gain consistency and visibility, too. There is also the issue that these meta-assumptions now appear more as complementary, rather than as competitive. Regarding the edited papers in this volume, the summary seems to indicate that efforts to keep the direction of study on the formality and informality interlinkage has inevitably led us to discover various, analyzable and conceptualizable, manifestations within formality (see *meta-informality*, *hybrid legal system*, *deficient rule of law*, *charter*) on the one hand, and informality (see *trans-informality*, *formalized informality*, *instrumental informality*, *multidimensional informality*) on the other. Although this is obviously related to the recent changes in these domains as well, one cannot help but notice that initial efforts to strictly delineate the formal and informal "sectors" took us in the direction of seeing the boundaries not as clearly defined, but instead as blurred.

References

Beckert, Jens. 2006. "*Interpenetration Versus Embeddedness*: The Premature Dismissal of Talcott Parsons in the New Economic Sociology." *American Journal of Economics and Sociology* 65: 161–188.

Böröcz, József. 2000. "Informality Rules." *East European Politics and Society* 14: 348–380.

Bromley, Patricia, and Walter W. Powell. 2012. "From Smoke and Mirrors to Walking the Talk: Decoupling in the Contemporary World." *The Academy of Management Annals* 6: 1–48.

Carruthers, Bruce G. 2010. "Institutional Dynamics: When is Change «Real Change»?" Comparative-Historical Social Science (CHSS), Working Paper No. 12–004, February.

Centeno, Miguel Ángel, and Alejandro Portes. 2006. "The Informal Economy in the Shadow of the State." In *Out of the Shadows: Political Action and the Informal Economy in Latin America*, edited by Patricia Fernández-Kelly and Jon Shefner, 23–48. University Park: Pennsylvania State University Press.

Dingwall, Robert, and Phil M. Strong. 1985. "The Interactional Study of Organizations: A Critique and Reformulation." *Journal of Contemporary Ethnography* 14: 205–231. Accessed November 5, 2014. doi:10.1177/089124168501400204.

Fernández-Kelly, M. Patricia, and Anna M. Garcia. 1991. "Informalization at the Core: Hispanic Women, Homework, and the Advanced Capitalist State." In *The Informal Economy: Studies in Advanced and Less Developed Countries*, edited by Alejandro Portes, Manuel Castells and Lauren A. Benton, 247–264. Baltimore and London: The Johns Hopkins University Press.

Grzymala-Busse, Anna. 2010. "The Best Laid Plans: The Impact of Informal Rules on Formal Institutions in Transitional Regimes." *Studies in Comparative International Development* 45: 311–333.

Guha-Khasnobis, Basudeb, Ravi Kanbur, and Elinor Ostrom, eds. 2006. *Linking the Formal and Informal Economy. Concepts and Policies*. Oxford: Oxford University Press.

Haldar, Antara, and Joseph E. Stiglitz. 2013. "Analyzing Legal Formality and Informality: Lessons from Land-Titling and Microfinance Programs." In *Law and Economics with Chinese Characteristics. Institutions for Promoting Development in the Twenty-First Century*, edited by David Kennedy and Joseph E. Stiglitz, 112–148. Oxford: Oxford University Press.

Helmke, Gretchen, and Steven Levitsky. 2006. "Introduction." In *Informal Institutions and Democracy*, edited by Gretchen Helmke and Steven Levitsky, 1–30. Baltimore: The Johns Hopkins University Press.

Kanbur, Ravi. 2012. "The Informality Lens." Plenary Presentation, GDN Annual Conference, Budapest, June 17, 2012. Accessed April 14, 2014. http://www.gdn.int/admin/uploads/editor/files/2012Conf_Papers/GDN2012_RaviKanburPPT.pdf.

Knight, Jack. 1992. *Institutions and Social Conflict*. Cambridge: Cambridge University Press.

Kus, Basak. 2006. "State and Economic Informality in a Comparative Perspective." Center for Culture, Organization, and Politics, Working Paper #2006–11.

Lauth, Hans-Joachim. 2000. "Informal Institutions and Democracy." *Democratization* 7: 21–50.

Lauth, Hans-Joachim. 2004. "Formal and Informal Institutions: On Structuring their Mutual Co-existence." *Romanian Journal of Political Science* 4: 66–88.

Meyer, John W., and Brian Rowan. 1977. "Institutionalized Organization: Formal Structure as Myth and Ceremony." *American Journal of Sociology* 83: 340–363.

Misztal, Barbara A. 2000. *Informality. Social Theory and Contemporary Practice*. London: Routledge.

Misztal, Barbara. 2005. "The New Importance of the Relationship between Formality and Informality." *Feminist Theory* 6: 173–194.

Nee, Victor, and Paul Ingram. 2001. "Embeddedness and Beyond: Institutions, Exchange, and Social Structure." In *The New Institutionalism in Sociology*, edited by Mary C. Brinton and Victor Nee, 19–45. Stanford: Stanford University Press.

North, Douglass C. 1990. *Institutions, Institutional Change and Economic Performance*. Cambridge: Cambridge University Press.

Pejovich, Svetozar. 1999. "The Effects of the Interaction of Formal and Informal Institutions on Social Stability and Economic Development." *Journal of Markets and Morality* 2: 164–181.

Pejovich, Svetozar with contributions from Enrico Colombatto. 2008. *Law, Informal Rules and Economic Performance. The Case for Common Law.* Cheltenham (UK) and Northampton (MA): Edward Elgar.

Portes, Alejandro, and William Haller. 2005. "The Informal Economy." In *Handbook of Economic Sociology*, edited by Neil J. Smelser and Richard Swedberg (2nd edition), 403–425. New York: Russell Sage Foundation; Princeton (NJ): Princeton University Press.

Portes, Alejandro. 2010. *Economic Sociology: A Systematic Inquiry*. Princeton and Oxford: Princeton University Press.

Sassen, Saskia. 1997. "Informalization in Advanced Market Economies." *Issues in Development*, Discussion Paper 20, Development Policies Department, International Labour Office Geneva.

Stinchcombe, Arthur. L. 2001. *When Formality Works: Authority and Abstraction in Law and Organizations*. Chicago: University of Chicago Press.

Tsai, Kellee S. 2006. "Adaptive Informal Institutions and Endogenous Institutional Change in China." *World Politics* 59: 116–141.

Van Assche, Kristof, Anastasiya Shtaltovna, and Anna-Katharina Hornidge. 2013. "Visible and Invisible Informalities and Institutional Transformation in the Transition Countries of Georgia, Romania, and Uzbekistan." In *Informality in Eastern Europe: Structures, Political Cultures and Social Practices*, edited by Christian Giordano and Nicolas Hayoz, 89–118. Bern, Berlin, Bruxelles, Frankfurt am Main, New York, Oxford, Wien: Peter Lang.

Part I: The Sociology of Informally Embedded Formality

Robert Dingwall

Formality in the Interactional Study of Organizations

> [...] Dingwall and Strong recognized the interpretative divide between interactionist and mainstream definitions of organizations...despite the promise of their insight, the work of Dingwall and Strong did not gain much attention in terms of promoting theoretical or methodological innovation in either organizational analysis or interactionist theorizing [...] they were simply ahead of their time (McGinty 2014).

In 1985, Phil Strong and I published a paper in what is now the *Journal of Contemporary Ethnography*, examining the condition of the interactionist approach to the study of organizations (Dingwall and Strong 1985). The paper had had a long gestation, beginning in lectures given by Strong to undergraduates at the University of Aberdeen in the early 1970s and developed through his ethnography of a clinic serving children with learning disabilities (Strong 1979). Strong supervised my PhD thesis on a professional school training public health nurses (Dingwall 1977) and the paper was also informed by my experiences in a research group in socio-legal studies, studying the inter-organizational system for child protection in England (Dingwall, Eekelaar and Murray 1983). Strong died in 1995, at the age of 49, and I have given little thought to the paper since. However, McGinty's advocacy persuades me that it is worth revisiting, particularly to explore the path not taken by the new institutionalists. Where they read Meyer and Rowan's (1977) seminal paper on organizational structures as myth and ceremony and turned to Weber for a future direction (DiMaggio and Powell 1983), we incorporated the same paper into Strong's scholarship on Goffman, and my reading of Everett Hughes, to develop an interactionist vision for the study of organizations. In the marketplace of ideas, we lost out. However, as McGinty notes, we anticipated problems with the institutionalist account that were never quite resolved and which have recently been underlined by the proposals for an "inhabited institutionalism" (Hallett and Ventresca 2006).

This chapter, then, has three elements. First, it revisits the arguments of the 1985 paper, at least as I understand them, and its debate with the negotiated order approach that dominated interactionist studies at the time. Why did we insist that there was more to organization than the informal relations stressed by negotiated order writers? Second, it considers the history of the new institutionalism and its partial reading of Weber. Finally, it examines the recent work in "inhabited institutionalism" and discusses whether this is a satisfactory solution to the problems that it identifies – or whether it remains unduly constrained by the interactionist tradition that Strong and I challenged.

The interactional study of organizations

The 1985 paper begins by positioning the negotiated order approach to the study of organizations, represented by the work of Anselm Strauss and his collaborators, in contrast to the rational, comparative-structural, approaches that had dominated the literature since World War II. The latter focussed on the study of formal organizations, which was separated out from the study of social organization, the orderliness of everyday life. Formal organizations were those established for an explicit purpose to achieve specified goals. Systematic research on the effectiveness of various organizational forms for achieving particular goals could inform the rational design of structures to deliver preferred outcomes. However, this rational paradigm had struggled with the accumulation of empirical anomalies. Four were identified: members did not observably act in the rational and rule-governed ways assumed by the model; members did not share clear and unified goals; rules were found to rest on bargaining power rather than rational planning and, moreover, could not be followed unreflectively; power and hierarchy were not associated in a simple fashion. During the 1950s and 1960s, various attempts were made to accommodate these findings by acknowledging that organizations had both formal and informal aspects. The latter, however, were regarded mainly as a random source of noise and subversion that evidence-based organizational design would eventually eliminate.

Strauss turned this thinking on its head. The negotiated order programme abolished the distinction between formal and social organization by treating the former merely as an ecologically bounded version of the latter. Formality was an epiphenomenon, a transient creation for certain purposes in certain contexts. He summarized the approach in six axioms: all social order is negotiated; these negotiations take place in a patterned and systematic fashion; their outcomes are temporally limited; the negotiated order constantly has to be reconstituted as a basis for concerted action; the negotiated order on any day consists of the sum total of the organization's rules, policies, and local working understandings or agreements; and, finally, any change arising within or imposed on the order will require renegotiation to occur (Strauss 1978, 5–6). While Strauss and his collaborators created a very substantial body of work, it had limited impact upon the mainstream of organizational studies, although it attracted the attention of some Marxist writers who saw it as offering a way to investigate praxis. As they pointed out, in criticisms echoed by some interactionists, discarding the notion of organizational structure made it difficult to deal with issues of coercion or constraint. Strauss (1978, 247–258) responded by elaborating two subsidiary concepts: the "negotiation context" and the "structural context". The negotiation context referred to the properties of a local situation that constituted conditions in the course of negotiation. The structural context was the overall framework of conditions within which all local negotiations occurred. Strauss gives the examples of production technology, the size and differentiation of firms in an industrial sector and the balance between fixed and variable investment. Neither of these completely solved the problem: other investigators found it hard to determine what would actually count as limiting conditions and which conditions were practically relevant to actors. Nevertheless, they did constitute an acknowledgement that some account of extrasituational constraints was necessary. Strong and I concluded that negotiated order had inverted the original error of formal organization theorists. By asserting that there was only informality, scholars in this tradition were forced into awkward accommodations to reintroduce formality.

The accumulation of empirical problems had also prompted a parallel response among specialists in the study of organizations, the so-called "new organization theory". Meyer and Rowan (1977) argued that organizational structures should be treated as legitimating myths rather than literal descriptions of relationships between actors. While the mainstream tradition had focussed on those elements of Weber's work that examined the rational design of bureaucracies as the hallmark of modernity, they noted that Weber's own analysis of organizational forms stressed their embedding in particular types of society. The legitimacy of bureaucracies did not derive from their rationality *per se* but from the expectation of other actors in their environment that these organizations would adopt a rational "vocabulary of structure". Formal rationality was only important in a society that valued formal rationality. The actions of organization members were constrained by the prospect of having to account for them in terms that could be reconciled with the established cultural expectations of organizations. The formality of these expectations – in laws, regulations, bookkeeping, etc. – was what distinguished formal and social organization. Strong and I noted that, although this made both language and its use into central topics of inquiry for the study of organizations, the "new organization theory" did not seem to have a way to deal with this, beyond conventional ethnographic methods.

I shall discuss later the way in which the "new organization theory" evolved into the "new institutionalism". Strong and I struck out in a different direction. In particular, we noted the way in which Meyer and Rowan's discussion echoed Garfinkel's remarks on the concept of organization:

> [T]he term "an organization" is an abbreviation of the full term "an organization of social actions". The term "organization" does not itself designate a palpable phenomenon. It refers instead to a related set of ideas that a sociologist invokes to aid him in collecting his thoughts about the ways in which patterns of social action are related (Garfinkel 1956, 181).

While Strauss and his associates had used Garfinkel's analysis of the indeterminacy of rules to make them irrelevant (Strauss et al. 1964, 313) ethnomethodologists always underlined their constraining nature. Although they accepted Wittgenstein's argument that the meaning of a rule depended on the context in which it was used, they recognized that the context could require that the rule be acknowledged and action justified by reference to it. Bittner, for example, proposed that the official descriptions of formal organizations could be treated as:

> a *generalized formula to which all sorts of problems can be brought for solution…* [acquiring] *through this reference a distinctive meaning that they would not otherwise have.* Thus the formal organizational designs are schemes of interpretation that competent and entitled users can invoke in yet unknown ways whenever it suits their purposes (Bittner 1965, 249–250).

He went on to identify three practices that both members and observers (including researchers) used to make this generalized formula visible: compliance, stylistic unity and corroborative reference. These have "determining power", "discipline" and "prohibitions" in relation to actions that take place within the jurisdiction of the generalized formula. While broadly endorsing this approach, Strong and I pointed to some limitations. Firstly, the properties of the generalized formula were excessively vague. Minimally, we thought, it had to be able to characterize action as intended to achieve some end – which might be why conventional organizational analysis was so preoccupied with goals. Secondly, we were not convinced that it provided an adequate basis for distinguishing between social and formal organization, which was clearly a distinction that members of our society thought to be important. We went on to propose our own conceptual terminology for the analysis of formal organizations, of "charters" and "missions", paralleling Hughes's (1971) use of "license" and "mandate" to characterize occupations. (Occupations may be thought of as another distinctive kind of organization in modern societies.) This terminology was then linked to a detailed set of methodological and research design proposals that constituted a programme for investigation.

The synthesis proposed by Strong and myself was not unique within its context: the original paper has extensive citations of the work that we built on and other authors were making similar proposals, particularly Silverman (1975). As a movement, however, the impact was relatively limited. McGinty (2014) could only find seven identifiable citations to our work, although Silverman's certainly had more impact than this. It is probably fair to say that neither Strong nor I were great advocates for our approach, mainly because of our career contingencies. Strong went to the Open University and was heavily involved in writing course texts and then in a big project on the social history of AIDS, left unfinished at his death. For me, the ideas were formulated at the end of two projects on organizational ethnography and I did not become involved in such projects again, particularly as the kind of access that I had enjoyed became more difficult with the rise of ethical regulation regimes. I only fully used the approach once, in a re-analysis of some of my PhD data (Dingwall 1986). Neither interactionists nor ethnomethodologists and conversation analysts were particularly interested in the kind of synthesis we were proposing – these fields were less sharply distinguished in the UK in the 1980s than they were in the US or became later in the UK (Strong 1988). In effect, the work was something of a dead-end. This does not, however, necessarily mean that it has become irrelevant.

The new institutionalism

The mainstream inheritance from Meyer and Rowan (1977) lies in what has become known as the "new institutionalism". The connection is explicitly made – Powell and DiMaggio's (1991) edited collection that serves as a manifesto for this approach reprints Meyer and Rowan's paper as its first chapter, before their own seminal paper, *The Iron Cage Revisited* (DiMaggio and Powell 1983). The authors begin by anchoring their approach in the Weberian agenda for the study of organizations and its depiction of the spread of bureaucracy as the expression of formal rationality in a system of control that is both efficient and powerful. The iron cage is a metaphor for the suppression of human individuality so that rational production techniques can assure that all goods and services will be created and delivered in an optimum fashion without discrimination, except perhaps in relation to one's ability to pay for them. While formal bureaucratic rationality had come to dominate large corporations and the administration of the state, its drivers had changed in other sectors. DiMaggio and Powell suggested that all organizations are being forced into the same mould by the concerns for legitimacy that were identified by Meyer and Rowan's reading of Weber. While the early entrants to any particular institutional field might have some scope for variation, their competitors were pressed to adopt the same models, which, in turn, become constraints on the first movers. Legitimacy is crucial to the perception, by those other organizations that constitute its wider environment, that any particular organization is robust, stable, well-functioning and a credible partner for the exchange or supply of material or symbolic resources. DiMaggio and Powell describe this process as isomorphism, occurring through three types of mechanism: coercive, mimetic and normative.

What has not, perhaps, been recognized with sufficient clarity is that Meyer and Rowan, on the one hand, and DiMaggio and Powell, on the other, are not addressing exactly the same problem. Meyer and Rowan's question might be characterized as "how do we know a formal organization when we see one?" DiMaggio and Powell's, however, might be phrased as "why do formal organizations tend to look the same?" Both are entirely proper questions for organizational scholars to ask but they do lead in different directions. Strong and I argued that it was possible to look at Meyer and Rowan's work as promoting a programme of research into the interactional construction of organizational formality, repairing some of the problems created by the negotiated order emphasis on informality. This would be consistent with the historic legacy of institutionalism, which, as Stinchcombe (1997, 2) notes, was traditionally a study of organizations "created by purposive people". In the new version promoted by DiMaggio and Powell, however, organizations lose their grounding in action. In Stinchcombe's (1997, 2) words, "collective representations manufacture themselves by opaque processes, are implemented by diffusion, are exterior and constraining without exterior people doing the creation or the constraining". He goes on to criticise both of these key papers for failing to ask why formality might be demanded of an organization. Consider, for example, the well-documented tensions when Silicon Valley companies have sought stock market listings to raise additional capital – and potential investors have demanded that the companies acquire structures that look more familiar to them as a condition of supplying funds. If ceremonial compliance with environmental expectations is all there is, why do people take the ceremonies seriously? Formality needs to be enacted by organization members who acknowledge what is at stake. Ceremony must be accompanied by moral commitment, not just lip-service. There is a difference between a trial and a show trial: although both may share the same ceremonial form, only one of them seeks to achieve substantive justice.

In effect, new institutionalism ended up reifying organizations as actors on their own account rather than as the outcome of action by people pursuing their own strategies and logics in response to an environment. This allowed it to advance a set of hypotheses about isomorphic change that are potentially testable by various kinds of quantitative study (Greenwood and Meyer 2008). DiMaggio and Powell draw on a good deal of ethnographic work, and note its consistency with their approach (DiMaggio 1998). DiMaggio (1988) also complained about the neglect of agency by other institutionalists. Nevertheless, their programme left space for the new institutionalism to be domesticated by established approaches in the field.

> Much of the narrowness in modern institutionalism...is explained by the lack of detail in the conceptions of institutions. A narrow conception is easier to mathematize. This in turn is due to ignoring the work of people who put the detail into institutions and who constrain people and organizations to conform to institution's exteriority. But if the guts of the causal process of institutional influence are left out of the model, then we successfully mathematize abstract empiricism, an empiricism without the complexity of real life (Stinchcombe 1997, 6).

DiMaggio and Powell's account has worn less well in some other respects too. Firstly, there is the translation error at the heart of their key metaphor. As Baehr (2001) has pointed out, the phrase "iron cage" was introduced by Parsons in his translation of *The Protestant Ethic* and is not consistent with Weber's original text. The German phrase expressed the idea of a shell "as hard as steel", into which an individual can withdraw in retreat from the harsh world of formal rationality. This does not necessarily undercut DiMaggio and Powell's argument about the ways in which organizations may be imprisoned by isomorphic pressures. However, it does affect the legitimation of the argument and goes to some of Stinchcombe's points about its neglect of individual agency. There is a difference between an individual who has formal rationality imposed on them and one who chooses to retreat into some private space to escape from it. Secondly, the representation of the driver for organizational rationalization as deriving from strategies of control on the part of governments or large corporations may well have been true in Weber's day but looks less credible with the advance of technology and the stripping out of middle management. While these were barely visible at the time when the new institutionalism was proposed, the implied path dependency in these sectors has certainly been disrupted. Call centres need minimal supervision because the computer logs individual performances. Functionally, this may still operate like a Prussian bureaucracy – but the army of supervising bureaucrats has disappeared.

Recent writers in the tradition of organization studies have, then, been calling increasingly for a new synthesis that retains some of the insights of neo-institutionalism but grounds these in stronger accounts of the practical construction of organizations by people dealing, in a historical context, with the everyday contingencies of technology, resources and culture.

> [...] when we only count the outcomes of institutional processes we overlook everything that is interesting in the "institutional story." Institutional theory has largely failed to retain methodologies that are consistent with their need to attend to meanings systems,

> symbols, myths and the processes by which organizations interpret their institutional environments (Suddaby 2010, 16).

Barley (2008) notes that, while neo-institutionalism is the only macro-sociology of organizations to have developed out of micro-sociological roots, its research programme has largely neglected them. He returns, as Strong and I did, to Hughes and Strauss to lay out their claims to have offered a substantial theory of institutions and organizations that has been obscured by some of the sectarian conflicts of US sociology. Although Barley does not put it in quite this fashion, the fact that Hughes and Strauss did not write theory that looked like "theory" in the writings of Marx, Adorno, Foucault or Deleuze did not mean they did not write theory. Within the Chicago tradition, theory was a second-order generalization from empirical work that allowed case studies to be placed within a wider frame, rather than an armchair exercise in *a priori* critique. While Barley presents arguments from within organization studies for taking the interactionist tradition more seriously, a parallel development within interactionism has sought to re-embrace institutionalism.

Inhabited interactionism

Although the phrase "inhabited institutionalism" has a longer history, its most systematic exposition is in the work of Hallett and various collaborators or associates. While the approach has been used in a number of empirical studies, its programme is best articulated in Hallett and Ventresca's (2006) discussion of Gouldner's classic study, *Patterns of Industrial Bureaucracy* (1954). They argue that this text stands at a critical juncture in studies of organizations before symbolic interactionism was set apart, or set itself apart, from the mainstream. Moreover, it is unquestionably acknowledged as institutionalist work by leading figures like DiMaggio (1988) and Perrow (1986). Gouldner's study of a gypsum mine occurred at a moment of transition with the death of a long-serving manager, whose approach had been highly informal, and his replacement by an incomer, who was expected by the company's management to use a more formal and bureaucratically rational style. The study provides a detailed account of the external sources that pressed on the local manager for the formalization of controls, target-oriented production and the adoption of rational bureaucratic models for the mine's organization. In this respect, Hallett and Ventresca note, Gouldner's work is clearly a precursor to neo-institutionalism. However, his ethnographic approach added a dimension on the enactment of the supposed institutional logic of formalization. Its encounter with a prior order, characterized by informality and personal authority, blunted the force of rationality, generating a complex of responses from mock compliance and partial engagement to active conflict. The expression of the apparently disembodied social forms of institutionalism was accomplished through interactions that changed their practical meaning. Specifically, Gouldner's analysis placed the organization that he studied within a context of history and community, within the encounters of specific actors, and within a spirit of scepticism about the claims made on behalf of abstracted models of organization and their effects. The result was to treat the mine as an institution inhabited by people whose interactions generated the observable order, while being constrained by a variety of external audiences. In this it effected a reconciliation between structure and agency.

Hallett (2010) himself has published an account of a similar moment in the life of an elementary school, where a new principal was appointed by the school district to introduce a new accountability regime. This was designed to standardize the practice of an established group of teachers accustomed to a high degree of diversity and autonomy in their pedagogy. It became the focus of an intense struggle between teachers and principal, with casualties on both sides. As with other studies in this line, it is based on skilful ethnography. However, it is not clear that it fully resolves the problems that Strong and I identified with the negotiated order approach, as opposed to simply reinvigorating it for a new generation. As Hallett and Ventresca (2006) pointed out, there was a sub-text to Gouldner's work in his interest in the politics of class struggle, which could not be openly expressed in the early 1950s when McCarthyism was in full flow. *Patterns of Industrial Bureaucracy* (1954) is a celebration of the workers' fight back against the corporation. In the same way, Hallett's elementary school teachers are the true heroes of his paper in their defence of classroom autonomy against target-driven management. This romantic strain permeates a good deal of interactionist work (Strong and Dingwall 1989). As Strong (1988, 18) once commented, qualitative sociologists had a preference for being "right-on" rather than right – in the sense of correct rather than neo-conservative! The elementary school is a site for a study of change, conflict, professional autonomy and the like – but the organization is still missing. These comments on the negotiated order programme seem equally applicable to inhabited institutionalism (Dingwall and Strong 1985, 208).

> Formal organizations [are] merely ecologically bounded social organizations in which actions [are] united only by territorial or temporal coincidence. A hospital, for instance, [is] simply "a professionalised locale, a geographical site where persons drawn from different professions come together to carry out their respective purposes" (Strauss et al. 1963, 150).

To be sure, Hallett (2010) discusses the environmental context and something of the history of the school that he studied but his main focus is on the struggle for control rather than on the constitution of the organization, "the study of politics in the fundamental sense" (Hughes 1971, 291).

This is not to criticise Hallett for doing something different from what I or, I imagine, Strong would have done – but it is to say that inhabited institutionalism continues to leave an empty space when it comes to the question of formality. What does it mean for social organization to become formalized? How is this accomplished?

Why does formality matter?

As Stinchcombe (2001, 1) observes, sociologists have come to view formality as a fraud. The study of formal organizations has been abandoned because formality is thought to be purely a matter of myth, ceremony and ritual. Against this, he argues that formalization occurs when substantively important stuff needs to get done. This process may be hijacked to add importance to stuff that is not really all that important – but we should not be misled by this. The availability of formality makes it possible for specific matters to be referred to general and abstract principles, which then provide a basis for the unification of those matters into coherent and consistent patterns of action. Although Stinchcombe comes from a very different starting point, this is, in essence, the same argument that Bittner (1965, 249–250) makes in the passage quoted earlier. Formality means that we do not need constantly to stabilize meaning by reference to the fundamental substance that it encodes. It is an inscription device (Latour and Woolgar 1979) that defines events for all practical purposes within the limits of its application. That process of application is available for empirical study and we may from time to time wish to open up the black box and explore its workings. However, practical actors do not need to do this. So, for example, Stinchcombe notes that most of the time we can assume that courts do justice because there are special times and places where that process is examined by both internal and external actors, either to fill in gaps in the application of law or to question whether the law itself achieves the purported goal of delivering justice as a substantive good. The fact that courts may rest on a great deal of ad hoc and informal organizational actions (Feeley 1979; Church Jr. 1985) does not mean that formality is irrelevant, because it provides the very framework that makes those actions possible. This is not Stinchcombe's example, but Lynch's (1997) discussion of the way in which references to "the judge" integrate and make sense of action in a courthouse might exemplify what this kind of approach can achieve. Reference to the judge transforms interaction into *organizational* interaction, governed by a distinctive reference that is not wholly determined by the parties. As Lynch shows, the judge need not be physically present for this influence to be evident. As such, the interaction is no longer informal and self-governed by the parties but operates within the shadow of whatever it is that the office of judge represents. The elevation of informality risks destroying features of social life that we consider to be substantively important (Atkinson 1982). As such, it is critical for sociologists of organizations not to approach formality as if it were a sideshow, a mere veneer on the reality of informal organization.

Studying formality

Strong and I proposed that the formal dimension of organizational life could be captured by the notion of a charter, which we modelled on Hughes's concept of professional, or occupational, licence.

> An occupation consists in part in the implied or explicit license that some people claim and are given to carry out certain activities rather different from those of other people and to do so in exchange for money, goods, and services (Hughes 1971, 287).

A charter is the concept to which organization members orient in their dealings with one another and with non-members to establish the limits of action that can be legitimately considered to be organizational. It refers to the organization's notional contract with other institutions for the coordination of a certain area of human action for an agreed or specified purpose. In some sense, a charter can be said to represent the constraints on a member's freedom of action that he or she experiences or depicts as exterior, objective and given. As Goffman had discussed this phenomenon,

> [E]ach of these official goals or charters seems admirably suited to provide a key to meaning – a language of explanation that the staff, and sometimes the inmates, can bring to every crevice of action in the institution (Goffman 1968, 81).

Formal organizational action describes those elements that operationalize the charter and make it visible or hearable to both members and non-members.

Where should we look to understand formality? Strong and I suggested that interactionists had been somewhat misled by their tendency to focus on the underlife of organizations, what later became known as the "street-level" (Lipsky 1980). As I showed (Dingwall 1986), professional education looked very different from the perspective of faculty trying to decide who they could certify as competent compared with the traditional approach of focussing on students trying to get through. Formality was critical to the construction of an account of the grounds for failing students that would be defensible under unknown future circumstances of possible challenge or review. Like other senior organizational personnel, faculty make a set of decisions that resolve and embed organizational charters in ways that provide a framework for the actions of lower-level personnel, students or clients. While one could identify implicit charter references in everyday interactions at this level, it was unusual to find them regularly and actively invoked.

There are, however, occasions when the organization's charter is placed on display or called into question. Some of these have been treated as uninteresting by interactionists, such as official ceremonies, which may, indeed, be full of platitudes – but these are platitudes that tell the acute observer something important about how members of the organization choose to represent the organization to its audiences. The induction of new members or the staging of public launches for new initiatives may also have something of this character. Other events may place the meaning of the charter in question. In some contexts there are formal policymaking discussions on boards and at management retreats: here is a change in the legal regime that surrounds us – how can we respond in order to survive and prosper while retaining our collective identity? In the same category, we might consider occasions like audits, assessments, evaluations or complaints handling. All of these require the people involved to articulate their understanding of the charter and demonstrate their compliance with it. This list is not exhaustive – nor could it be: charter discussions may occur anywhere at any time. Our concern in drawing attention to such occasions was to encourage efficiency in fieldwork. Here are places where it might be profitable to start looking, if we are to get a proper appreciation of the role of formality in organizations.

Conclusion

The study of organizations has yet fully to resolve the relationship between formal and social organization. What is it that distinguishes some specific set of purposive and co-ordinated actions from the general production of orderliness in society? This chapter does not fully answer that question. However, it has established that it is still a question worth asking, that it has not been satisfactorily answered and that there might be relatively straightforward ways to produce an answer. The first step, though, is to take formality more seriously. It is not mere show, an epiphenomenon to be pushed aside in search of the "real life" of informal organization. Formality creates the conditions under which informality can occur and plays a key role in their integration. It is the fulcrum for sustaining legitimacy in the eyes of various environmental actors. As such, it requires equal weight and attention in organizational analyses – we need to examine the suits as well as the sandals.

References

Atkinson, J. Maxwell. 1982. "Understanding Formality: The Categorization and Production of «Formal» Interaction." *The British Journal of Sociology* 33: 86–117. Accessed November 5, 2014. doi:10.2307/589338.

Baehr, Peter. 2001. "The «Iron Cage» and the «Shell as Hard as Steel»: Parsons, Weber, and the Stahlhartes Gehäuse Metaphor in the Protestant Ethic and the Spirit of Capitalism." *History and Theory* 40: 153–169. Accessed November 5, 2014. doi:10.1111/0018-2656.00160.

Barley, Stephen R. 2008. "Coalface Institutionalism." In *The SAGE Handbook of Organizational Institutionalism*, edited by Royston Greenwood, Christine Oliver, Roy Suddaby and Kerstin Sahlin-Andersson, 491–518. London: Sage.

Bittner, Egon. 1965. "The Concept of Organization." *Social Research* 32: 239–255.

Church, Jr., Thomas W. 1985. "Examining Local Legal Culture." *Law & Social Inquiry* 10: 449–510. Accessed November 5, 2014. doi:10.1111/j.1747-4469.1985.tb00508.x.

DiMaggio, Paul J. 1988. "Interest and Agency in Institutional Theory." In *Institutional Patterns and Organizations: Culture and Environment*, edited by Lynne G. Zucker, 3–22. Cambridge (MA): Ballinger.

DiMaggio, Paul. 1998. "The New Institutionalisms: Avenues of Collaboration." *Journal of Institutional and Theoretical Economics (JITE)/Zeitschrift Für Die Gesamte Staatswissenschaft* 154: 696–705.

DiMaggio, Paul J., and Walter W. Powell. 1983. "The Iron Cage Revisited: Institutional Isomorphism and Collective Rationality in Organizational Fields." *American Sociological Review* 48: 147–160.

Dingwall, Robert. 1977. *The Social Organisation of Health Visitor Training*. London: Croom Helm.

Dingwall, Robert. 1986. "The Certification of Competence: Assessment in Occupational Socialization." *Urban Life* 15: 367–393.

Dingwall, Robert, John Eekelaar, and Topsy Murray. 1983. *The Protection of Children: State Intervention and Family Life*. Oxford: Basil Blackwell.

Dingwall, Robert, and P. M. Strong. 1985. "The Interactional Study of Organizations: A Critique and Reformulation." *Journal of Contemporary Ethnography* 14: 205–231. Accessed November 5, 2014. doi:10.1177/089124168501400204.

Feeley, Malcolm M. 1979. *The Process Is the Punishment: Handling Cases in a Lower Criminal Court*. New York: Russell Sage Foundation.

Garfinkel, Harold. 1956. "Some Sociological Concepts and Methods for Psychiatrists." *Psychiatric Research Reports* 6: 181–195.

Gouldner, Alvin W. 1954. *Patterns of Industrial Bureaucracy*. Glencoe (IL): Free Press.

Goffman, Erving 1968. *Asylums*. Harmondsworth: Penguin.

Greenwood, Royston, and Renate E. Meyer. 2008. "Influencing Ideas: A Celebration of DiMaggio and Powell (1983)." *Journal of Management Inquiry* 17: 258–264. doi:10.1177/1056492608326693.

Hallett, Tim. 2010. "The Myth Incarnate: Recoupling Processes, Turmoil, and Inhabited Institutions in an Urban Elementary School." *American Sociological Review* 75: 52–74.

Hallett, Tim, and Marc J. Ventresca. 2006. "Inhabited Institutions: Social Interactions and Organizational Forms in Gouldner's Patterns of Industrial Bureaucracy." *Theory and Society* 35: 213–236. Accessed November 5, 2014. doi:10.1007/s11186-006-9003-z.

Hughes, Everett C. 1971. *The Sociological Eye: Selected Papers on Work, Self and the Study of Society*. Chicago: Aldine-Atherton.

Latour, Bruno, and Steve Woolgar. 1979. *Laboratory Life: The Social Construction of Scientific Facts*. London: Sage.

Lipsky, Michael. 1980. *Street-Level Bureaucracy: The Dilemmas of the Individual in Public Service*. New York: Russell Sage Foundation.

Lynch, Michael. 1997. "Preliminary Notes on Judges' Work: The Judge as a Constituent of Courtroom »Hearings.«" In *Law in Action: Ethnomethodological and Conversation Analytic Approaches to Law*, edited by Max Travers and John F. Manzo, 99–132. Aldershot: Ashgate.

McGinty, Patrick J. W. 2014. "Divided and Drifting: Interactionism and the Neglect of Social Organizational Analyses in Organization Studies." *Symbolic Interaction* 35: 155–186. Accessed November 5, 2014. doi:10.1002/symb.101.

Meyer, John W., and Brian Rowan. 1977. "Institutionalized Organizations: Formal Structure as Myth and Ceremony." *American Journal of Sociology* 83: 340–363.

Perrow, Charles. 1986. *Complex Organizations: A Critical Essay*. New York: Random House.

Powell, Walter W., and Paul J. DiMaggio, eds. 1991. *The New Institutionalism in Organizational Analysis*. Chicago: University of Chicago Press.

Silverman, David. 1975. "Accounts of Organizations. Organizational »Structures« and the Accounting Process." In *Processing People: Cases in Organisational Behaviour*, edited by John B. McKinlay, 269–302. New York: Holt, Rinehart and Winston.

Stinchcombe, Arthur L. 1997. "On the Virtues of the Old Institutionalism." *Annual Review of Sociology* 23: 1–18.

Stinchcombe. Arthur L. 2001. *When Formality Works: Authority and Abstraction in Law and Organizations*. Chicago: University of Chicago Press.

Strauss, Anselm. 1978. *Negotiations: Varieties, Processes, Contexts, and Social Order*. San Francisco: Jossey Bass.

Strauss, Anselm L., Leonard Schatzman, Rue Bucher, Danuta Ehrlich, and Melvin Sabshin. 1963. "The Hospital and Its Negotiated Order." In *The Hospital in Modern Society*, edited by Eliot Freidson, 147–169. New York: Free Press.

Strauss, Anselm L., Leonard Schatzman, Rue Bucher, Danuta Ehrlich, and Melvin Sabshin. 1964. *Psychiatric Ideologies and Institutions*. New York: Free Press.

Strong, P.M. 1979. *Ceremonial Order of the Clinic: Parents, Doctors and Medical Bureaucracies* (1st edition). London: Routledge & Kegan Paul.

Strong, P. M. 1988. "Qualitative Sociology in the UK." *Qualitative Sociology* 11: 13–28. Accessed November 5, 2014. doi:10.1007/BF00988686.

Strong, P. M., and Robert Dingwall. 1989. "Romantics and Stoics." In *The Politics of Field Research: Sociology Beyond Enlightenment*, edited by Jaber F. Gubrium and David Silverman, 49–69. London: Sage.

Suddaby, Roy. 2010. "Challenges for Institutional Theory." *Journal of Management Inquiry* 19: 14–20. Accessed November 5, 2014. doi:10.1177/1056492609347564.

Grażyna Skąpska and Grzegorz Bryda

Empirically Grounded Rule of Law

This paper is based on the following data:

(1) We have drawn first on the data regarding the level of trust in the judiciary in Poland, collected in 2009 in a survey sponsored by the Polish Council of the Judiciary – *Public Confidence in the Judiciary in Poland. Report from Quantitative Research Prepared by CBM Indicator* (CBM Indicator 2009). This research project was conducted from April 25th to May 7th, 2009, with the use of individual standardized questionnaire interviews. Research tools consisted of 71 basic questions and 35 questions related to socio-demographic data. The research project was conducted on a representative sample of adult Poles aged 18–75. It was a nation-wide representative sample regarding gender, age, education, size of locality and region (voivodship). 1500 interviews were conducted. The research project was prepared by Grażyna Skąpska, and the survey questionnaire by Grażyna Skąpska and Grzegorz Bryda. In this paper, this research will be referred to as the *Public confidence in the judiciary in Poland* (CBM Indicator 2009) project.

(2) The paper is also based on the general data from evaluations of the courts and other political and legal institutions collected in public opinion surveys conducted in Poland in the years 1997–2011 (see Skąpska and Bryda 2013).

Introduction

Even a not particularly keen observer of the public debate in Poland could easily assess that there is a great discrepancy between opinions concerning the implementation of the rule of law between lawyers and non-lawyers. According to lawyers, the rule of law, or rather in the European continental tradition, the law-governed state principle, is fully protected in Poland, it presents one of the Polish constitution's opening norms (The Constitution of the Republic of Poland of 2 April 1997, Art. 2; The Constitution of the Polish People's Republic of 22 July 1952 (amended), former Art.1), and it is supported by other constitutional provisions (notably by Art. 7). Further, this crucial principle is developed in constitutional provisions on the protection of fundamental rights, especially regarding penal law, and the protection of private property, as well as in other constitutional provisions, including the Constitution of the Republic of Poland of 2 April 1997,

Chapter VIII, which defines the autonomy of the system of justice and Art. 178, which defines the independence of judges. Special attention should be given to the impartiality and independence of judges. The rule of law is also protected by the Constitutional Tribunal, and the system of justice as a whole. Moreover, its realization is protected by the European Convention on Human Rights and Fundamental Freedoms, by the decisions of the European Tribunal of Human Rights, and of the European Tribunal of Justice.

On the other hand, according to many publicly expressed opinions, the rule of law is severely violated in Poland. Such critical opinions concern those, who are engaged in the making of law, mainly parliamentarians, and those, who are responsible for law enforcement: judges, prosecutors, police and public functionaries. Such critical opinions were recently expressed in Poland in two popular movies released in 2013: Ryszard Bugajski's *Układ zamknięty* [Closed Network] and Wojciech Smarzowski's *Drogówka* [Road Police]. In those movies – of which one is based on a true story – there are many examples given of drastic violations of the rule of law, the corruption of the police officers and the corruption schemes of the involved prosecutors, police and the high circles of authority in the prosecutor's office and politics.

The discrepancy of opinions could be easily explained. Lawyers think about the rule of law in terms of the "written law", that is, constitutional provisions, statues and procedures, whereas non-lawyers debate the rule of law in terms of the "living law", with their personal experiences with the execution of laws in courts and governmental institutions, and the news about the application of law they find in the mass media, predominantly on TV and newspapers. The latter are characterized by a high degree of criticism. The critical point, and the "junction" of the promise expressed in the written law, and the reality of this principle's real implementation and protection, presents the anatomy of a functioning system of justice as it is broadly understood: the decisions of courts and the functioning of the prosecution and the police. Courts and judges have an especially salient importance. They connect the legal expectations of an abstract principle with an actual, real case, or a conflict, and the social expectations and experiences linked with it. Hence the formation of an empirically grounded and not merely abstract concept of the rule of law is rooted in the decisions taken by judges, and other public functionaries. Consequently, the measure of the popular perception of the rule of law is the level of trust in courts and the judiciary[1].

---

1 The results of the international comparative research on the courts and the judiciary are also worthy of notice. The research conducted by two leading organisations,

Concerning the abovementioned discrepancy of opinions, we are going to develop a twofold argument in this paper. Firstly, we will argue that the very concept of the rule of law, which was invented in the XIX century, has been both reconsidered and conceptualized anew in the XXI century, following the dramatic social changes, and in view of the entirely new challenges faced by law making, and law enforcement. Such reconsideration is particularly important during times of fundamental social change and democratic reconstruction following the collapse of dictatorial or totalitarian regimes. It is the "account" of the rule of law not only with reference to important and rudimentary legal values – namely the classical values of legal certainty and predictability, but also, in its more contemporary version, justice and protection of human rights. Secondly, as it was already argued, sociologists would stress an empirical account of the rule of law grounded in social experiences, in local memory and local knowledge, reflected in popular opinions on the application of law, and in the trust in the judiciary, discussed earlier.

It must be particularly emphasized that the level of trust in public institutions, similar to the level of trust within personal relationships, is strictly connected to social cohesion, stability and orderly changes of a transitional nature, or abrupt changes characteristic of transformation. The situation of Eastern European courts is outstanding because the law and judiciary have been subject to deep transformation. According to editors of a book on the court system in East-Central Europe, despite some profoundly transitional features and the huge and complex system

Freedom House and the World Bank, indicate a decrease in the quality of the rule of law in Poland. In 2007 an assessment of justice in Poland by Freedom House gave the country a score of 2.25 (with 1 as the best and 7 as the worst mark), which suggests a decrease in comparison with the 2003–2004 period, when the mark was 1.5 (see Freedom House (2003; 2004; 2007) *Nations in Transit* reports on Poland). Emphasised was the excessive number of cases in relation to the possible efficiency of the court, and the resulting delays in court proceedings, low trust in courts on the part of society and the insufficient number of places in penal institutions. The World Bank data on the rule of law show a similar assessment of the functioning of the rule of law in Poland. The result in 2005 was 60, in 2006 and 2007 it was 59, and in 2008 the grade was 65. This means that in the opinion of the World Bank, on the basis of several dozen research projects, 65% of the countries have worse functioning legal systems than the one in Poland (the higher the index, the higher the grade; the index fluctuating between 0–100). It should be added that in previous years the assessments of the rule of law in Poland conducted by the World Bank were better, yet since 1998 they have been regularly decreasing (in 1998, 70% of the countries worldwide had legal systems functioning worse than ours, in 2002 – 66%, in 2006/2007 – only 59%) (see Worldwide Governance Indicators (WGI, n.d.) on Poland 1996–2013).

of written law inherited from the past, these developments in Eastern Europe are discontinuous and include rather revolutionary features of the abrupt changes of the political system followed by complex social and economic transformations, and finally, of the legal system (Přibáň and Roberts 2003, 1).

The differences between lawyers and non-lawyers regarding opinions on the practical implementation of rule of law is to be debated in the particular context of the transformation. After the "round table agreements" these countries reconstructed their political, economic and legal systems. They enshrined the rule of law in their constitutions, but they also had to deal with the legacies of the communist past. Those legacies, and the innovative "transitory justice" (Teitel 2000) approach to law contributed in an important way to the discrepancy between the formal, legal understanding of the rule of law, especially in its classical entourage, and the empirical reality of the "living law" after the collapse of the former system. Therefore in this paper, after the short description of the difficult issue of the past, the legacies of "lawful revolutions" and the approach to the rule of law they promoted, and the empirical reality with which they were confronted, we will debate the contemporary accounts of the "living rule of law" in popular opinion, on the basis of empirical data on trust in the judiciary.

Brief remarks on the important issue of the past: The socialist rule of law

The historical experiences of Eastern European countries, in particular those of Poland, have determined a special approach to law, treated as an instrument of a foreign, often hostile power. These historical experiences – and in the Polish case also the experiences of the XIXth century when Polish territory was divided and the Polish state ceased to exist – have an obvious impact on the legal consciousness and legal cultures of the societies in question. The deciding factor, namely the imposition of communist governments, was the catalyst for fundamental political, economic and social changes that were introduced in this region after 1945. Thus, in 1945, the Soviet Army entered Poland, and other East-Central European countries. That was followed by a series of political and legal acts. First, prior to the Soviet Army entering Polish territories, the provisional government was created, which was already forging a new Polish state in accordance with communist principles. It worked closely with the Soviet Union, and was under the particular control of Stalin. Its Department of Justice ensured that politically reliable judicial officials were recruited in the future administration of justice in the Polish People's Republic. The provisional government's aims in Poland, similar to other new communist countries, were threefold: to legitimize the quasi-judicial functions of the secret police, to control citizens' activities and to assume control over private property. After the first "elections", in the early 1950s Soviet-style constitutions were introduced and the councils-of-state – the primary executive organs of government – were created and were granted full legislative powers. Moreover, the legal system imposed on the societies of East-Central Europe was highly bureaucratized to ensure complete control over decision-making. In Poland and elsewhere, the concept of "socialist rule of law" was invented, and the judiciary, the bar and police were restructured under the tight supervision of the communist party. In accordance with the "socialist rule of law" principle, the whole legal system and the judiciary as its functionaries were transformed firstly into instruments of the class struggle and, secondly into door-keepers for the new system of power.

As is stressed, the significant feature of communist law and its application was a predominance of public law over private law. The public interest was, as one reads, "[…] translated into state interest, i.e., the Communist Party's interest" (Fijałkowski 1999, 247). The judges were trained according to this framework, and received legal indoctrination in Marxist-Leninist theory (to attend the so-called evening university of Marxism-Leninism was obligatory for legal practitioners until the 1990s). All of that led to the formation of a particular culture of lawyers, characterized by political conformity, formalistically understood legal positivism and the absence of an individual rights culture. The judges, as it is argued, were primarily seen as law enforcement agents and not as defenders of individual rights (Wagnerová 2003, 177, 178). They were also strict adherents to the formalistically applied law and subjects to the political power of the state.

To these remarks on the general political context in which the judiciary was functioning and the socialist rule of law was practised, one should additionally note the contempt for law and judges, so characteristic of Leninist-Marxist ideology (Krygier 1994, 137ff; Czarnota and Krygier 2007, 164ff). The low wages of judges, the very poor equipment of courts, and the general organizational chaos contributed strongly to the decline of legal culture, and influenced the decline of the prestige of courts and the judiciary.

Legal certainty and predictability characteristic of the juristic version of the rule of law versus "transitory justice"

As it is also observed, the break with communism and the building of new regimes has become to an important extent a legal problem; and the East European transitions took a legal and often also a constitutional form (Arato and Sajó 1991, 101; Király 1995; Sólyom 2003; Skąpska 2011). The rule of law, in its classical and juristic version, indeed at those times almost a civil religion, legitimized profound political change in the eyes of the outside world. Legality, i.e. the strict adherence to the juristic version of the rule of law, presented also a key component of the semantics promoted by members of the democratic opposition in East-Central Europe.

The initial description of these events as "lawful revolutions" came from Hungary – notably from a prominent intellectual and participant of the Hungarian roundtable, János Kis, as well as from Kálmán Kulcsár, the then Minister of Justice in the still communist government. For the former author, the parties seated at the Hungarian roundtable were mostly interested in securing stability and continuity for the legal order in Hungary (Kis 1995). This position was later strongly supported by the actions of the Hungarian Constitutional Court, especially by its then President, Justice László Sólyom, who often stressed legal stability and certainty as dominant features of the Hungarian transition (Sólyom 2003, 20, 21, 39) having described it as a "revolution under the law" (Sólyom 2003, 39).

This attitude had also been highly evident in the writings and activities of persons like Václav Havel; it was practiced by Charter 77 in Czechoslovakia, by the Committee for the Defense of Workers in Poland, and by the oppositionists of the Polish Round Table even at the price of losing their credentials among the society (Skąpska 1990). Thus, according to a prominent Czech author, "[…] the rule of law became one of the most important revolutionary demands" (Přibáň 1999, 45).

The concept of lawful revolutions refers to the unquestioned formal legality of the political transformation limited by the existing law and its procedures. Thanks to not only the classical notion of rule of law, but also legal certainty and predictability, citizens could reasonably arrange their affairs, and the law preserved its autonomy, that is, its distance from political pressures. Presently, as it was mentioned, such a classical concept of the rule of law is supplemented by the protection of human rights and the promise of justice (The Constitution of the Republic of Poland of 2 April 1997, Art. 2). This modern, albeit still juristic version of the rule of law appeals to important moral values and to citizens' expectations. It also has an important political dimension. The second article of the Polish Constitution declares Poland to be a "democratic, law governed state", stressing in this way the democratic law-making procedures. It stirs great expectations, put to the test by the challenges inherent in the difficult legacies of the past and dilemmas created by the post-communist transformation, as well as the results of the introduction of the free market economy. First and foremost among those results was the promise to simultaneously protect individual property rights, and to create rights for the new owners of privatized national enterprises (Skąpska 2011, 185–218).

Moreover, in the light of the previously mentioned argument of this paper, the classical and juristic notion of the rule of law should be supplemented by the daily life experiences of citizens, their contacts with law and their perceptions of law, linking the liberal ideal with the locally rooted interpretations of it.

Hence, to understand the discrepancy of opinions between the lawyers and non-lawyers on the installment of the rule of law after the collapse of the former regimes in Eastern Europe, one has to take into account the broader axiological and sociological understandings of this concept. Above all, one has to remember that the new governments, courts and judges were confronted with tasks that required more than simply a formal understanding of this concept and which went far beyond the already mentioned rudimentary values of legal certainty and predictability. These courts had to deal with new issues and cases having only the old, or the slowly changing system of law at their disposal, not compatible with the legacies of the past that were "greater than the existing law", such as the mass scale human rights violations, or the mass scale deprivation of property rights. They were also confronted with quite novel tasks and issues brought about by rapid economic change, or the accession to the European Union.

A burning question emerged regarding how to deal with past atrocities, and the grand scale human rights violations committed by the functionaries of the former regimes, how to punish perpetrators and instigators as well as how to compensate the victims and still protect the rule of law principle. Equally important were the tasks linked with the restitution of the property rights of the owners deprived of their property by the communists, and the simultaneous protection of the newly created property rights of the new owners of the privatized state property.

There is a vast literature already devoted to the human rights violations committed by the dictatorships, and the ways of dealing with such a legacy have been undertaken by the particular governments. Here we would only briefly outline three ways in which the courts dealt with such issues. All of them are to be evaluated in the political context of East-Central Europe in which the new governments were to act: the delicate balance which resulted from the initial agreements between the communist functionaries engaged in dismantling the old system and the democratic opposition on the one hand, and the emerging networks and interconnections between the interests groups involved in the post-communist reconstruction on the other.

The first approach to those tasks was compatible with the classical and juristic understanding of the rule of law. It consisted of a strict adherence to the existing law and procedures, based on the mentioned values of legal certainty and predictability. Because of this characteristic, the approach turned out to be rather "perpetrator oriented". This approach to past human rights violations could be summarized in the following way: "we had expected justice, but we got the rule of law". The second could be described as a way led by moral standards, especially in the "hard cases" of transformation. It was closer to the daily life experiences of people living under the past regime and especially to those of its victims: persons whose fundamental rights were violated, political prisoners, persons fired from their jobs or from universities because of their political convictions, persons deprived of their property, or those defined as "capitalists", "kulaks" or "bloodsuckers". It was later described as a "transitory justice" (Teitel 2000). It consisted of an innovative, even ingenuous, approach to state law based on the international regulations and practices of dealing with human rights violations, and also on the standards set by the Nuremberg Trial of the top Nazi functionaries responsible for the Holocaust, and by the trial of Rudolf Eichmann in Jerusalem. In contrast to the former, classical understanding of the rule of law, this approach was rather victim oriented. To be legally valid, it was supported by the defining of the human rights violations committed by the communist functionaries as genocide, and of the communist party as a criminal organization. Such an approach required strong political support by a politically strong leadership, such as it was at that time of the presidency of Václav Havel in the Czech Republic.

The third approach to the legacies of the past could be described as opportunistic, and even cynical. It consisted of a formal adherence to the existing state law, in order to avoid the punishment of perpetrators, as well as of policies aimed at the protection of the former regime functionaries because of the deep interconnections, nepotistic or corruptive relations between the judiciary, public prosecution, parliamentarians, government officials and later, the "newly rich". These relations could have very old roots, based in former friendships or collaboration between functionaries of the former regime (Skąpska 2009). In the event of its application to the "hard cases", its main effect was a manifestation of corruption in all possible meanings of the concept.

All those difficult issues became the subject of popular opinions on law and justice. The growing criticism in the actual realization of the rule of law principle reflects dissatisfaction with its fulfillment with regard to the crimes committed in the past. Thus, in the year 2014, 25 years since the beginning of democratization in Poland, more than one-third of the Polish population is of the opinion that the Polish government and courts did not resolve satisfactorily the difficult issues of the past[2]. The importance of such difficult issues is indicated by court cases, which are discussed and remembered as symptomatic, and are indicators of the failure of rule of law. According to the already mentioned research conducted in Poland in 2009 and published in the report *Public confidence in the Judiciary in Poland* (CBM Indicator 2009) – 20 years after the Round Table Talks and the revolutionary changes they initiated – still among the most important and unsatisfactorily resolved court cases mentioned spontaneously by the respondents answering an open question of the questionnaire were "the Wujek coal mine pacification by the military in 1981", and "the events of December in 1970". Among the cases from the list, the respondents indicated the Wujek coal mine pacification, the case of Grzegorz Przemyk, who was murdered by the political police, and the events of December 1970, next to economic transformation related cases such as the coal scandal and the death of Barbara Blida, the Fund for Foreign Debt Servicing scandal, the case of Roman Kluska, the *Żywiec* brewery trademark case, and the case related to the property rights of the Wedel trademark[3]. Needless to say, these cases are among the most often reported and debated in the mass media and are also the most often cited sources of information about law and its application[4]. Other cases broadly reported by the media, and that are important for the formation of opinions on the application of law in courts, include the homicide of Wojciech Olewnik, the so-called "sex-for-employment affair" of the political party *Samoobrona*, the Rywin case, where a grand scale corruption attempt led to governmental change, the Gen. Papała

2 These are the newest data from the survey conducted by CBOS in February 2014 (see Pankowski 2014).

3 Cases including the Wujek Coal Mine Pacification, when several people were killed by the police forces, following the order of the highest ranking officials, above all Gen. Kiszczak, as well as the events of December 1970, where many persons were also killed following orders from Gen. Jaruzelski, the homicide by the police of student Grzegorz Przemyk in 1983, and the still unpunished criminal activities of the high ranking officials in Poland, as well as the criminal activities of the ordinary policemen (Grzegorz Przemyk case). The case of the *Żywiec* brewery trademark from 1993, the Roman Kluska case of 2002, and the case of Barbara Blida from 2007 present cases related to economic transformation and of a grand- scale corruption and nepotism in which the new authorities were involved. Roman Kluska was a very successful businessman accused of tax evasion and arrested, but was ultimately cleared by the court; Barbara Blida, a former deputy minister of construction, was accused of corruption and committed suicide as police searched her house.

4 It is noteworthy that the media, the TV and the press present a constant, most important source of knowledge, and opinions on law and the judiciary (Daniel 2007, 72, 73).

case, and the case concerning the deaths in medical ambulances in Łódź, as well as the unresolved criminal cases and the case illustrating the methods used when attempting to corrupt the law making process[5].

Factors important for the popular perception of the rule of law

All these difficult issues help to explain the relatively low level of trust in courts and the judiciary in post-communist Poland.

Aside from history, popular opinion on courts and judges is also formed from various sources of knowledge regarding the execution of laws as well as personal contacts with courts.

As already mentioned, the main source of knowledge about law and the application of laws are presentations given by mass media – TV and newspapers. They play an enormous role in the formation of popular images of law, the judiciary and the court system; therefore they have a salient impact on the opinions about the fulfilment of the rule of law in popular consciousness. Opinions communicated by the media are of particular importance also because they are accessible to everybody, and are subject to informal exchange. They awaken emotions, because they are focused on the most controversial, political and criminal cases. Therefore, they are not only informative, but also performative.

The media fulfil this important role in a very peculiar way.

In light of the research conducted in Poland, the readers are not only wrongly informed about law and court decisions, but the media often create a negative image of the judicial system, and impair its authority. The opinions propagated in the press on law and courts – even by some prominent journalists – are mostly one-sided, i.e. critical, highly emotional; they are concentrated on cases that awaken serious doubts, and they are often accompanied by sensational "news" on the corruption of the system of justice, accusations of nepotism, and the ties of judges with the old regime (Daniel 2003, 134; 2007, 73). The contacts with courts and court proceedings present another important factor in shaping the

5 Wojciech Olewnik was the son of a Polish businessman. He was kidnapped and murdered in 2001. After years of investigation the case is still not resolved, and allegations that high ranking police officers were somehow involved in it have not been falsified. The assassination of General Marek Papała – the then Police Chief – in 1998 bore all the signs of a contract killing, with evidence pointing to business and political circles of that period. The attempt of Lew Rywin in 2003 to corrupt one of the most known dissidents and journalists had riveted Poland like no other corruption scandal. It led to the resignation of the Prime Minister and a change of government.

opinions about the judiciary. This factor underwent a significant change after the collapse of the former regime. Thus, in the year 2009, 30% of Poles declared personal contacts with courts to be the most influential factor in forming their views (Skąpska and Bryda 2013, 87).

The formation of trust in the judiciary is affected by features that determine differences between law-applying institutions, especially courts, and other public institutions. American research on satisfaction with court decisions (Tyler 2006) indicates the particular importance of procedural fairness (fulfilment of the fair trial principle). It means the observance of procedural standards in each particular case, including especially the independence and objectivity of judges, the equal treatment of parties, the right to defence, understood as a right to present one's own argument and to respond to the opponent's argument, transparency of proceedings, and a right to legal support, to decide on the personal, subjective feelings that one was treated fairly, independently of the final decision (verdict) of the court (Tyler 2006, 37). Thus, in this broader conceptualization, procedural justice means not only the fulfilment of the strictly legal standards, but also the observance of some cultural standards, i.e. the cultural norms responsible for the "civility" of the proceeding's participants treatment, and the communicative standards: the communicative rationality of the court's proceeding, i.e., the truthfulness, the intelligibility and argumentative validity of the communication in the court. Here the particular responsibility rests on the judge, his or her way of communication with the parties and the broadly defined "court trial culture", including in particular respecting the authority of the court. Aesthetics and dramaturgy are not to be disregarded here – the architecture of buildings, arrangement of court rooms, outfits of the parties and the "judicial etiquette" on the one hand, and the substantive competence of the judges on the other. Cultural factors overlap or cross with the mentioned factors comprising a "fair trial" and constitute an important context for the implementation of this principle. Yet another important context for the popular perception of the rule of law, and in effect, the popular trust in judges and courts consists of the organizational standards and the actual law enforcement. It should be emphasised that even the most suitable formal guarantees of a fair trial may be ineffective if participants in the proceedings will experience mediocrity, disorganization and arrogance.

Hence, the level of trust placed in courts and judges, and the opinions on the fulfilment of the rule of law principle differ depending on historical experience, the approach to the violation of human rights committed by the functionaries of the former regime, and the actual fulfilment of the fair trial principle: the broadly understood procedural justice in the process of executing the laws in the context of the communication, culture and organisation of the court proceedings and the enforcement of court decisions.

Let's look closer into data from the *Public confidence in the Judiciary in Poland* (CBM Indicator 2009) report.

As it was already mentioned, according to the results collected in the survey on the public confidence in the judiciary in 2009, in Poland the number of persons who have participated in proceedings before the court is rapidly increasing. The respondents' declarations shed light on this matter, revealing that this percentage had reached 30% in 2009. The data confirm the hypothesis that the system change results in an increase in contacts with instances of the actual application of laws. Another set of data indicates that Poles participate in court proceedings mainly in the role of witnesses (59.1%) and claimants (27.5%).

It should be emphasised that Poles most often appear before civil courts. Thus, the opinions on the courts and trust in courts among the participants in the proceedings are mostly shaped by contacts with civil courts.

As the data indicate, only 3% of the Poles surveyed have absolutely no trust in courts and 3.5% declare complete trust. The prevailing opinion, however, indicates high and rather high trust in the courts' ability to serve justice. In fact, on a 7-step scale, the respondents assessed their level of trust in courts mainly as 4 (29.9%), 5 (25.3%) and 6 (14.9%). Thus, in the light of the survey, over one-third of the Poles, by assigning a digit corresponding to their level of trust, indicate a somewhat ambivalent attitude, whereas 40% of the Poles assess their level of trust in courts as predominantly positive (that is, a total grading of one's own level of trust as 5, 6 or 7).

The idea that Poles believe in a significant role for the courts and the law in regulating disputes is confirmed in the data on the institutions to which the Poles would turn when faced with a disputable matter regarding ownership or financial liabilities. Most often it would be either attorneys at law (lawyers) to whom one should turn for assistance (25.5%) or to the court directly (16.8%). The respondents hope for an accurate and reliable resolution and the chance of retrieving the disputed item or money. Therefore, over 40% of adult Poles notice and appreciate the importance of professional advice and the significance of proceedings before the court in civil cases. What is characteristic is the low percentage of opinions indicating professional mediation as a method of regulating disputes over property and financial cases (6.1%). The survey research leads to one more conclusion: Despite the strong emphasis placed on the so-called alternative methods of regulating disputes as well as the relatively high costs of proceedings before the court, financial and otherwise (e.g. in the form of wasted time, the need to prepare for the proceedings, stress and uncertainty as to the outcome), Poles are somewhat more oriented towards fighting and a willingness to prove their rights rather than adopt an amicable attitude and willingness to reach a compromise. Without a doubt, the chances for the enforcement of judicial decisions play an important role here.

An attempt was also made to examine the assessment of courts and judges with regard to their independence and impartiality, independence from the media, resistance to the influence of other participants in the proceedings (attorneys at law, prosecutors, experts and other participants) as well as the fairness of judgments and overall professionalism.

The gathered data indicate that on a five-point scale the independence of courts was evaluated favourably, when asked about in general. Specifically, 46.1% of the respondents strongly agree or agree with the declaration that nobody has influence on judicial decisions and judgments, 24.4% disagree and strongly disagree. The opinions concerning the professionalism of judges are also high. 55.8% of the respondents agree or strongly agree with the opinions that judges demonstrate high professionalism and experience in hearing cases, 9.6% disagree or strongly disagree. Further questions about the independence of courts, however, indicate inconsistencies and contradictions in the opinions of the respondents.

Thus, 25% of the respondents disagree or strongly disagree with the opinions that judges succumb to undue influence, whereas 42.8% agree or strongly agree with such opinions. 40.6% of the respondents agree or strongly agree with the opinion that courts are free from any political influence, especially the influence of political parties, whereas 27.4% disagree or strongly disagree; 25% disagree or strongly disagree with the statement that courts are influenced by the media, and 42.8% of the respondents agree or strongly agree with this statement. In the light of the opinion of the majority of the respondents, courts succumb to the influence of lawyers, prosecutors and experts, and the pressures from the participants in the proceedings. 18.1% of the respondents disagree or strongly disagree with the statement that judges are impartial in relation to the participants in the proceedings, whereas 47% agree and strongly agree with it.

The respondents differed rather considerably in their opinions as to whether courts pass fair judgments: 29.5% of the respondents disagree or strongly disagree with the statement that courts always pass fair judgments, whereas 36.5% of the respondents agree or strongly agree with it (CBM Indicator 2009).

In conclusion, a major part of Polish society believes that in general nobody can influence judicial decisions and that courts are under no influence of political parties, yet a considerable part of the respondents is of the opinion that judges are not impartial, but instead they succumb to the influence of the media and other participants in the proceedings – which affects the realisation of the fair trial principles.

Moreover, as indicated by the data, whether one participates in proceedings before a court has no direct influence on the opinions of the respondents. This finding is confirmed by the comparison of the declared level of trust in supranational and international courts (the European Tribunal of Human Rights, the European Tribunal of Justice, and even the Tribunal of The Hague), domestic courts of the highest level (the Supreme Court and Constitutional Tribunal) and various types of domestic courts. The respondents declare higher trust in European courts and supranational courts than in the Polish courts (CBM Indicator 2009).

Supranational courts and domestic courts of the highest level – with which only few respondents had any direct contact and about which they know almost nothing – enjoy the highest level of trust in the opinion of Poles. Thus, 7.7% of adult Poles declare full trust in the European courts and international courts, only 0.7% declare a complete lack of trust in such courts, whereas nearly 50% of the Poles surveyed declare that their level of trust is 5 and 6, which indicates above-average trust (an increase of 10% in comparison with the abovementioned declared average level of trust in Polish courts).

The survey was focused on the assessment of the work of the courts, the manner in which cases are heard, and on the functioning of the courts. Moreover, the respondents were asked about their level of satisfaction with the judicial decision with regard to their own case. The respondents were requested to assess the work of courts on a 1 to 5 scale with regard to the availability of information at court secretariats, the speed, professionalism and efficiency of service at the secretariats, the openness of the proceedings, the reliability of the evidence assessment, the efficiency of the court proceedings and the politeness of the court secretariats' personnel. On the basis of the findings it is possible to formulate a general conclusion that the respondents rather highly assess the work of the courts with regard to the abovementioned features. As the data indicate, the highest grade was given to the politeness of the personnel and to the professional service at court secretariats, and also, respectively, the openness of the proceedings and reliable assessments of evidence. The lowest scored was the assessment of the efficiency of the proceedings. The last of the mentioned opinions is reflected in the data on the duration of proceedings regarding the cases in which the respondents participated. In fact, in the opinion of a vast majority (58.7%), the proceedings last too long. In general, however, the assessment of the work of courts was rather high: the mean assessment on the 7-point scale is 4.27.

It should also be noted that the respondents participating in the proceedings were mostly satisfied with the judgment (68% of the respondents); one-third of the respondents were dissatisfied with the judgment as it did not satisfy the respondent or was unfair, or the case was lost. The majority, over 70% of the respondents who participated in the proceedings, declared satisfaction with the court's argumentation, 29.1% of the respondents were not satisfied with it, assessing it as unclear, biased or unfair.

Summing up, the majority of the participants in the proceedings before the courts in Poland assess the work of courts positively. This includes persons who lost their cases. It is a significant and beneficial difference with the popular images of law and the application of law.

In order to examine the culture of the trial and atmosphere of the courtroom, the respondents were also asked about the features describing the conduct of judges. A list of 14 various features describing the conduct of judges during proceedings were presented to the respondents. The features included: chaotic – organised, emotional – logical, lenient – strict, unworthy of respect – respectable, unprofessional – professional, unfriendly – friendly, impolite – polite, careless – careful, unbalanced – composed, hasty – calm, biased – impartial, submissive – firm, superior – approachable, jocular/witty – serious. As the results of the analysis indicate, the mean assessment of each of these features is high – about 5 on a 7-step scale. Moreover, a vast majority (80.5%) of the participants in the proceedings before the court stated that the judge was in control of the proceedings, and the authority of the court was observed.

The personal culture of a judge and the outcome of court proceedings play an important role in the shaping of trust in courts and the judiciary. The level of trust is definitely higher when a judge conducted a trial politely and professionally rather than coldly/impolitely and with an air of superiority, and when a respondent declares that he or she won a case in a court and was pleased with the judgment and the statement of reasons. From the respondent's perspective, the outcome of the proceedings is more important than the personal culture of a judge. A positive outcome increases the level of contentment (satisfaction and the feeling that a fair trial was conducted).

Finally, the duration of court proceedings – in particular the length of the proceedings and waiting time for the court's judgment – is very important with respect to the level of trust placed in courts and the judiciary. Trust in courts and the judiciary increases as the duration of court proceedings decreases, in accordance with the principle that courts should judge quickly, effectively and justly. According to the data analysis, trust in courts and the judiciary is thus a direct reflection of personal experience connected with judges, the institutional efficiency of courts, the duration of court proceedings and the effectiveness of the enforcement of decisions/verdicts.

The level of trust in public institutions (including courts) is diversified, and the institutions assessed by the respondents have different locations/positions in social space. In order to present this mutual position of institutions in relation to the public trust survey, a distance matrix between the assessed public institutions was constructed on the basis of multidimensional scaling. Multidimensional Scaling, MDS, is a statistical technique enabling detection of so-called latent variables that, although not directly observed, explain similarities and differences between the studied objects (in this case, public institutions). At the beginning of the procedure there is usually the distance or similarity matrix between the objects. This may be, for example, a correlation matrix. Multidimensional scaling aims at the distribution of objects as points in n-dimensional space, so that similar objects are closer to one another. The result of the analysis is for each object *n* of real numbers forming the Cartesian coordinates[6].

The distance matrix of the studied public institutions was created on the basis of the arithmetic means of the evaluations of trust in these institutions. The obtained results were subject to multidimensional scaling with the use of the ALSCAL algorithm (Alternating Least Square Scaling). This algorithm enables the analysis of data measured in order, interval and quotient scales; it also ensures the possibility of analysis on discrete and constant, symmetric and non-symmetric data.

The application of multidimensional scaling for the data from the trust in courts and the judiciary survey would enable the creation of a geometrical representation of the analysed objects, i.e. public institutions and the presentation of similarity between them.[7] The positioning of public institutions in the context of trust using the ALSCAL algorithm is presented in the chart below.

6 Multidimensional scaling is used for finding structure in the set of distances between particular objects. It is possible by attributing observations to particular places in conceptual space (usually two- or three-dimensional) so that the distances between the points in space, as close as possible, correspond to particular measures of non-similarity. In many cases dimensions of this conceptual space may be interpreted and used for better understanding of the data. If n=<3 (n number of dimensions) the results may be presented in a chart. Rotation of the coordinates system and the mirror reflection do not change the distance between the items so that the result of scaling may be subject to rotation or reflection. Most often it is done when the discovered dimensions correspond to geographical coordinates. Multidimensional scaling is an alternative method with respect to factor analysis.

7 In the analysed example the index of adjustment of the received model to input data (STRESS) amounts to 0.148 and the RSQ measure = 0.895. If the coefficient STRESS =< 0.2, the received configuration is in monotonic relation to the input data.

Chart 1: Social perception of distances between public institutions

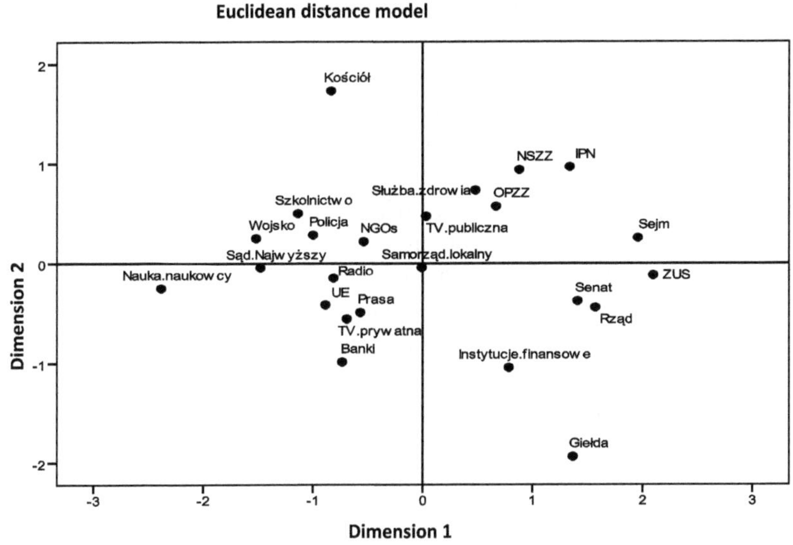

Legend:

Kościół	Church
NSZZ	NSZZ (Independent and Self-Governing Trade Union
IPN	IPN (Institute of National Remembrance)
Służba zdrowia	Health Services
OPZZ	OPZZ (All Poland Alliance of *Trade Unions*)
Szkolnictwo	Education
TV publiczna	Public TV
Wojsko	Army
Policja	Police
NGOs	NGOs
Sejm	Sejm
Sąd Najwyższy	Supreme Court
Samorząd lokalny	Local self-government
ZUS	ZUS (Social Insurance Company)
Nauka, naukowcy	Science, scientists
Radio	Radio
UE	EU
Prasa	Press
TV prywatna	Private TV
Banki	Banks
Senat	Senate
Rząd	Government
Instytucje finansowe	Financial Institutions
Giełda	Stock exchange

The analysis of the chart makes it possible to state that the directly observed similarities between public institutions (the social positioning of institutions) may be presented as the resultant of two dimensions of social perception. Dimension 1 can be interpreted in the context of the implementation of private – public interests by institutions, whereas dimension 2 can be interpreted as the power – knowledge dominance[8]. The two-dimensional systematisation space distinguished under the analysis suggests that the original criterion for the shaping of trust in public institutions is the relation with public interest, whereas the secondary criterion is political neutrality.

According to these two criteria, the Supreme Court, in comparison with other public institutions, is positioned in the social consciousness as a politically neutral institution engaged on the one hand in the protection of the public interests/good, and on the other hand in the protection of private interests. The Supreme Court is perceived as a mediator of the interests of various social actors, a mediator independent of political pressure. Moreover, the Supreme Court is located among the institutions which are not involved either with politics or the market, and are rather connected with the protection of the public good. This finding, presented above in the distance model, supplemented by the data on the level of trust in the supranational law-applying institutions (European and international courts) focused on the protection of human rights, justify an argument that the protection of the public good and human rights protection are deciding factors for the level of trust in courts and judges, and for an empirical legitimation of the rule of law.

Concluding remarks

The rule of law presents a salient characteristic of liberal democracy; it is enshrined in constitutional and state law. However, this principle is to be reconsidered, especially after the collapse of authoritarian/totalitarian regimes. At those times, in order to establish an empirically rooted rule of law, i.e., to form an empirically valid popular conviction that the rule of law is protected, the legal principle should be understood broadly, including social knowledge about past human rights violations and social experiences with the injustices of the past.

The instalment of liberal democracy and the introduction of a market economy also constitute important contributions to the formation of popular opinions on the rule of law, measured by the level of trust in the judiciary.

8 If we make a rotation (transformation) of the coordinate system for a better interpretation of data, the trust in public institutions appears to be shaped in the following dimensions: religion – economy and knowledge/law – tradition.

The findings of the public confidence in the judiciary survey indicate a significant increase in the number of contacts within Polish society with applications of the law, which is closely related to the consolidation of the market economy. They clearly indicate the trend, observed within the last years, consisting of the slow improvement of the opinion concerning the execution of laws. It thus appears that we approach the reversal of trends and the overcoming of the "transformation trauma".

In the light of the survey data, trust in courts is diversified and arranged in a hierarchy in social consciousness. The courts most trusted by the Poles are transnational courts, including European courts, as well as the supreme national courts (the Supreme Court and Constitutional Tribunal). Contact with a court and the respondents' experience connected with it are also important factors. Also, the final effect and respondents' satisfaction with their contact with courts are significant. The obtained results enable the formulation of further conclusions in that respect.

Compliance with the rules governing a fair trial, i.e. the application of procedures by courts and the fulfilment of procedural guarantees, are important for the level of trust. Moreover, the "soft" components of a fair trial appear to be significant for the shaping of trust in courts and judges, such as the appearance/architecture of court buildings, courtrooms' arrangement, judges behaviour in the courtroom and mutual respect shown by participants before a court. The influence of the said "soft factors" related to the execution of a fair trial should become the subject of further sociological research. It is worth indicating here other interesting relations, revealed by the analysis of survey data. What is particularly interesting is the assessment of the Supreme Court as a politically independent institution, and an institution positioned among those that are focused on the protection of the public good and human rights.

References

Arato, Andrew, and András Sajó. 1991. "Editors Introduction." *Law and Policy* 13: 101–106.

CBM Indicator. 2009. *Public Confidence in the Judiciary in Poland. Report from Quantitative Research Prepared by CBM INDICATOR*, Warsaw, May.

Czarnota, Adam, and Martin Krygier. 2007. "Po postkomuniżmie – następny etap? Rozważania nad rolą I miejscem prawa." *Studia Socjologiczne* 2: 145–198.

Daniel, Krystyna. 2003. "Normatywny i społeczny obraz sędziego." In *Sądy w opinii społeczeństwa polskiego*, edited by Maria Borucka-Arctowa and Krzysztof Palecki, 99–148. Kraków: Ratio.

Daniel, Krystyna. 2007. "Kryzys społecznego zaufania do sądów." *Studia Socjologiczne* 2: 61–82.

Fijałkowski, Agata. 1999. "The Judiciary's Struggle towards the Rule of Law in Poland." In *The Rule of Law in Central Europe*, edited by Jiří Přibáň and James Young, 242–256. Dartmouth: Ashgate.

Freedom House. 2003. "Poland." *Nations in Transit 2003*. Accessed October 27, 2014. http://www.freedomhouse.org/report/nations-transit/2003/poland#.VE6luvmG-W4.

Freedom House. 2004. "Poland." *Nations in Transit 2004*. Accessed October 27, 2014. http://www.freedomhouse.org/report/nations-transit/2004/poland#.VE6l-fmG-W4.

Freedom House. 2007. "Poland." *Nations in Transit 2007*. Accessed October 27, 2014. http://www.freedomhouse.org/report/nations-transit/2007/poland#.VE6mFfmG-W4.

Király, Béla K. 1995. "Dictatorships, Lawful Revolutions, and the Socialist Return to Power." In *Lawful Revolutions in Hungary 1989–1994*, edited by Béla K. Király and András Bozóki, 4–34. Boulder (Colorado): Atlantic Research and Publications.

Kis, János. 1995. "Between Reforms and Revolution: Three Hypotheses about the Nature of the Regime Change." In *Lawful Revolutions in Hungary 1989–1994*, edited by Béla K. Király and András Bozóki, 99–120. Boulder (Colorado): Atlantic Research and Publications.

Krygier, Martin. 1994. "Marxism, Communism and the Rule of Law." In *Marxism and Communism: Posthumous Reflections on Politics, Society, and Law*, edited by Martin Krygier, 237–172. Amsterdam and Atlanta (GA): Rodopi.

Pankowski, Krzysztof. 2014. "Dwudziesta piąta rocznica okrągłego stołu." *Raport CBOS* 16. Accessed October 27, 2014. http://cbos.pl/SPISKOM.POL/2014/K_016_14.PDF.

Přibáň, Jiří. 1999. "Legitimacy and Legality after the Velvet Revolution." In *The Rule of Law in Central Europe. The Reconstruction of Legality, Constitutionalism an Civil Society in the Post-Communist Countries*, edited by Jiří Přibáň and James Young, 29–56. Dartmouth: Ashgate.

Přibáň, Jiří, and Pauline Roberts. 2003. "The Central European Systems of Justice in Transition: Introductory Remarks." In *Systems of Justice in Transition. Central European Experience Since 1989*, edited by Jiří Přibáň, Pauline Roberts and James Young, 1–10. Dartmouth: Ashgate.

Skąpska, Grażyna. 1990. "The Rule of Law from the East Central European Perspective." *Law and Social Inquiry* 15: 699–701.

Skąpska, Grażyna. 2009. "Rule of Law, Economic Transformation and Corruption after the Fall of the Berlin Wall." *Contemporary Problems of the Rule of Law* 1: 59–72.

Skąpska, Grażyna. 2011. *From "Civil Society" to "Europe": A Sociological Study on Constitutionalism After Communism*. Leiden and Boston: Brill.

Skąpska, Grażyna, and Grzegorz Bryda. 2013. "Apolitycznosć czy sprawiedliwosć proceduralna? Zaufanie do sądowego wymiaru sprawieldiwosci na tle innych instytucji publicznych w Polsce." *Studia Socjologiczne* 1: 77–94.

Sólyom, László. 2003. "The Role of Constitutional Courts in the Transition to Democracy with Special Reference to Hungary." *International Sociology* 18: 133–161.

Teitel, Ruti G. 2000. *Transitional Justice*. Oxford: Oxford University Press.

The Constitution of the Polish People's Republic of 22 July 1952.

The Constitution of the Republic of Poland of 2 April 1997.

Tyler, Tom R. 2006. *Why People Obey the Law*. Princeton: Princeton University Press.

Wagnerová, Eliška. 2003. "Position of Judges in the Czech Republic." In *Systems of Justice in Transition: Central European Experience since 1989*, 163–179. Aldershot: Ashgate Publishing Limited.

Worlwide Governance Indicators. n.d. *Country Data for Poland, 1996–2013*. Accessed October 27, 2014. http://info.worldbank.org/governance/wgi/pdf/c177.pdf.

Part II: The Sociology of Formally Embedded Informality

Liela Groenewald

Understanding Informality: Conceptual Lessons from Informal Settlement in Southern Africa

Introduction

Urban Sociology and Urban Studies more broadly have provided an intellectual home for the empirical study and the conceptual development of informality, primarily as it manifests in the sectors of the economy, politics, and housing. Since different definitions of informality betray normative assumptions about informality, these definitions also have implications for social justice. Descriptions of informality in hegemonic discourses that portray informality as separate or decoupled from the formal tend to portray informality as deviant, betraying normative assumptions that the formal ideal should be pursued or protected (Groenewald et al. 2013). Instead of strategies that empower those who resort to informality, these assumptions can lead to strategies to regulate, control or repress informality. This is often to the detriment of marginalised communities.

In the context of South Africa, where three hundred years of colonialism and legislated racial discrimination lasting until the late twentieth century have produced large-scale, deep and chronic poverty coupled with extraordinary levels of inequality, researchers have emphasised the ethical obligation on social scientists to develop accounts that could resonate with the poor, rather than to reiterate and rationalise the perspectives of the well-off classes. Failing to heed this obligation leads not merely to ethical difficulties, but also to concerns about rigour, since significant knowledge gaps and biases arise from the overwhelming domination of social science by the accounts of the powerful (Nader 1972, 292–295; Connell 2007, 216). Taking account of the knowledge risks involved in generalising about informality based on empirical work in the comparatively privileged context of the global North, the purpose of this paper is to demonstrate that the knowledge and theories of informality can be advanced by taking account in particular of work from southern Africa, where informality is concentrated.

Mindful of the need to produce work that excavates the perspective of the marginalised, rather than to simply reinforce power relations, scholars working within the critical paradigm have embraced definitions that emphasise the particular precariousness of people who build livelihoods in the informal sector, the conceptual distance between hegemonic definitions of informality and the aspirations of people who employ informality in pursuit of incremental improvements to these precarious lives, and the need for intellectual interventions to shift dominant thinking in a direction that can resonate with and assist ordinary people.

The question of whether the formal and informal sectors are separate entities or are entangled with one another has therefore arisen as a central concern for critical scholars. In this regard, important lessons can be drawn from diverse experiences of urban informality. To do so, this paper draws on the theoretical foundations of a research project[1] focused on the interaction between the extremes on this formal-informal continuum, i.e. the response of the formal state to informal settlement in the context of post-apartheid South Africa. First, the paper draws attention to the plurality of experiences of urban informality that is available to social researchers. Next, public policy responses and their implications for informality are considered. Based on this, the paper concludes that a conceptual shift with regard to the definition of informality is necessary for the sake of accuracy and rigour.

Plural experiences of urban informality

At the beginning of the twenty-first century, for the first time since the dawn of humankind, most people are living in cities (Bekker 2006; Davis 2006). In about twenty years, city dwellers are expected to constitute two-thirds of the world's population. Almost all of this projected urbanisation will occur in less developed regions, so that not a single local region where the majority is rural should remain by 2030 (United Nations Population Fund 2005). In 2001, the UN estimated that nearly a third of the global urban population, numbering nearly a billion people, lived in slums (UN-Habitat 2003). Poor people constitute approximately half of the 7 billion people in the world (Population Reference Bureau 2011). In less developed countries, where population growth of about 76 million people per year is taking place, the proportion of people living in poverty is also escalating. As a result, the world's largest numbers of slum dwellers can be found on the continents of Asia and Africa as well as in Latin America (UN Habitat 2003).

The need for comparison across countries has led to the widespread use of the UN definition of urban areas as those areas with more than twenty thousand inhabitants within its boundaries and of cities as those with more than a million inhabitants (Kasarda and Crenshaw 1991, 470). The unprecedented

1 I wish to thank the National Research Foundation (NRF) of South Africa for funding this work.

pace of urbanisation and the concentration of poverty in informal settlements may be interpreted as the key challenges facing the local state in the developing world today. The pace of urbanisation is both new and unique to the developing world, but the process of urbanisation fits patterns of economic and social change that occurred in the West during its earlier period of industrialisation (Kasarda and Crenshaw 1991, 481–482). This leads to explanations of urbanisation that consider earlier urbanisation in the West and later urbanisation in the developing world as continuities of the single historical process of capitalist expansion (Kasarda and Crenshaw 1991, 483). Dependency or world systems interpretations, on the one hand, consider underdevelopment as the outcome of "plunder and exploitation of peripheral economies by … core areas", whereas interdependency theories regard urbanisation as the result of capitalism, which encourages the concentration of resources, infrastructure and production (Clark 1998, 88). Alternative explanations of urbanisation include modernisation theories, which consider industrialisation rather than capitalism to be the main driver of urbanisation in the developing world, and urban bias theories, which consider powerful political elites to be reliant on urban resources, leading them to implement policies that promote cities and neglect rural areas (Kasarda and Crenshaw 1991, 484). Whatever the explanation, this urbanisation has also seen a concentration of poor people in informal settlements, often called slums.

Although the terminology of "informal settlement" is new, academic concern with the phenomenon is not. Slums have been mentioned in the literature for as long as urbanisation has been a focus of sociological analyses. In his seminal 1845 rendition of *The Condition of the Working Class in England*, Engels (1845, 28) describes a settlement of makeshift dwellings, which in physical terms corresponds to a layman's understanding of contemporary informal settlements:

> Passing along a rough bank, among stakes and washing-lines, one penetrates into this chaos of small one-storied, one-roomed huts, in most of which there is no artificial floor; kitchen, living and sleeping-room all in one. In such a hole, scarcely five feet long by six broad, I found two beds – and such bedsteads and beds! – which, with a staircase and chimney-place, exactly filled the room. In several others I found absolutely nothing, while the door stood open, and the inhabitants leaned against it. Everywhere before the doors refuse and offal; that any sort of pavement lay underneath could not be seen but only felt, here and there, with the feet. This whole collection of cattle-sheds for human beings was surrounded on two sides by houses and a factory, and on the third by the river, and besides the narrow stair up the bank, a narrow doorway alone led out into another almost equally ill-built, ill-kept labyrinth of dwellings. … The whole side of the Irk is built in this way, a planless, knotted chaos of houses, more or less on the verge of uninhabitableness, whose unclean interiors fully correspond with their filthy external

> surroundings. … Privies are so rare here that they are either filled up every day, or are too remote for most of the inhabitants to use (Engels 1845, 28).

Engels (1845, 34) criticises the slum-clearing response of the local state for failing to provide a solution, but appears unconcerned about painting working class areas as areas of dirt, crime, disease, and early death. His description represents an early instance of the tendency to understand informal settlement from the perspective of its physical structures, social shortcomings, and deviance from the mainstream norm. It should be noted that even during these earliest recorded cases of urbanisation linked to capitalist industrialisation, informal settlement emerges as a salient concern closely associated with access to the city and to state-society relations or citizenship.

By half a century later, Simmel (2009 [1908]), who considers a lack of shelter to represent the most extreme kind of poverty, argues that the homeless *Penner* who find shelter for the night in seasonal haystacks in the Berlin area are pushed into hiding because their original communities cannot bear the sight of poverty, so that these individuals become a new community bound together primarily by poverty.

It is clear, then, that neither informal settlement nor repressive state responses are new, or particular to the Global South, although contemporary informal settlement is concentrated in those parts of the world that have high levels of inequality, is most prevalent in sub-Saharan Africa as compared to other regions (UN-Habitat 2003), and is manifested in the visible concentration of poverty in informal settlements. Existing theory relies heavily on evidence gathered in the North. In this regard, urban Africa is under-explored. More inclusive considerations of informality are therefore critical for increasing our current understanding.

Public policy and the production of informality

Despite broad similarities between contemporary urbanisation in the developing world and the urbanisation of the West from the mid-nineteenth century, important differences have also been identified. Greater degrees of development, access to more natural resources, and a context of far less competitive economic production had eased the difficulties associated with urbanisation at the time of industrialisation in the North (Kasarda and Crenshaw 1991, 468). The exponential increase of the urban population has resulted in the attainment of megacity status by some cities in the developing world (Kasarda and Crenshaw 1991, 469). Two factors contributing to the developing world's average urban growth rate of approximately double that experienced during European industrialisation, are its inability to "export its surplus population" to the colonies as Europe did, coupled with the higher natural population increase, which was estimated during the 1990s at roughly double the rate experienced by industrialising European cities (Kasarda and Crenshaw 1991, 468).

Cities or countries that have been occupied by foreign powers count the cost in the loss of human life, the decimation of indigenous architecture and culture, the repression of indigenous language and education, and trauma, all with crippling consequences for human capital and economic development. The legacy of over three hundred years of European colonial occupation and extraction – from the mid-seventeenth century until the late twentieth century – has left cities of southern Africa with severe backlogs in the most basic infrastructure development.

City growth in the developing world has contributed to the dilemma that the demand for state services has grown beyond what the local state has the capacity to provide (Curtis 1999; Smith 2000; Qomfo 2005). Economic changes have also informed the rationalisation of urban subsidies or grants, and the resulting inadequacy of state-provided services has contributed to conflict between citizens and the state (Roberts 1989, 673). Since "city-space remains the node where multiple identities and modernities emerge, are contested and refashioned in context to the way citizenship has been defined and organized" in Patel's (2006, 34) conception, informality can be understood as a particular articulation of modernity.

While local government in South Africa is required to consult communities, the ability of individuals or communities to participate in such formal consultation fora has proved to be particularly constrained by informality. The questionable legal status of informal settlements often means that the people who live there have access only to "clientelist politicians and bribe-seeking public employees" (Mohamed 2006, 36).

With regard to participation in formal local consultation processes, some writers argue that poor and voiceless people are excluded from being able to provide input into local policies by inadequate systems. The constitution obliges municipalities to consult their constituent communities, but not to establish ward committees (Steytler and Mettler 2001, 2). Even where ward committees are established, they may reinforce exclusion or fail to serve the interests of the excluded. Social movements that operate across ward boundaries may be represented on ward committees. But, based on a study conducted in Western Cape towns, Bekker and Leildé (2003, 144) show that it is largely middle-income people who maintain both loyal and critical local participation, to the extent that "the affluent as well as the poor had withdrawn from local civil society". This concern is echoed by Mohamed (2006, 45), who concludes that in South African cities, including Johannesburg, "disadvantaged sectors of the urban population, especially the informal settlement communities, are isolated from the processes of policy-making" due to their exclusion from ward committees.

The obstacles to the participation of informal settlement residents in ward committees include the sectoral constitution of these committees, and the lack of ward committees' capacity to function effectively and regularly meet with communities (Mohamed 2006, 39–40). This lack of capacity is illustrated by a Rustenburg case in which, on the one hand, ward councillors who chaired ward committees were ultimately able to control the agenda of ward committee meetings, and, on the other hand, politicians and residents in poor areas shared an inability due to limited education to deal meaningfully with technical aspects of development plans (Putu 2006, 29–30). Residents of informal settlements not only suffer from a lack of basic services, such as water, sanitation and electricity, but in most cases also from a lack of essential social services such as health centres, roads, drainage, schools, and market places (Mohamed 2006). In the absence of these facilities, life may consist of a daily drudgery in pursuit of bare necessities, and regular meeting attendance may simply be impossible. A broad range of factors thus contributes to the limited impact of the residents of informal settlements on local government.

On the one hand, public participation has been institutionalised as part of formal political processes, but on the other hand, citizens of the developing world have also played an active role "in shaping urban space" as they secure amenities by means of self-help and reciprocity rather than to rely on the state (Roberts 1989, 686–687, 672–673). Informal work and informal settlement are two significant self-help strategies that are employed by poor city residents. In the face of exponential population growth, informal work has generated substantial interest due to its potential role in labour absorption and job creation (Kasarda and Crenshaw 1991, 477). The considerable contribution of informal settlement in meeting the need for shelter has also been recognised (Kasarda and Crenshaw 1991, 480). A substantial difficulty in studying the informal sector has been the challenge of defining informality and drawing the boundaries of the informal sector (Kasarda and Crenshaw 1991, 494). Common mistakes have been to assume a concentration of recent migrants in informal work or in informal settlements and to assume an overlap between those who make a living by means of informal economic activity and those who live in informal settlements (Castells 1983; Kasarda and Crenshaw 1991, 478). For UN-Habitat, it will take time and concerted effort to turn the tide of informal settlement in sub-Saharan Africa (UN-Habitat 2010). For many academics and analysts, if population and urbanisation projections are anywhere near correct, then the growth of informal settlement is inescapable (Kasarda and Crenshaw 1991, 480; UN-Habitat 2008).

It has become clear that informal settlements represent different things to their residents and to those who look in on them from the outside. Rather than viewing informal settlement as a "last resort", residents gain privacy, autonomy, informal ownership, and integration into "vital, if oftentimes poor, communities" by moving from "squalid rental quarters or alternative living arrangements" to informal settlements even though their access to public health facilities, transport, and security of tenure remains limited (Kasarda and Crenshaw 1991, 480). A significant portion of urban residents rely on informal housing because they are unable to afford formal residential stock (Bredenoord and van Lindert 2010). As a result, only remedial schemes such as site-and-service programmes are likely to ameliorate the shortage of shelter (Kasarda and Crenshaw 1991).

Public policy responses have often seemed oblivious to this aspect of informal settlement. Based on a comparison of state responses to informal settlement in Brazil and South Africa, Huchzermeyer (2002, 98) identifies two possible founding interpretations that drive state responses to informal settlement. The first sees informal settlement as a result of class relations and seeks to support mobilisation, participatory intervention and pressure for broad-ranging policy changes, while the second sees informal settlement as a threat to conservative, middle-class interests and pursues relocation of informal settlement residents to segregated developments.

Apart from their lack of affordability for the poor, another problem with large-scale formal housing projects is their location at the periphery of urban areas where development is most affordable for states, thereby exacerbating residents' struggles to access employment, public transport and social services. Alongside residents, states also experienced problems with the peripheral concentration of low-cost housing: the cost of providing basic services increased with distance from the centre; states were obliged to include core housing units from where residents could begin the construction of their dwellings; and infrastructure development was associated with an increase in self-help informal housing at the urban periphery (Bredenoord and van Lindert 2010).

Even outside the global South, the insecure position of informal settlements in comparison to the residents and government of the formal, planned city has caused Yiftachel (2009, 90) to name these power relations "colonial", arguing that the term can be used to characterise city management that enables confiscation and annexation, and institutionalises in the urban political economy a code for expanding dominant spatial and other interests, exploiting marginalised groups, essentialising identities as "different and unequal", and for involuntary, hierarchical segregation. Despite the difficulties that contested definitions, the questionable legal status of slum dwellers and the fluidity and vulnerability of poor livelihoods present to states, social workers, researchers, and activists, it is clear that the high profile of the Millennium Development Goals and the target of improving the lives of 100 million slum dwellers have helped informality to emerge as a key consideration with regard to inclusion in citizenship and access to its benefits.

Contested definitions of informal settlement

The broad spectrum of theoretical approaches to the state and informal settlement provide no easy solutions to achieving the ideals of inclusive citizenship. Based on a variable-sum concept of power, consensus theories have interpreted the democratic state as a mediator of the various interests in society that ultimately promotes social cohesion in the common interest. Government, in this view, is based on consent. Consensus theories fly in the face of the plight endured by the impoverished majority of the urban population of present-day sub-Saharan Africa. Cities in Africa achieve the highest Gini coefficients on the planet (UN-Habitat 2008), indicating vulgar contrasts in standard of living. The gradual policy and power fluctuations of a pluralist democratic system offer only a crassly inadequate response to the welfare needs experienced urgently and constantly by the poorest residents of these cities.

Despite the inadequate attention that the functionalist perspective pays to conflicts of interest, its distinction between the functional and dysfunctional outcomes of a single event exposes a need to understand the continuities between everyday practices and oppressive or violent actions, an insight that is critical for effecting social change (Johnson 2000). This is helpful in the case of informality, where the constant iteration of the formal ideal arguably reproduces the precariousness of informality and thereby contributes to the reproduction of informality.

In contrast, conflict theorists have defined the state as a coercive institution that protects positions of privilege. Although a Weberian conception of the state as relying on the bureaucracy as well as a monopoly on the legitimate use of violence has been popular, more recent work indicates that the means by which the state controls the population has multiplied (Foucault 1975). The collusion between structures and agents confirms the need, identified by feminist and postcolonial theorists, to discover ways of escaping imposed taxonomies and ways of reading the world, also with regard to formality and informality.

In pursuit of more inclusive incarnations of citizenship, contemporary authors seek alternatives to those definitions of informality that pathologise informal settlements and their residents. These settlements often represent something very different to residents compared to what it represents to outsiders. While the phrase "informal settlement" is often used by authors in Africa and Latin America, the term "slums" is more common in the literature from the North. The two do not overlap exactly, and both are plagued by conceptual difficulties. Earlier scholarship commonly referred to freestanding informal settlements as "shantytowns", as in the work of Crankshaw (1993, 31, 50), who considers this form of housing to be one manifestation of homelessness. Such an interpretation may be sympathetic to the desperate plight of residents of informal settlements, but raises problems because it obscures the gains that these settlements represent to residents, including a foothold that gives them access to cities and which can easily be ripped out from under them by insensitive government policies. This problem persists in the definition of the term "slum" under the Millennium Development Goals in 2005 as "any area that met the following six criteria: lack of basic services, inadequate building structures, overcrowding, unhealthy and hazardous conditions, insecure tenure, and poverty and exclusion" (Huchzermeyer and Karam 2006, 2). Given these shortcomings of common-sense and dominant definitions, Huchzermeyer and Karam (2006) argue that the central, defining characteristic of informal settlements is that of tenure insecurity, even as they acknowledge that this insecurity may be exacerbated by a range of other hazards.

A central concern that arises with regard to the definition of informality is the question of whether informality can be understood as decoupled or separate from the formal sector. In this regard, critical scholars have argued that it is important *not* to understand the informal sector as a decoupled underclass of housing, work or politics. On the messy continuum from formal to informal, different instances of informality rather appear to be differentially entangled with the formal economy. Leaning towards the formal side of this continuum are those informal strategies that people use to access or manage formal services such as the way in which *Štela*, a local interpretation of social capital, is used to access services in Herzegovina (Koutkova 2013) or when judges informally present barriers to public access to courtrooms (Burdziej and Pilitowski 2013), while on the informal side one finds the collective savings organisations of poor communities (Tshoose 2009). In these intermediate or grey areas, it is not difficult to demonstrate that informality and formality are entangled. The most devastating implications for social justice have, however, emerged from definitions that crudely decouple the informal from the formal by pathologising informality as deviant based on descriptions of a physical or other inadequacy. An example of this occurs when dwellings in informal settlements are bulldozed by the state because they contradict the ideal of neat, formal homes, even when such homes are not available. It is critical, therefore, to consider the entanglement of formality and informality precisely where they occur further removed from each other on the extremes of this continuum. The description of global institutions and the formal, state response to informal settlement in cities of southern Africa present such an opportunity. The state response to informality betrays a hegemonic definition of carefully controlled formality as the superior option. This conception does not take adequate account of the experiences of ordinary people whose circumstances have led them to seek recourse in informal strategies.

The difficulties of living in informal settlements are substantial, but recent empirical work in five southern African cities indicates that the residents of informal settlements populate these homes and communities not only out of a need for shelter, but also to secure certain advantages. Rather, informal settlement residents are aware of contradictory dimensions of their home communities, including many different "deficiencies, gains, and aspirations" (Groenewald et al. 2013, 107). While the deficiencies are familiar and have been emphasised repeatedly in the hegemonic literature, the gains and aspirations include a place in the city versus no access to city life, the close proximity to jobs, and freedom of movement (Groenewald et al. 2013). This work therefore presents substantial challenges to hegemonic definitions of informality and the normative assumptions that they accompany. Such judgments contribute to precariousness and thereby reproduce informality, but they also ignore the conceptually distinct articulations of ordinary people about their own involvement in informality. This approach resonates with work emerging from the Latin American context, for example the concept of the kinetic city conceived by Hernández et al. (2010, xiii–19), which is characterised as spontaneous, dynamic, productive, and interlinked and responding to the biases of the formal economy. Like their southern African contemporaries, Hernández et al. (2010, xiii) recognise that it is far more complex to delimit informality than to identify the shortcomings of hegemonic definitions.

Emphasising agency and spontaneity as key traits of informal settlement successfully avoids the pathologising tendencies of other hegemonic discourses about informality, but carries an inherent risk in such an approach, which is that of obscuring class structure and conflict of interest, while romanticising informality. Recent empirical work on informal settlements in southern Africa demonstrates that residents seek better services and greater security, with their primary concern being their acute vulnerability to state-sanctioned evictions. In Luanda, the capital city of Angola, residents of the most central informal settlements indicated that they were the most vulnerable to violent eviction projects implemented by the state in order to make way for private development (Raposo 2008, 4–5). In interviews conducted in Cape Town, South Africa, residents described informal settlements as lacking a future, since they "do not know where they will live and how they will live wherever they will be" (Tredoux 2008, 3–4). Residents told Tredoux (2009) that they urgently needed essentials such as water, electricity and toilets. In the Gauteng province of South Africa, where the legacy of colonialism and apartheid includes an acute lack of adequate shelter for generations of families trapped in chronic poverty, residents of various informal settlements expressed a "process of frustration over many decades" (Huchzermeyer 2008, 3). This includes the tendency that the social and economic vulnerability of residents of informal settlements are worsened by being relocated to developments far from the urban centre (Huchzermeyer 2008, 5). Residents of informal settlements lack the means to access the formal rental or property market; but while the informal strategies they use to secure shelter serve to strengthen their own sense of security within their communities, they are not recognised by the state (Rubin 2008, 6). While residents of informal settlements therefore require services and other forms of support, the primary concern shaping their strategies and fears is that of their vulnerability to eviction. This self-reflective conceptualisation of the residents of informal settlements contributes to the argument that the key characteristic of informal settlements that researchers should recognise, is that of tenure insecurity (Huchzermeyer and Karam 2006, 2). Risk and insecurity are contributed primarily by the state inclination to protect the property market and private development opportunities at the expense of security for the poor, whose agency and innovation open up opportunities within the gap left by various regimes of state austerity or welfare.

While a conceptual binary between formality and informality is limiting, an analytical approach that obscures class similarly lacks rigour. The growing group of authors who work within this framework criticise the dominant construction of the formal and informal sectors as a binary, but avoid the fallacy of universalising or generalising across distinct interest groups who are in competition with one another by recognising both the acute vulnerability of people in informal settlements and their agency with regard to forging a place for themselves in the city despite a hostile social structure. The interconnections and interdependency of the formal and informal sectors must be foregrounded in an adequate account of contemporary cities, since an analytical approach and distinct policy formulations are required for understanding and responding to people whose lives are constantly exposed to risk and those whose livelihoods are more stable and secure.

Conclusions

This paper set out to demonstrate that attention to the experiences of ordinary, poor people living in informal settlements in the global South and in particular in southern Africa, where informality is concentrated, can strengthen both the conceptual rigour and the broader legitimacy of social science conducted in the urban spaces of the twenty-first century. The paper showed that urban experiences vary widely, and that theorising that draws on the global North alone inherently runs the risk of erroneously generalising parochial findings. The pathologising effect of dominant policy responses further demonstrated the potential for knowledge and legitimacy gaps within contemporary analyses of informality. Despite significant differences in the context and texture of urban life across the globe, the recent global economic recession has created conditions that increase the relevance of the informality literature based on empirical work in the global South for the global North as well. It has been argued that the empirical work emerging from southern Africa and elsewhere in the global South provides a foundation for a conceptual shift with regard to informality.

Informality is a broad idea that can be difficult to pin down. Together with a growing international interest in slums, the growth of the South African informal sector in the last decade of the twentieth century has given rise to an extended literature on informal housing and land tenure, informal trade and labour, and informal politics. Partly because informal settlement occurs in plural forms across the world, it is a hotly contested and questioned concept. Recent research from southern Africa, however, opens up the possibility of integrating ordinary people's conceptions of informality into academic definitions, with the promise of gains both for the residents of informal settlements and for scholars. Specifically, a shift of definition to recognise the particular precariousness contributed primarily by the state, and to balance this challenge with the gains and aspirations of those who resort to informality, creates conceptual problems for a purely repressive state response. Such a shift not only facilitates scholarship oriented to the interests of marginalised, informal communities, but is also more comprehensive than hegemonic and pathologising definitions, which has the benefit of not only strengthening the conceptual rigour and accuracy of social research, but also the broad legitimacy of social science in the international context of austerity and high levels of unemployment.

References

Bekker, Simon. 2006. "Introduction." In *Reflections on Identity in Four African Cities*, edited by Simon Bekker and Anne Leildé, 1–10. Cape Town: African Minds.

Bekker, Simon, and Anne Leildé. 2003. "Residents' Perceptions of Developmental Local Government: Exit, Voice and Loyalty in South African Towns." *Politea* 22: 144–165.

Bredenoord, Jan, and Paul van Lindert. 2010. "Pro-Poor Housing Policies: Rethinking the Potential of Assisted Self-Help Housing." *Habitat International* 34: 278–287.

Burdziej, Stanisław, and Bartosz Pilitowski. 2013. "Informal Restrictions on Public Access to Open Court Sessions in Poland." Paper presented at "Formality and Informality: From Decoupling to Entanglement. Workshop on the Formal and the Informal in Law, Institutions and Economy", Warsaw, Poland, 13–14 May.

Castells, Manuel. 1983. *The City and the Grassroots: A Cross-cultural Theory of Urban Social Movements*. London: Edward Arnold.

Clark, David. 1998. "Interdependent Urbanization in an Urban World: An Historical Overview." *The Geographical Journal* 164: 85–95.

Connell, Raewyn. 2007. *Southern Theory*. Cambridge: Polity Press.

Crankshaw, Owen. 1993. "Squatting, Apartheid and Urbanisation on the Southern Witwatersrand." *African Affairs* 92: 31–51.

Curtis, Donald. 1999. "Performance Management for Participatory Democracy: Thoughts on the Transformation Process in South African Local Government." *International Journal of Public Sector Management* 12: 260–272.

Davis, Mike. 2006. *Planet of Slums*. London: Verso.

Engels, Friedrich. 1845. *The Condition of the Working Class in England*. London: Dodo Press.

Foucault, Michel. 1975. *Discipline and Punish: The Birth of the Prison*. London: Vintage Books.

Groenewald, Liela, Marie Huchzermeyer, Kristen Kornienko, Marius Tredoux, Margot Rubin, and Isabel Raposo. 2013. "Breaking Down the Binary: Meanings of Informal Settlement in Southern African Cities." In *Governing Cities in Africa: Politics and Policies*, edited by Simon Bekker and Laurent Fourchard, 93–116. Cape Town: HSRC Press.

Hernández, Felipe, Peter Kellet, and Lea K. Allen, eds. 2010. *Rethinking the Informal City: Critical Perspectives from Latin America*. New York: Berghahn Books.

Huchzermeyer, Marie. 2002. "Informal Settlements: Production and Intervention in Twentieth-Century Brazil and South Africa." *Latin American Perspectives* 29: 83–105.

Huchzermeyer, Marie. 2008. "Settlement Informality: The Importance of Understanding Change, Formality and Land and the Informal Economy." Paper presented at Informality Workshop of the GDRI on Cities in Africa, Johannesburg, South Africa, 3–4 July.

Huchzermeyer, Marie, and Aly Karam. 2006. "The Continuing Challenge of Informal Settlement: An Introduction." In *Informal Settlements: A Perpetual Challenge?*, edited by Marie Huchzermeyer and Aly Karam, 1–18. Cape Town: UCT Press.

Johnson, Allan G., ed. 2000. *The Blackwell Dictionary of Sociology: A User's Guide to Sociological Language* (2nd edition). Oxford: Blackwell Publishers Ltd.

Kasarda, John D, and Edward M Crenshaw. 1991. "Third World Urbanization: Dimensions, Theories, and Determinants." *Annual Review of Sociology* 17: 467–501.

Koutkova, Karla. 2013. "The Importance of Having Štela: Reproduction of Informality through Democracy Assistance in Bosnia and Herzegovina." Paper presented at "Formality and Informality: From Decoupling to Entanglement. Workshop on the Formal and the Informal in Law, Institutions and Economy", Warsaw, Poland, 13–14 May.

Mohamed, Salah Elzein. 2006. "From Ideas to Practice: The Involvement of Informal Settlement Communities in Policy-Making at City Level in South Africa." *South African Review of Sociology* 37: 35–47.

Nader, Laura. 1972. "Up the Anthropologist: Perspectives Gained From Studying Up." In *Reinventing Anthropology*, edited by Dell Hymes, 284–311. New York: Pantheon Books.

Patel, Sujata. 2006. "Urban Studies: An Exploration in Theories and Practices." In *Urban Studies*, edited by Sujata Patel and Kushal Deb, 1–38. New Delhi: Oxford University Press India.

Population Reference Bureau. 2011. "2011 World Population Data Sheet." Accessed May 14, 2012. http://www.prb.org/pdf11/2011population-data-sheet_eng.pdf.

Putu, Mpho. 2006. "The Role of Ward Committees in Enhancing Public Participation in Rustenburg Municipality: A Critical Evaluation." Accessed January 10, 2009. http://www.idasa.org.za/gbOutputFiles.asp?WriteContent=Y&RID=1667.

Qomfo, Luyando. 2005. "Local Government Responsibilities." *Idasa Budget Brief* 153: 1–8.

Raposo, Isabel. 2008. "Urban Planning and Management Instruments and Institutions versus Socio Spatial Cohesion in Maputo and Luanda: Resettlement or Upgrading and Rehabilitation of the Informal Peri-urban Areas." Paper presented at the Informality Workshop of the GDRI on Cities in Africa, Johannesburg, South Africa, 3–4 July.

Roberts, Bryan R. 1989. "Urbanization, Migration, and Development." *Sociological Forum* 4: 665–691.

Rubin, Margot. 2008. "Beyond the Informality/Formality Dualism – Findings from South Africa." Paper presented at the Informality Workshop of the GDRI on Cities in Africa, Johannesburg, South Africa, 3–4 July.

Simmel, Georg. 2009 [1908]. *Inquiries into the Construction of Social Forms*. Leiden: Koninklijke Brill NV.

Smith, Laïla. 2000. "The Urban Political Ecology of Water in Cape Town, South Africa." Accessed June 4, 2007. http://www.queensu.ca/msp/pages/Project_Publications/Papers/Water.pdf.

Steytler, Nico, and Johan Mettler. 2001. "Proposals Towards the Establishment of Ward Committees." Accessed January 14, 2009, http://www.community-lawcentre.org.za/local-government/working-papers03/wp004-2001-steytler-mettler-proposals-towards-the-establishment-of-ward-committees.pdf/attachment_download/file.

Tredoux, Marius. 2008. "Views Expressed by Stakeholders in the Housing Delivery Process on Informal Settlements and Their Formalization: A Cape Town Case Study." Paper presented at the Informality Workshop of the GDRI on Cities in Africa, Johannesburg, South Africa, 3–4 July.

Tshoose, Clarence. 2009. "The Emerging Role of the Constitutional Value of Ubuntu for Informal Security in South Africa." *African Journal of Legal Studies* 3: 12–19.

UN-Habitat. 2003. *The Challenge of Slums: Global Report on Human Settlements 2003*. Nairobi: United Nations Human Settlements Programme.

UN-Habitat. 2008. *State of the World's Cities 2010/2011: Bridging the Urban Divide*. London and Sterling (VA): Earthscan.

UN-Habitat. 2010. *The State of African Cities: Governance, Inequality and Urban Land Markets*. Nairobi: UN-Habitat.

United Nations Population Fund. 2007. "Fast Facts." Accessed February 19, 2007. http://www.unfpa.org/pds/facts.

Yiftachel, Oren. 2009. "Theoretical Notes on "Gray Cities": The Coming of Urban Apartheid?" *Planning Theory* 8: 88–100.

Stef Adriaenssens, Dieter Verhaest and Jef Hendrickx

Lineland and the Underground Economy: The Multidimensionality of Informal Work by Secondary Education Students

> His subjects [...] were all alike confined in motion and eye-sight to that single Straight Line, which was their World. It need scarcely be added that the whole of their horizon was limited to a Point; nor could anyone ever see anything but a Point. Man, woman, child, thing – each was a Point to the eye of a Linelander (Abbott 1884, 55).

Whenever states impose formal rules on people in a territory, noncompliance will occur. There is no empty set of underground activities in any given modern society. This is not to say that underground activities have a comparable frequency in different societies. The width and breadth of informal activities depends upon a number of socio-economic and personal characteristics. Apart from their *pure* sociological relevance, underground activities have a number of economic, political and social effects. For all these reasons, it is essential that the phenomenon is researched in a conceptually consistent and valid way.

This text primarily aims to contribute to the methodological knowledge of informal activities. It attempts to do justice to the true nature of informal activities by taking their multidimensional and gradual nature into account. The analysis of this phenomenon is applied to the part-time paid work often performed by students. The evidence shows a substantial activity rate of students in labour markets in virtually every country. As is the case in every other category of employment, not all of this work is declared to the proper authorities, is regulated by the formal institutions of society or is protected by government regulations and enforcement. In short, a substantial part of the paid work carried out by students takes place within the informal economy.

However, the existing literature does not excel in a careful dissection of the informal nature of student work. The basis of informality may lie in the violation of different types of rules: taxes, social security regulation, permits, health and safety legislation, labour law and so forth. One should therefore not speak of *informality*, but rather of *informalities*. At the same time, the transgression of most rules is not of a binary nature either: in reality, transgression is an ordered feature. Indeed, the central task of this paper is to show that different dimensions, measures and categorizations of informal work may have a significant impact on the outcome of any study.

In the first part we build our case starting from a careful definition of what "informal activities" entail. This phenomenon then is connected to the world of student work. The second section introduces the data collection from the Belgian *Student Employment Survey* and how these data will be put to use. The third section then presents the results of the analyses.

Problem: Student work and the informal economy

The introductory section explains the theoretical rationale of this contribution. Therefore we begin by building upon our understanding of the literature regarding informal activities in the first two subsections. We start by developing the agreement that seems to exist about the concept and the main paths into informal work and consumption. Most informal economy studies reduce the subject to a one-dimensional and binary phenomenon.

In reality it is multidimensional and ordered. The problem is *not* that this is introduced as a stylized fact in order to make measurement possible; that is wholly understandable and a legitimate approach. The problem is that most commentators do not acknowledge that a form of reductionism is taking place. The second subsection develops this general critique.

The final subsection introduces the specific problem of informal work by students. Students are among what might be called the "usual suspects" of informal work. In the third subsection we briefly look into the literature on part-time work done by adolescents during their education.

What "informality" refers to

In this subsection we explore the concept of informality in the literature on underground and informal activities. Our focus is mainly on work, although the general discussion is also applicable to underground aspects of consumption. One of the general problems of this literature is its conceptual disarray. Nevertheless, underneath the apparent lack of agreement, some conceptual consensus exists. We build upon this latent concurrence to propose a definition of informal activities in relation to formal, underground, criminal and self-provisioning activities. Within the informal sector, some agreement exists regarding the main paths into informal activities. This general model is also explained.

As in other uses of the word "informal", its meaning can only be understood in relation to those things that are defined as "formal". In the economic realm of work and consumption, formality and informality primarily refer to the control and protection, as well as the meddling in, of activities by the polity. This starting point excludes from the world of the formal economy all activities that escape in one way or another from the supervision, regulation, taxation, enforcement and interference of the government at large. Feige (1990) used the phrase "underground economy" for this largest set of non-formal activities and transactions. The problem is, however, that the drug baron, the housewife, the beggar, the street vendor and the citizen paying the plumber in cash for the heating repair, are in this way all in the same "underground" boat. It requires little imagination and even less social-scientific rigidity to see that such a diverse company and set of activities is difficult to describe or analyse. Nevertheless, this is the starting point of most conceptualisations of the economy that is deemed not to be formal.

One of the problems with research into the underground economy is that many of its subdivisions are used with slightly different meanings, and there is no agreement as to which type of activity refers to which concept. However, underneath the Babel-like confusion of words and meanings (a good overview is given in Gërxhani 2004), there is considerable conceptual agreement about a further distinction of underground activities into three relevant subtypes. The agreement goes as far as the distinction between three types of activities that somehow elude government involvement (or attempt to do so). Most conceptual explorations distinguish between the non-monetary, the criminal, and the informal area in underground economic life (Feige 1990; Schneider and Enste 2002; Portes and Haller 2005). In some cases, these activities are hardly regulated by the state at large, and no exchange of money in return for commodities is involved. This is the world of self-provisioning in the household or other primary groups, neighbourhood reciprocal relationships and voluntary engagement. The "underground" nature is mainly due to the fact that the formal registration of the economy, as captured, for example, in national accounts or labour market statistics, does not count the added value or activities comprised in this non-market sphere (Chadeau 1985). The second domain of underground activities involves the production, distribution and consumption of commodities for the marketplace, where these commodities are illegal in nature (Losby et al. 2002, 6). This of course depends on the regulation applicable to the territory at hand. Illicit drugs are more or less forbidden commodities in all countries of the world, but in some countries the ban also applies to diverse things such as the purchase of physical sexual services, alcohol, some sorts of meat for consumption (e.g. pork or dog) or the use of certain additives. This second economy of illicit commodities is usually defined as the criminal or illegal economy. This contribution, however, also deals with a third type of underground activities, which we define as "informal economic activities" (following Feige 1990, 992; Portes 2010, 134). The informal economy comprises those activities that are legal by law, but some regulation is transgressed in the process of producing or distributing the commodity.

This may refer to the transgression of rules such as labour regulations, migration status, taxation, licensing or credit. For the remainder of this contribution, we will confine ourselves to these informal economic activities.

Within the informal sector there is one more distinction between two possible motivators of informal activities that seems to have considerable support from different students of the informal economy (in the sense defined above). There is some agreement regarding the way people are motivated or pushed into informal activities (see e.g. Portes and Haller 2005; Williams and Round 2008; Oviedo, Thomas and Karakurum-Özdemir 2009), although, once again, the criterion and ensuing types get different names. For reasons of clarity and concision, we will make use of the binary distinction developed by Perry et al. (2007). This typology distinguishes between two ideal typical paths that lead to engagement in the informal economy: exclusion and exit.

The first logic of *exclusion* refers to people without access to formal employment (or also: consumption) due to a number of pressures or barriers. Their lack of entry opportunities into the official job market, leads to segmented labour markets with little mobility between the official job market and the informal. The relative inaccessibility of formal labour markets pushes workers to engage in informal activities, often forced by the sheer drive for survival. Some have argued that the emergence of a dual labour market follows from the increased (international) competition between large formal firms, pushing them towards informal employment through subcontracting or direct hiring at a lower cost (Portes, Castells and Benton 1989). Notwithstanding the relevance of increased competition, the entry barriers to the formal labour market are all too often of a regulatory or fiscal nature. For instance, when the polity is not able to effectively enforce its immigration regulation, a relevant number of inhabitants may be forced to provide for themselves through informal income. High tax levels and social security contributions may also lead to the exclusion of a part of the labour supply from the labour market of formal jobs, pushing them into the second sector of informal employment.

The second lens for highlighting, or approach to, informal activities, *exit*, prioritizes the positive choice workers or consumers may make for the informal option. The basic logic here is that, for instance, workers have a choice between formal and informal employment, and that their preferences and the given supply of jobs makes them better-off in the informal sector. Possible advantages are the exemption from cost-increasing taxation, the reduction of administrative burdens and flexibility. At the same time, the possible advantages of formal employment may be insignificant, depending on the relative quality of the institutional architecture of the polity (compared to the cost of compulsory contributions, that is). Indeed, the empirical evidence indicates that the quality of the institutions and the legitimacy of the government predict the level of engagement in informal activities quite well, both at the individual and the societal level (Kus 2010; Adriaenssens and Hendrickx 2012).

The multiple dimensions and gradations of informality

The empirical literature dealing with informal economic activities often presumes that there is a clear boundary between work that is deemed formal and work that is not. Also, many studies deal with only one aspect of informality, in particular some sort of tax evasion. Both these versions have a different rationale, but at the very least they are stylized facts. Two elements should be brought into the equation: there are many forms of informal transactions, and most of them are in reality multi-categorical, and maybe even ordered or continuous phenomena. We will briefly go into both issues, and indicate how we will test the relevance of both arguments in the empirical part of this contribution.

First of all, the informal sector is not confined to one type of regulation. The possible set of transgressions indeed depends on the proliferation of rules by the polity. As most contributions define informal activities as noncompliant behaviour occurring outside of the formal institutions of society and neither adhering to nor comprised in formal regulation (compare Feige 1990, 990; 1999, 17), the foundations of informality are manifold. The set of possible informal activities is a function of the formal institutions imposing and enforcing rules: "the formal economy creates its own informality" (Lomnitz 1988, 54). To be clear, Lomnitz' citation takes a short-cut that is rightly criticized (e.g. Ruhs and Anderson 2010). It is not formal rules *producing* informal activities; they merely create the opportunity to transgress. Regulation thus provides the base of an opportunity set of feasible actions. The behavioural outcome from rules therefore is not linearly associated to the extent of regulation.

Comparatively and historically speaking, the extent of regulation varies hugely. It is the breadth of regulation within a state or other polity that defines the number of possible transgressions, and hence the scope of the informal economy (Portes 2010). In this contribution, we will focus on two forms of possible informal transactions in student work. Students and their employers in Belgium have the legal obligation to enter into an employment contract and employers have to set up a pay slip documenting the total wage paid and send that to the employee and the tax administration. Of course, distinguishing only two dimensions is only slightly less of a reduction of the multidimensional world of informal work. It will nevertheless allow us to illustrate the conjecture that dimensions of informality do not necessarily overlap, and therefore adding well-chosen dimensions will reveal non-redundant information.

A second problem is that many studies depict the informal sector as an all-or-nothing choice: either persons or activities are in or they are out. Especially in empirical studies, one usually dichotomizes between formal and informal. This is fairly clear in country studies, where each activity is supposed to be either informal or formal (e.g. Loayza, Oviedo and Serven 2006; D'Hernoncourt and Méon 2012). Individual-level quantitative studies directly measuring informal activities are not particularly numerous. Nevertheless, the examples tend to dichotomize between those in the informal world and those in the formal, e.g. in consumption (Lindström 2008) or tax evasion (an overview in Alm and Torgler 2011).

The way the exclusion literature conceptualizes informality is strongly conducive to this binary depiction of the informal sector. The strongest objection to this binary depiction is, of course, that careful observation of informal activities shows that the pathways into, and the logic and effects of, informal activities may differ according to the level of engagement (see also Davidov 2006). Another problem is that in reality many informal exchanges are only partly so, and many workers engaging in the informal sector also have a formal job in the labour market. Hence families depend on income from both informal and formal markets. Part of the informal activities are somewhere in a grey area between the formal and the underground. Take for instance the well-known problem of multiple jobholding or moonlighting (Dickey, Watson and Zangelidis 2011). Many moonlighters hold one job in the formal sector, and have another informal part-time occupation. Another illustration of the thesis that the distinction between formal and informal practices is a continuum rather than a boundary lies in the existence of envelope wages. This means that part of the wage of formal employees consists of an undeclared envelope wage, a practice that has been documented to exist frequently (Williams 2009).

Are these reductions a problem? There may be good empirical, methodological and efficiency reasons to reduce informality to one decision with a binary distinction. For one thing, it is already hard to measure one aspect in a valid and reliable way. Doing justice to the multidimensional nature of informal activities may very well be an exaggerated expectation. In survey research, nonresponse and social desirability bias probably increase in longer questionnaires and when more items are devoted to sensitive matters (Tourangeau and Yan 2007). Also, the distribution of, for instance, the frequency of transgressions tends to be extremely skewed, so that a dichotomization may be a defensible strategy (as in Adriaenssens and Hendrickx 2009). Finally, in comparative research it is already hard to obtain equivalent data for one dimension of informality. Doing so for more dimensions may become a sheer impossibility. In short, methodologically it may be defensible to measure informal activities as just one binary variable. The cost in terms of higher non-response and loss of comparative possibilities may outweigh the benefits. The problem is that *not* researching the feasibility and the possible costs of a multidimensional approach makes it hard to assess the appropriateness of these possible objections. Theoretical laziness may be more than a peccadillo here. One of the problems is that we do not know whether dichotomizing captures the most important aspects of the phenomenon.

Students as the "usual suspects" of participation in informal work

Students are usually defined as mainly young people whose daily task is attending school and studying. Overall, a significant proportion of school-aged adolescents is involved in part-time work (e.g. Canny 2002; Warren and Forrest Cataldi 2006; Wolbers 2008; Staff and Schulenberg 2010). Adolescents in Belgium are in compulsory education until 18, and are legally allowed to work on the side starting at 15. There is only scarce direct research into student work in Belgium. The existing statistics, based on the Eurostat Labour Force Survey (Eurostat 2014), seem to suggest that student work in Belgium happens less than in comparable countries (Tielens and Vermandere 2007). However, there is reason to believe that underreporting is affected by the prevalence of underground and informal work (see further). In the case that there is significant informal employment, the comparison thus may be flawed. Also the data collection through peer response leads to underreporting (Freeman and Medoff 1982). Estimates of the proportion of Belgian students that work regularly strongly depends on the source: between one in twenty according to the Labour Force Survey (Tielens and Vermandere 2007) to around two-thirds according to another survey-based estimate (Randstad 2012). One study in the French speaking part of the country estimates that one in five students are at work (Demeulemeester and Rochat 2000).

People attending education are probably the only category among the economically inactive population that have a relatively high chance of working in an underground job. The evidence shows that the labour market of student employment, indeed, has an exceptionally high prevalence of informal employment. Being a working student often implies being an off-the-books worker. For instance, in a Eurobarometer survey in 2007, 9% of the questioned students admitted to having carried out undeclared activities during the past year (European Commission 2007). That is almost twice as high as the overall average (5%). The study therefore concluded that informal work among students scores at a comparable level to often-cited groups strongly prevalent in the informal sector such as the unemployed (9%) and self-employed (8%) (European Commission 2007, 25). Pedersen's (2003) research in Germany, Great Britain and several Scandinavian countries concludes that students are at the top of all the studied groups where informal work is concerned. This conclusion is in line with a recent Russian survey, where students seem to be "the most involved in informal employment" (Gorisov 2005, 9), and with a set of older surveys in the Netherlands (Van Eck and Kazemier 1988).

What are the probable reasons for this high prevalence of informality among student jobs? Certain incentives to go informal seem to exist in the demand and supply characteristics of this market. From the perspective of the demand side, student employment is a price competitive and flexible means to meet sudden increases in demand or gaps in production capacity. This function is strongly reinforced by the advantageous fiscal and social regulation for student work in Belgium. Students (in Belgium) enjoy a quite favourable legal position with regard to taxation and social security. If they do not work too many days, they are exempted from most of the fiscal and social security contributions due for adult workers. In itself, this exemption would be expected to lead to low levels of informality. However, labour market rigidities might explain the high prevalence of informal work. Students have little experience and are expected to be less productive than mainstream workers. If, as in Belgium, there are compulsory minimum wages, employers will be inclined to offer informal jobs with lower wages to students. This makes it possible to pay students according to their productivity. Renooy (2004, 26, 120), working on the tradition of informal work as a result of labour market segmentation, rephrases the argument: students, together with women and the unemployed, have to take the less favourable positions in the informal labour market, as opposed to skilled men between 25 and 45. The demand for such less attractive tasks may also imply that students are often driven to jobs without legal commitments such as a written labour contract or the compulsory reporting of working hours to the labour inspectorate, for example.

There are probably also supply side factors fuelling the undeclared employment of students. Some students may prefer off-the-books jobs because of the legal constraints of paid student employment. If students perform more work than the limit, this may lead to the loss of benefits, for instance the child benefits for the parents, and advantages, for instance the exemption from income and social security taxation. Another factor is the legal minimum age of 15, pushing under-age workers into informal jobs.

The central question of this contribution is: will distinct phenomena be revealed when one makes use of multiple measures of informality that also take the gradual nature of unofficial employment into account? Starting from the exit and exclusion perspectives on informal work, we expect that the transgression of rules regarding tax records and pay slips benefits both parties. Employers save administrative costs and some taxation and win flexibility; the latter two advantages also apply to the student. The absence of a written labour contract on the other hand, may be closer to an exclusionary logic. Once again employers save administrative costs and win flexibility. Conversely, a lack of knowledge about their legal position may increase the student's enforcement costs in the case of a breach of the agreement with the employer. The benefits for the students, on the other hand, are unclear here.

Data

Data collection

The analysis is based on data collected in a large scale data survey in secondary schools in the Dutch-speaking northern part of Belgium: the Student Employment Survey (SES). The *SES* aims for a representative picture of the work done by adolescent students who are in the second and third grade of secondary education (i.e. year 3 to 6) and attending school full-time. Most of these pupils are between 14 and 18 years old (except for some who repeated a class). Students in part-time education thus fall outside the sample. Adolescents in special needs programmes or in the very small Arts track (2.2% of the school population) have not been included in the sampling frame.

In line with most educational research, respondents were indirectly sampled through their school. Practically and economically, it is more feasible to follow an indirect sampling strategy. An extra quality of indirect samples in schools is that unit non-response (of the higher unit) is not dependent on the characteristics of the respondent. The definite disadvantage is that respondents are clustered in schools. This problem becomes more serious when schools are larger than average (Lavallée 2007). We reduced clustering by decreasing the basic unit size in the sampling frame by dividing schools into smaller parts and making those the elementary units for sampling. The sampling frame was provided by the Flemish educational administration, and is based on data from the 2009–10 school year. Initially 60 units and 60 replacements were selected with a "probability proportional per size" sampling. The odds of a unit being selected thus depend on the number of students, so that every student has the same chance of being sampled. Of the sample, 26 initial units and 10 replacements agreed to participate.

The data were collected through written questionnaires from November 2010 to January 2011. Usually, the questionnaires were filled out in the classroom under the supervision of a school teacher, but some schools chose to assemble all the students in one place. Eventually 4,018 respondents filled out the questionnaire. Because the distribution of this response does not entirely reflect the population, a weighting was carried out on the two central variables of class year and educational track (for more information about these matters, see Adriaenssens et al. 2014).

Measures

Those respondents who indicated that they had worked in a position that provided financial compensation at some point since the previous summer received a distinct questionnaire. This inquired into

(1) aspects of the job carried out during the past summer holiday, and
(2) during the school year, both the job performed on a fixed schedule and work without a fixed day or regularity,
(3) personal characteristics, study experience and study results, and
(4) personal and family background.

The questions looking into the nature of the job relevant to informality inquired whether a contract was signed, whether a pay slip (a legally compulsory tax record) was handed over, and whether the contract and the tax record covered all of the work carried out. A job is classified as "white" when the signed contract regulated all the work carried out, or when the pay slip mentioned the total wage earned. When no contract was signed or no pay slip was handed over at all, the job was classified as black. The intermediate "grey" jobs did have a contract or tax record, but part of the work or wage was not included in them (envelope wage or non-contracted extra hours). Because we are mainly interested in the formal-informal distinction in the firms and firm-like organizations, all descriptive data and analyses only refer to work performed for organizations and firms, and not for households. All in all, work for households is virtually always informal work according to the definitions used here. Contracts and tax records are hardly ever used in household work by students. In a way, this type of work represents its own form of informality, or at least of underground economic activities, one that is close to self-provisioning.

Results

In the results section, three types of results are presented. In the first subsection, descriptive and bivariate output regarding the prevalence of informal work is discussed. "Informal" is measured according to both labour contract and pay slip regulation. These data are represented per student, but also in terms of wage and weekly hours worked. Finally, we give an estimate of the integral socio-economic weight of each type of work. This first subsection concludes by discussing the association between both dimensions of informal activities: the absence of a written contracting and non-declared wages.

In the second subsection we test whether a model explaining the engagement in informal work differs according to the dimension used. The independent variables introduced in the model have been shown to be relevant antecedents for access to work and the intensity of work. Because of the strong indications that the relation between some of the independent variables and informality may be U-shaped (see the next section), we have chosen not to estimate these models with the help of an ordinal technique, but with techniques for unordered categorical variables.

Prevalence and consistency

In the *Student Employment Survey*, the respondents were asked to give information about three types of jobs: the job they held during the past summer holiday, the job they perform regularly during the school year, and a possible odd job they work on an occasional basis. The last job type was documented in a limited way, so that we can only report whether a full pay slip was handed over for the wage paid. Therefore we will only report the prevalence of informal work in these odd jobs in *tables 1* and *2* reporting specific aspects (participants, wages and hours worked).

Table 1: Distribution of respondents engaging in informal activities per job type (weighted)

	Holiday job		Regular job		Irregular job
	Contract	Record	Contract	Record	Record
Formal (white)	50.5%	47.2%	41.5%	37.3%	22.2%
Informal					77.8%
Grey	22.8%	13.2%	17.9%	9.0%	
Black	26.7%	39.6%	40.6%	53.3%	
N	1594	1529	686	649	190

Data source: SES, weighted data, own calculations.

We begin by showing the distribution of informal activities according to the particular law which has been broken or ignored: the labour law imposing a written contract on the one hand (*Table* **1**, columns "Contract"), and the fiscal obligation to hand over a pay slip with all the wages paid (as a written record) on the other (*Table* **1**, columns "Record"). These descriptive results may already give an indication as to whether contract or record informality differs in prevalence. Formal summer jobs count for around half of the total employment reported in the study. The share of formal jobs during the school year is smaller. Overall, the prevalence of grey jobs is higher in the labour contract measure than in the tax record measure. Paying envelope wages on top of the official wage is thus a less prevalent practice than working extra hours on top of the work schedule agreed upon in the contract.

These results represent the proportion of students engaged in one type of informal activity or another. More precisely, it is a count of jobs. If these jobs differ in terms of wages earned or working hours, this could lead to a biased estimation of the social, economic and also personal relevance of informal work. Therefore we also documented the average working time and hourly wages (*Table 2*)[2]. All averages are based on geometric means, so that distortion from remaining outliers remains limited. This seems appropriate in the light of the skewed distribution of wages and working hours.

In summer there is a wage differential between white and grey jobs on the one hand, and black jobs on the other. The lowest pay is reserved for black summer jobs. This wage difference between black and grey jobs does not occur during the school year, however. The relevance of a three-tier distinction between white, grey and black is hereby illustrated. It is also clear that they are more than just higher dimensions of informality. The dividing line in the wage category lies between black and the others, not between formal (white) and informal (black and grey). That conclusion is consistent with the fact that evasion of the minimum wage is effected through tax evasion, which does not necessitate contract informality. It is striking that the best paid jobs are grey summer jobs (in tax records).

The differences in working time are even more pronounced. In most instances the working time in grey jobs is higher than in white and black jobs. The latter two have approximately the same levels of working time. Here too the binary distinction between informal and formal would have concealed a lot. The real distinction in working times runs between grey jobs and the others (white and black).

---

2 Summer job wages were surveyed in several ways (e.g. wage per hour and total wage for the whole job), so a composite has been computed. Priority has been given to the direct measure, unless this was an extreme or missing value.
Extreme positive outliers in the wages have been omitted, with the help of an analysis of a histogram based on a logarithm. For the working hours, a maximum of 16 hours per day has been assumed.

Table 2: Average hourly wages and weekly hours (geometric means)

	HOLIDAY JOB		REGULAR JOB		IRREGULAR JOB (RECORD)
	Contract	Record	Contract	Record	
HOURLY WAGES					
Formal (white)	€ 7.80	€ 7.69	€ 7.92	€ 7.82	€ 7.78
Informal	€ 6.67	€ 6.77	€ 7.66	€ 7.73	€ 7.57
Grey	€ 7.39	€ 8.38	€ 7.82	€ 7.77	
Black	€ 6.11	€ 6.30	€ 7.58	€ 7.72	
F-test formal-informal	30.60***	19.47***	0.830	0.089	0.091
F-test 3 categories	26.96***	29.77***	0.601	0.049	
WEEKLY HOURS WORKED					
Formal (white)	11h36	12h30	7h14	7h50	4h44
Informal	12h40	11h41	6h53	6h42	3h54
Grey	15h54	14h43	8h36	8h42	
Black	10h27	10h48	6h15	6h26	
F-test formal-informal	3.85**	2.22	0.847	8.145***	1.15
F-test 3 categories	24.46***	10.37***	10.14***	9.11***	
N	1722	1658	716	677	608
*p < 0.1 **p < 0.05 ***p <0.01					

Data source: SES, weighted data, own calculations.

Finally we estimated the aggregated share of the different forms of work. This means that we weighed summer jobs and work during the year, taking the number of workers, the total hours worked and the wages (for total income) into account. This results in an estimate of the share of informal work in the total hours worked and the total income throughout the year (*Table 3*). This estimate takes into account the length of the summer holiday and of the school year. The conclusion does not change – namely that more than half of the work performed and the wages earned come from informal activities. In considering tax evasion, the black market represents by far the largest share in the informal market segment. From the perspective of contract informality, however, both markets are more or less evenly important.

Table 3: Aggregated share of informal activities by work time and income

	White	Grey	Black
Work time			
Contract	44.8%	25.3%	29.8%
Record	44.7%	13.2%	42.1%
Total income			
Contract	46.6%	25.5%	27.9%
Record	45.8%	14.2%	40.0%

Data source: SES, weighted data, own calculations.

In order to conclude this first exploration, we look for the relationship between both measures of informality (*Table 4*). This means that we test whether a job characterized as being black, white or grey in one dimension (tax or contract) is a good predictor of whether the job is also performed in the same status in the other dimension. Because the first results gave us no clear indication of an ordered structure in the white, grey and black types of work, we present both nominal and ordinal measures of association. All measures suggest that there is quite a strong association between both measures, but that it is far from perfect. The strength of the measure is more or less similar between the holiday job category and that of the work performed during the school year. The measurement of both concepts thus overlaps, but at the same time each measure provides considerable information that the other measure does not provide.

Table 4: Association between contract and tax record informality (weighted)[3]

	Holiday job	Regular job
Nominal		
Chi^2	594.7***	155.0***
Goodman & Kruskal's τ		
Contract dependent	0.200***	0.241***
Record dependent	0.244***	0.241***
Ordinal		
Kendall's τ_b	0.489***	0.539***
* $p < 0.10$, ** $p < 0.05$, *** $p < 0.01$		

3 All presented measures have a significance under 0.001.

Predicting informality

In this section, we test a model aiming to account for the choice between white, grey or black work. The variables entered in the model are adopted from earlier research into the determinants of access to student work, the intensity and the level of the job held (Adriaenssens et al. 2014). The individual-level factors introduced are gender, age and migration background. The job characteristics are the type, skill level and sector of the job performed. We briefly explicate these factors and their measurement.

In terms of general participation, some studies see a higher (Hirschman and Voloshin 2007; Howieson et al. 2012) or equal level of participation of girls in student work (Lucas 1997; Wolbers 2008). For informal work, the literature generally asserts that informal markets are strongly gendered, with higher levels of participation by women (Nelson 1999; International Labour Office 2002, 12–14).

For adolescents at school, age has a strong positive impact on participation (Hodgson and Spours 2000; Howieson et al. 2012). For participation in informal work, though, we expect a negative relation for two reasons. On the one hand, the legal limit of 15 years forces younger students to work off-the-books. Also, productivity is expected to grow with age. Due to this expectation as well as minimum wages, younger workers may be pushed into informal work.

Studies consistently show that students with an ethnic minority background, and in particular those from minorities with a precarious economic position, participate less in part-time work (Porterfield and Winkler 2007; Howieson et al. 2012). Migration background has been measured through the nationality of the grandmothers. We distinguish between three groups: respondents with two Belgian grandmothers, those with at least one grandmother of a Turkish or Moroccan nationality (individuals with these backgrounds together represent a group in Belgium with a well-documented overall precarious socio-economic background) and those with a non-Belgian background of other nationalities. In our own study we found both a lower participation rate as well as participation in jobs at a lower skill level for students with a Moroccan or Turkish migration background (Adriaenssens et al. 2014).

In terms of skill level, informal jobs are often depicted as low-level jobs (e.g. Rinehart 2004; Katungi, Neal and Barbour 2006). We thus expect that lower-skilled jobs tend to be informal more often. The job level in our study was coded by the research team on the basis of the Dutch occupational SBC-classification (the Standaard Beroepen Classificatie, see Centraal Bureau voor de Statistiek 2010). In the questionnaire, the respondents stated their employer, job and tasks performed in three open questions. These open variables were coded into occupational categories. This resulted in a variable of the occupational category according to the SBC-classification, a job level code of five ordered categories, that is. They refer to the (theorized) educational level needed to perform the job, starting from "elementary" up to "scientific" jobs. As almost all student jobs are of the lowest levels (elementary), we recoded the higher levels (intermediate level and up) into one group.

As for the sector, most research indicates that student jobs are strongly clustered in retail, the food service industry (usually bars and restaurants) as well as in households (van der Meer and Wielers 2001; Staff, Messersmith and Schulenberg 2009; Staff and Schulenberg 2010). As paid work in households has been omitted from this study, we entered three sectorial categories into the equation: retail, bars and restaurants, and other sectors. It so happens that the sectors of bars and restaurants and retail are often cited as hotbeds of informal activities as well, and are particularly targeted by social inspections services (SIOD 2013, 28). The same information that was used to code the job level provided the information to code the economic sector as well. This coding is based on the Belgian NACE-classification (Algemene Directie Statistiek 2011)[4].

Finally we also make use of the distinction between summer jobs and work throughout the school year. Although most studies on student work only document one type of work, there are good reasons to expect that there are strong seasonal variations in the demand for student work. For example, if one looks at the first years of the 21st century in the US, one longitudinal study found stable adolescent participation in jobs during the school year (Staff, Messersmith and Schulenberg 2009), while another one measured declining employment in summer jobs (Morisi 2010). In summer jobs, students often substitute for semi- and unskilled workers taking a holiday; jobs during the school year more often serve to meet with needs of a flexible work organisation (Canny 2002).

Our data have a two-level hierarchical structure, as up to two jobs are performed by each student (a holiday job and a job during the school year). In such a case a multilevel model is required. The problem at hand is usually referred to as a *choice model*. Whether a job is official, black or grey is a mutually exclusive matter. The standard set of methods studying this type of problem is multinomial logit and tobit models (Borooah 2001). The former is more popular due to its more economical modelling, so it has been chosen here too. A possible problem with multinomial logit analysis may be the "independence of irrelevant

4 The Belgian NACE is derived from the European NACE, which in turn is based on the global ISIC classification of economic sectors.

alternatives" assumption (IIA) – the problem that (quasi-)perfect substitutes preclude choice. However, more recent work shows that multinomial logit is robust against the breach of the IIA assumption (Hedeker 2008).

We first estimated the multilevel multinomial logit model with a first order marginal quasi-likelihood (MQL) algorithm. However, MQL estimates may be strongly biased downwards (Browne and Draper 2006; Rodriguez 2008). This is expected to be the case especially when level 1 (the jobs in our case) only consists of one or two observations per unit of level 2 (students). In order to remedy this potential bias, we also estimated the same parameters with the help of two techniques that should be more reliable in this case: Bayesian and Gauss-quadrature estimations. We applied the Bayesian estimation method with a diffuse prior (in MLwiN, Rasbash et al. 2009), using a Markov Chain Monte Carlo (MCMC) procedure with a burn-in length of 50,000 iterations and 2,000,000 simulations, with a thinning factor of 10 (so only every 10th simulation is kept). The Gauss-quadrature estimation (with 20 quadrature points) was estimated with the help of the GLLAMM-routine in Stata (Rabe-Hesketh, Skrondal and Pickles 2005). Overall the comparison between the Gauss-quadrature and the MCMC models shows a quite consistent estimation of parameters. In short, the GLLAMM-procedure and the Bayesian estimation lead more or less to the same conclusions. All the estimations are presented in the appendix (*Table 6* and *Table 7*), and show the changes over the different estimation techniques. In the text we confine ourselves to the final Bayesian model (*Table 5*). We should warn the readers that comparison between the estimates is to be confined to significance levels. In logit models the estimators are rescaled, so that comparison between the estimates in order to assess their relative strength makes no sense (Karlson, Holm and Breen 2012).

Table 5: Bayesian MCMC multilevel multinomial logit on contract and record (reference: white)

		INFORMAL CONTRACT		INFORMAL RECORD	
		b	SE	b	SE
Grey	Constant	4.972	3.665	8.447	5.781
	Gender (male)	1.110**	0.287	1.690**	0.536
	Age	–0.490**	0.213	–0.884**	0.334
	Moroccan or Turkish descent[13]	–0.652	0.686	–2.287*	1.321
	Other migration background	1.085**	0.326	0.115	0.551
	Job during school year[14]	–0.357	0.236	–0.961**	0.409
	Elementary level[15]	–0.532	0.616	0.243	1.439
	Lower level	0.352	0.599	1.468	1.419
	Retail[16]	0.599**	0.313	0.405	0.557
	Bars & restaurants	2.040**	0.381	2.569**	0.677
Black	Constant	29.833**	4.204	34.875**	5.698
	Gender (male)	1.067**	0.325	0.746*	0.404
	Age	-1.951**	0.262	-2.035**	0.336
	Moroccan or Turkish descent	0.200	0.736	-1.065	0.965
	Other migration background	0.787**	0.374	0.316	0.463
	Job during school year	1.411**	0.266	1.315**	0.297
	Elementary level	-1.657**	0.668	-3.751**	0.938
	Lower level	-0.703	0.627	-2.422**	0.866
	Retail	-1.321**	0.385	-1.532**	0.453
	Bars & restaurants	2.283**	0.420	2.210**	0.526
Random part	s^2_{const1}	10.206**	2.850	30.697**	11.737
	s^2_{const2}	15.272**	3.991	30.989**	8.135
	$s_{const1,const2}$	4.986**	2.428	17.691**	6.684
* p < 0.10, ** p < 0.05					

5 Reference: no migration background.

6 Reference: job during summer holiday.

7 Reference: intermediate level or higher.

8 Reference: all other sectors.

The personal characteristics of age and gender have fairly consistent effects on both forms of informality. Boys and younger adolescents work more often in grey and black jobs. That gender effect is contrary to the general expectation in the literature.

The effects of migration background depend on the group and the type and level of informality. Considering contract informality, students with a migratory background that is not Moroccan or Turkish are more often in informal jobs (both grey and black). This effect is absent if one consults tax records. Students with a Moroccan or Turkish background, on the other hand, seem to be less prevalent in tax evasion, at least as far as grey jobs go. The effect, however, is only significant at the 10% level.

However, factors such as the skill level of a job as well as the sector in which it is classified show a stronger effect on one's choice for informal work when compared to the effects of a worker's personal characteristics.

First of all, compared to jobs held during summer holidays, the school year jobs are more often black, while grey employment is less prevalent in tax records during the same time frame. This contrasting trend illustrates the relevance of the distinction between grey and black jobs.

With regard to skill levels, there seems to be little discrimination between white and grey jobs. The effect is quite different for black jobs. Overall there is less black employment in elementary jobs, both for contract and record informality. This is a surprising finding, contradicting the overall expectation that low skilled workers would be sought for black jobs.

With reference to sector, restaurants and bars consistently employ more students in grey and black jobs. Restaurants and bars thus specialize both in contract and tax record noncompliance. The evidence regarding the retail sector is paradoxical. Black jobs are consistently *less* prevalent in the retail sector. At first sight the retail sector thus complies very well with both tax record and labour contract regulation. The strange conclusion, however, is that the retail sector seems to specialize in grey contracts. This means that, while shops and other retail firms generally set up a labour contract more often than other sectors, they are quite flexible and not very compliant when it comes to working overtime outside of the contracted work schedule.

Conclusion and discussion

Informal work is defined as those economic activities that are legal in themselves, but at the same time they evade some regulation in the transaction. Quite often this transgression refers to rules regarding labour regulation or taxation. The central question of this contribution is whether it is relevant to measure multiple dimensions and the gradual nature of informality. The central question is answered through the analysis of empirical data that allow for the measurement of informal work in two dimensions (tax record and labour contract evasion) divided into three categories (formal or *white*, totally informal or *black*, and partially informal work, called *grey*). The data contain information about a representative sample of students between 14 and 18 years in secondary education in Flanders, Belgium. This group is particularly suited for this type of research, because the evidence shows that students are quite active in the informal sector.

This contribution attempts to investigate whether a multidimensional analysis of informality is a feasible and rewarding strategy. The problem thus falls into two questions: is it feasible; is it opportune? The answer to the first question follows quite straightforwardly from the analyses. The answer to the latter question is dealt with in the subsequent paragraph.

So what about the question of feasibility? The simple answer would be: yes, it is feasible. Both measurements of more than a binary distinction between informal and formal, and of more than one dimension of informality have proven to be possible.

Measuring more than one dimension of informality was exemplified by focusing on two forms of informal engagement. On the one hand we measured the levels of noncompliance with tax rules in the form of the compulsory pay slip: either no pay slip was set up at all (black), or part of the wage was paid on the side as an envelope wage (grey). The dimension of the labour contract was measured by the question of whether a labour contract had been set up (black), or extra hours were worked that were not in the contractually agreed work schedule (grey). Both dimensions correlate quite well, but in a far from perfect manner. In the regressions the effects usually point in the same direction, but different factors have an effect on both dimensions. In particular this is the case for migration background. Students with a Moroccan or Turkish background evade *fewer* taxes, while students with another migration background comply *less often* with the labour contract rules.

In abandoning the binary distinction between formal and informal and making a more subtle three-category framework, this division between white, black and grey jobs proved particularly instructive. The relevance of this distinction is documented in more than one way. For one thing, grey jobs are more work-intensive than both white and black jobs. If one just compared formal with (black and grey) informal jobs, one would totally overlook this difference. Also, there is a marked seasonal difference: for jobs during the school year there is more tax evasion through black, and less tax evasion through grey jobs, compared to jobs employment during the summer holidays. In the retail sector there is a lower level of the absence of labour contracts (black), but at the same time this sector specializes in noncompliance with the working schedule stipulated in the contract (grey).

If feasibility indeed poses no fundamental problem, the second question is what is gained and what is lost when one uses a multidimensional and ordered measurement of informality. That which may be gained is exemplified by some of the empirical conclusions. Accounting for six possible positions vis-à-vis the informal economy (2*3) shows which polymorphic causal priors informality can have. For local studies, this should be a strong argument in favour of multidimensional strategies. The possible objection is that equivalence and comparability become more difficult in comparative research. As in all matters regarding governance, policy and the law, the problems of equivalence and comparability are real. Because theoretical possibilities of transgressions depend on the prevailing rules, every survey into informality will have to check whether the type of transgression investigated exists, and which levels exist in these transgressions. For multidimensional surveys of informality, the complexity of documenting equivalence increases quickly. Whether it is worthwhile, depends on the research goals at hand. Notwithstanding the possible problems, documenting multidimensionality in comparative research may prove even more relevant, as it allows for the differential impact of institutional arrangements on different dimensions of informality.

In short, this contribution shows strong evidence that simple binary and one-dimensional measures of informal work may hide real differences between types and grades of informality. Without a doubt, the extra cost and complexity of collecting data about informal transactions with multiple categories and more than one dimension is considerable. This is all the more the case in comparative research. Notwithstanding this objection, it would prove useful and rewarding to consider a broader approach in data collection and the analysis of informal work.

Annex: full tables from the multinomial logit regressions

Table 6: Multilevel multinomial logit regressions on informal labour contracts (reference: white)

		Without multilevel ML		Multilevel MQL		Multilevel Gauss–quadrature		Multilevel MCMC			
		b	SE	b	SE	b	SE	b	SE	95% credible interval Lower	Upper
Grey	Constant	1.968	1.481	2.074	1.393	4.452	3.373	4.972	3.665	–2.116	12.232
	Gender (male)	0.544**	.121	0.508**	0.122	1.018**	0.259	1.110**	0.287	0.572	1.702
	Age	–0.210**	.088	–0.216**	0.082	–0.442**	0.197	–0.490**	0.213	–0.919	–0.082
	Moroccan or Turkish descent[17]	–0.276	.321	–0.252	0.319	–0.569	0.619	–0.652	0.686	–2.052	0.660
	Other migration background[17]	0.520**	.136	0.470**	0.137	0.997**	0.296	1.085**	0.326	0.472	1.752
	Job during school year[18]	–0.099	.137	–0.183	0.131	–0.320	0.223	–0.357	0.236	–0.828	0.097
	Elementary level[19]	–0.242	.317	–0.163	0.316	–0.514	0.570	–0.532	0.616	–1.726	0.685
	Lower level[19]	0.174	.308	0.193	0.307	0.289	0.554	0.352	0.599	–0.783	1.558
	Retail[20]	0.290**	.144	0.302**	0.148	0.551*	0.286	0.599**	0.313	0.001	1.233
	Bars & restaurants[20]	1.016**	.156	0.944**	0.151	1.878**	0.346	2.040**	0.381	1.332	2.826
Black	Constant	12.134**	1.284	12.257**	1.231	26.980**	3.897	29.833**	4.204	21.840	38.452
	Gender (male)	0.457**	.121	0.435**	0.122	0.969**	0.292	1.067**	0.325	0.455	1.735
	Age	–0.793**	.077	–0.800**	0.073	–1.765**	0.242	–1.951**	0.262	–2.494	–1.457
	Moroccan or Turkish descent[17]	0.115	.283	0.066	0.289	0.201	0.662	0.200	0.736	–1.257	1.644
	Other migration background[17]	0.314**	.140	0.260*	0.140	0.715**	0.334	0.787**	0.374	0.074	1.542
	Job during school year[18]	0.739**	.127	0.667**	0.121	1.321**	0.249	1.411**	0.266	0.909	1.953
	Elementary level[19]	–0.645**	.267	–0.568**	0.271	–1.496**	–2.450	–1.657**	0.668	–3.000	–0.381
	Lower level[19]	–0.213	.257	–0.199	0.261	–0.626	0.276	–0.703	0.627	–1.952	0.508
	Retail[20]	–0.717**	.162	–0.689**	0.164	–1.241**	0.000	–1.321**	0.385	–2.097	–0.584
	Bars & restaurants[20]	0.992**	.147	0.914**	0.143	2.093**	0.379	2.283**	0.420	1.505	3.151
Random part		s^2_{const1}		0.746	0.183	7.931	2.157	10.206	2.850	5.629	16.689
		s^2_{const2}		0.765	0.177	11.732	2.961	15.272	3.991	8.874	24.428
		$s_{const1,const2}$		–0.296	0.138	3.807	1.863	4.986	2.428	0.752	10.311

* < 0.1 ** < 0.05 ***<0.01

9 Reference: no migration background.
10 Reference: job during summer holiday.
11 Reference: intermediate level or higher.
12 Reference: all other sectors.

Table 7: Multilevel multinomial logit regressions on informal tax records (pay slip) (reference: white)

		Without multilevel ML		Multilevel MQL		Multilevel Gauss–quadrature		Multilevel MCMC			
		b	SE	b	SE	b	SE	b	SE	95% credible interval Lower	Upper
Grey	Constant	1.598	1.832	1.504	1.783	7.762	4.984	8.447	5.781	−2.436	20.422
	Gender (male)	0.568**	0.153	0.549**	0.155	1.313**	0.408	1.690**	0.536	0.735	2.851
	Age	−0.241**	0.106	−0.234**	0.103	−0.737**	0.283	−0.884**	0.334	−1.604	−0.279
	Moroccan or Turkish descent[21]	−0.723	0.447	−0.761*	0.453	−1.671*	0.986	−2.287*	1.321	−5.123	0.088
	Other migration background[21]	0.002	0.176	−0.044	0.180	0.089	0.421	0.115	0.551	−0.978	1.207
	Job during school year[22]	−0.430**	0.184	−0.456**	0.178	−0.739**	0.351	−0.961**	0.409	−1.813	−0.202
	Elementary level[2319]	0.366	0.551	0.393	0.546	0.005	1.079	0.243	1.439	−2.361	3.318
	Lower level[23]	0.839	0.543	0.835	0.538	1.023	1.059	1.468	1.419	−1.033	4.552
	Retail[24]	0.012	0.181	0.046	0.185	0.164	0.438	0.405	0.557	−0.619	1.577
	Bars & restaurants[24]	0.756**	0.189	0.763**	0.188	2.024**	0.519	2.569**	0.677	1.367	4.017
Black	Constant	10.107**	1.194	10.289**	1.200	29.635**	4.784	34.875**	5.698	24.700	47.127
	Gender (male)	0.212**	0.104	0.206*	0.110	0.618*	0.340	0.746*	0.404	−0.023	1.569
	Age	−0.592**	0.071	−0.603**	0.071	−1.732**	0.282	−2.035**	0.336	−2.758	−1.435
	Moroccan or Turkish descent[21]	−0.222	0.254	−0.308	0.272	−0.863	0.816	−1.065	0.965	−3.017	0.790
	Other migration background[21]	−0.011	0.123	−0.019	0.129	0.260	0.394	0.316	0.463	−0.585	1.246
	Job during school year[22]	0.473**	0.113	0.456**	0.111	1.179**	0.271	1.315**	0.297	0.755	1.918
	Elementary level[23]	−1.023**	0.241	−1.026**	0.253	−3.161**	0.792	−3.751**	0.938	−5.733	−2.039
	Lower level[23]	−0.583**	0.233	−0.592**	0.245	−2.000**	0.734	−2.422**	0.866	−4.225	−0.813
	Retail[24]	−0.673**	0.132	−0.650**	0.138	−1.415**	0.397	−1.532**	0.453	−2.444	−0.665
	Bars & restaurants[24]	0.564**	0.132	0.588**	0.134	1.860**	0.441	2.210**	0.526	1.244	3.308
Random part		s^2_{const1}		1.046	0.289	16.399**	5.703	30.697**	11.737	13.982	59.218
		s^2_{const2}		0.747	0.150	21.354**	5.395	30.989**	8.135	18.182	49.935
		$s_{const1,const2}$		−0.207	0.156	11.479**	3.879	17.691**	6.684	7.055	33.280

* < 0.1 ** < 0.05 ***<0.01

13 Reference: no migration background.
14 Reference: job during summer holiday.
15 Reference: intermediate level or higher.
16 Reference: all other sectors.

References

Abbott, Edwin Abbott. 1884. *Flatland. A Romance with Many Dimensions, by A. Square*. London: Seeley & co.

Adriaenssens, Stef, and Jef Hendrickx. 2009. "Modes of Production in Home Maintenance: Accounting for the Choice between Formality, off the Books and Self-provisioning." *International Journal of Consumer Studies* 33: 596–603. Accessed January 25, 2014. doi: 10.1111/j.1470-6431.2009.00807.x.

Adriaenssens, Stef, and Jef Hendrickx. 2012. *Institutions, Trust and Relations: A Comparative Analysis Explaining Informal Economic Activities*. Edited by HUB. *HUB Research Papers* Brussels: HUB – University College Brussels.

Adriaenssens, Stef, Dieter Verhaest, Anja Van den Broeck, Karin Proost, and Dries Berings. 2014. "De arbeidsparticipatie van Vlaamse scholieren." *Tijdschrift voor Arbeidsvraagstukken* 30: 281–301.

Algemene Directie Statistiek. 2011. *NACE-BEL activiteitennomenclatuur*. Brussels: FOD Economie.

Alm, James, and Benno Torgler. 2011. "Do Ethics Matter? Tax Compliance and Morality." *Journal of Business Ethics* 101: 635–651. Accessed January 25, 2014. doi: 10.1007/s10551-011-0761-9.

Borooah, Vani K. 2001. *Logit and Probit: Ordered and Multinomial Models*, vol. 138, *Sage University Papers Series on Quantitative Applications in the Social Sciences*. Thousand Oaks: Sage.

Browne, William J., and David Draper. 2006. "A Comparison of Bayesian and Likelihood-based Methods for Fitting Multilevel Models." *Bayesian Analysis* 1: 473–514.

Canny, Angela. 2002. "Flexible Labour? The Growth of Student Employment in the UK." *Journal of Education and Work* 15: 277–301.

Centraal Bureau voor de Statistiek. 2010. *Standaard Beroepenclassificatie 2010*. Den Haag: CBS.

Chadeau, Ann. 1985. "Measuring Household Activities: Some International Comparisons." *Review of Income and Wealth* 31: 237–253.

D'Hernoncourt, Johanna, and Pierre-Guillaume Méon. 2012. "The Not So Dark Side of Trust: Does Trust Increase the Size of the Shadow Economy?" *Journal of Economic Behavior & Organization* 81: 97–121. Accessed January 25, 2014. doi: 10.1016/j.jebo.2011.09.010.

Davidov, Guy. 2006. “Enforcement Problems in »Informal« Labor Markets: A View from Israel.” *Comparative Labor Law and Policy Journal* 27: 3–25.

Demeulemeester, Jean-Luc, and Denis Rochat. 2000. “Labour Participation of Higher Education Students.” *Labour* 14: 503–522.

Dickey, Heather, Verity Watson, and Alexandros Zangelidis. 2011. “Is It All about Money? An Examination of the Motives behind Moonlighting.” *Applied Economics* 43: 3767–3774. Accessed January 25, 2014. doi: 10.1080/00036841003724403.

European Commission. 2007. *Undeclared work in the European Union*. Brussels: Social Affairs and Equal Opportunities Directorate General Employment.

Eurostat. 2014. “EU Labour Force Survey – Methodology.” *Statistics Explained.*

Feige, Edgar L. 1990. “Defining and Estimating Underground and Informal Economies: The New Institutional Economics Approach.” *World Development* 18: 989–1002.

Feige, Edgar L. 1999. “Underground Economies in Transition: Noncompliance and Institutional Change.” In *Underground Economies in Transition. Unrecorded Activity, Tax Evasion, Corruption and Organized Crime*, edited by Edgar L. Feige and Katharina Ott, 11–27. Brookfield: Ashgate.

Freeman, Richard B., and James L. Medoff. 1982. “Why Does the Rate of Youth Labor Force Activity Differ across Surveys?” In *The Youth Labor Market Problem: Its Nature, Causes, and Consequences*, edited by Richard B. Freeman and David A. Wise, 75–114. Chicago: University of Chicago Press.

Gërxhani, Klarita. 2004. “The Informal Sector in Developed and Less Developed Countries: A Literature Survey” *Public Choice* 120: 267–300.

Gorisov, S. 2005. “The Scale and Structure of Informal Employment in the Russian Economy.” *Russian Social Science Review* 46: 4–22.

Hedeker, Donald. 2008. “Multilevel Models for Ordinal and Nominal Variables.” In *Handbook of Multilevel Analysis*, edited by Jan de Leeuw and Erik Meijer, 237–274. New York: Springer.

Hirschman, Charles, and Irina Voloshin. 2007. “The Structure of Teenage Employment: Social Background and the Jobs Held by High School Seniors.” *Research in Social Stratification and Mobility* 25: 189–203.

Hodgson, Ann, and Ken Spours. 2000. *Earning and Learning: A Local Study of Part-time Paid Work among 14–19 Year Olds*. London: University of London.

Howieson, Cathy, Jim McKechnie, Sandy Hobbs, and Sheila Semple. 2012. “New Perspectives on School Students’ Part-time Work.” *Sociology* 46: 322–338. Accessed January 25, 2014. doi: 10.1177/0038038511419183.

International Labour Office. 2002. *Decent Work and the Informal Economy*. Geneva: ILO.

Karlson, Kristian Bernt, Anders Holm, and Richard Breen. 2012. "Comparing Regression Coefficients between Same-sample Nested Models Using Logit and Probit: A New Method." *Sociological Methodology* 42: 286–313. Accessed January 25, 2014. doi: 10.1177/0081175012444861.

Katungi, Dennis, Emma Neal, and Aaron Barbour. 2006. *People in Low-paid Informal Work: »Need Not Greed«*. Bristol: Policy Press.

Kus, Basak. 2010. "Regulatory Governance and the Informal Economy: Cross-national Comparisons." *Socio-Economic Review* 8: 487–510. Accessed January 25, 2014. doi: 10.1093/ser/mwq005.

Lavallée, Pierre. 2007. *Indirect Sampling*. New York: Springer.

Lindström, Martin. 2008. "Social Capital, Political Trust and Purchase of Illegal Liquor: A Population-based Study in Southern Sweden." *Health Policy* 86: 266–275.

Loayza, Norman V., Ana Maria Oviedo, and Luis Serven. 2006. "The Impact of Regulation on Growth and Informality: Cross-country Evidence." In *Linking the Formal and Informal Economy. Concepts and Policies*, edited by Basudeb Guha-Khasnobis, Ravi Kanbur and Elinor Ostrom, 121–144. Oxford: Oxford University Press.

Lomnitz, Larissa Adler. 1988. "Informal Exchange Networks in Formal Systems: A Theoretical Model." *American Anthropologist* 90: 42–55.

Losby, Jan L., John F. Else, Marcia E. Kingslow, Elaine L. Edgcomb, Erika T. Malm, and Vivian Kao. 2002. *Informal Economy: Literature Review*. Newark: ISED.

Lucas, Rosemary. 1997. "Youth, Gender and Part-time Work – Students in the Labour Process." *Work, Employment & Society* 11: 595–614.

Morisi, Teresa L. 2010. "The Early 2000s: A Period of Declining Teen Summer Employment Rates." *Monthly Labor Review* 133: 23–35.

Nelson, Margaret K. 1999. "Between Paid and Unpaid Work. Gender Patterns in Supplemental Economic Activities among White, Rural Families." *Gender & Society* 13: 518–539.

Oviedo, Ana Maria, Mark R. Thomas, and Kamer Karakurum-Özdemir. 2009. *Economic Informality: Causes, Costs, and Policies – A Literature Survey, World Bank Working Paper*. Washington: World Bank.

Pedersen, Søren. 2003. *The Shadow Economy in Germany, Great Britain and Scandinavia. A Measurement Based on Questionnaire Surveys*. Copenhagen: Rockwool Foundation.

Perry, Guillermo E., William F. Maloney, Omar S. Arias, Pablo Fajnzylber, Andrew D. Mason, and Jaime Saavedra-Chanduvi. 2007. *Informality: Exit and Exclusion*. Washington: World Bank.

Porterfield, Shirley L., and Anne E. Winkler. 2007. "Teen Time Use and Parental Education: Evidence from the CPS, MTF, and ATUS." *Monthly Labor Review* 130: 37–56.

Portes, Alejandro. 2010. *Economic Sociology: A Systematic Inquiry*. Princeton: Princeton University Press.

Portes, Alejandro, Manuel Castells, and Lauren A. Benton. 1989. "The Policy Implications of Informality." In *The Informal Economy. Studies in Advanced and Less Developed Countries*, edited by Alejandro Portes, Manuel Castells and Lauren A. Benton, 298–311. Baltimore: John Hopkins University Press.

Portes, Alejandro, and William Haller. 2005. "The Informal Economy." In *The Handbook of Economic Sociology*, edited by Neil J. Smelser and Richard Swedberg, 403–425. Princeton: Princeton University Press.

Rabe-Hesketh, Sophia, Anders Skrondal, and Andrew Pickles. 2005. "Maximum Likelihood Estimation of Limited and Discrete Dependent Variable Models with Nested Random Effects." *Journal of Econometrics* 128: 301–323. Accessed January 25, 2014. doi: 10.1016/j.jeconom.2004.08.017.

Randstad. 2012. *Studenten aan het werk. Studentenstudie Randstad 2012*. Brussels: Randstad Belgium.

Rasbash, J., C. Charlton, W.J. Browne, M. Healy, and B. Cameron. 2009. *MLwiN Version 2.1*, vol. 2.1: Centre for Multilevel Modelling, University of Bristol.

Renooy, Piet H., Staffan Ivarsson, Olga van der Wusten-Gritsai, and Remco Meijer. 2004. *Undeclared Work in an Enlarged Union. An Analysis of Undeclared Work: an In-depth Study of Specific Items*. Brussels: European Commission - DG V.

Rinehart, Richard D. 2004. *Designing Programmes to Improve Working and Employment Conditions in the Informal Economy: A Literature Review*. Geneva: International Labour Office.

Rodriguez, Germán. 2008. "Multilevel Generalized Linear Models." In *Handbook of Multilevel Analysis*, edited by Jan de Leeuw and Erik Meijer, 335–376. New York: Springer.

Ruhs, Martin, and Bridget Anderson. 2010. "Semi-Compliance and Illegality in Migrant Labour Markets: An Analysis of Migrants, Employers and the State in the UK." *Population, Space and Place* 16: 195–211. Accessed January 25, 2014. doi: 10.1002/psp.588.

Schneider, Friedrich, and Dominik H. Enste. 2002. *The Shadow Economy: An International Survey*. Cambridge: Cambridge University Press.

SIOD. 2013. *Strijd tegen de sociale fraude: actieplan 2013 van de inspectiediensten*. Brussels: Sociale Inlichtingen- en Opsporingsdienst. Accessed January 28, 2015. http://www.siod.belgie.be/siodsirs/publicationDefault.aspx?id=40280.

Staff, Jeremy, Emilz E. Messersmith, and John E. Schulenberg. 2009. "Adolescents and the World of Work." In *Handbook of Adolescent Psychology*, edited by Richard M. Lerner and Laurence Steinberg, 270–313. New York: John Wiley & Sons.

Staff, Jeremy, and John E. Schulenberg. 2010. "Millennials and the World of Work: Experiences in Paid Work during Adolescence." *Journal of Business and Psychology* 25: 247–255.

Tielens, Maarten, and Caroline Vermandere. 2007. *Een doorlichting van de arbeidsmarktsituatie van jongeren in Vlaanderen en Europa*. Leuven: Steunpunt Werk en Sociale Economie.

Tourangeau, Roger, and Ting Yan. 2007. "Sensitive Questions in Surveys." *Psychological Bulletin* 133: 859–883. Accessed January 25, 2014. doi: 10.1037/0033-2909.133.5.859.

van der Meer, Peter, and Rudi Wielers. 2001. "The Increased Labour Market Participation of Dutch Students." *Work, Employment & Society* 15: 55–71.

Van Eck, Robert, and Brugt Kazemier. 1988. "Features of the Hidden Economy in the Netherlands." *Review of Income and Wealth* 34: 251–273.

Warren, John Robert, and Emily Forrest Cataldi. 2006. "A Historical Perspective on High School Students' Paid Employment and Its Association with High School Dropout." *Sociological Forum* 21: 113–143.

Williams, Colin C. 2009. "Formal and Informal Employment in Europe: Beyond Dualistic Representations." *European Urban and Regional Studies* 16: 147–159.

Williams, Colin C., and John Round. 2008. "A Critical Evaluation of Romantic Depictions of the Informal Economy." *Review of Social Economy* 66: 297–323.

Wolbers, Maarten H. J. 2008. "Scholieren met een bijbaantje: de gevolgen voor hun schoolprestaties." *Mens & Maatschappij* 83: 239–257.

Part III: The Sociology of the Interaction between Formality and Informality

Barbara A. Misztal

Configurations of Informality and Formality in Contemporary Society

Introduction: The relevance of informality

In contrast to many social scientists' prediction that forces of globalisation, by imposing legal forms of regulations on the world, would eliminate or undermine the significance of informality, the reliance on informality remains universally practised to facilitate the formal processes of business, politics and society. However, the first two decades of the new century's trends and new technologies have not only sustained the informality, they have also, at the same time, changed the scope of informality and its relations with formality. On the one hand, the expanding access to information and the process of globalisation, together with the growing level of complexity of the global system are seen as creating favourable conditions for less formal social encounters. Hence, with the loosening of formal hierarchies and the de-conventionalisation of organized practices, we observe the new importance of informality. On the other hand, in the context of the widespread unpredictability, deregulation and the complexity of the global economic system, there is an increase in the perception of informality being associated with corruption, nepotism and other malpractices, which prompts many to question the relevance of informality for democracy (Lauth 2000). Taken together, these two possible roles of informality, in the context of modern democracies becoming increasingly shaped by various types of "personalized governance and the profusion of modes of online provision" (Eriksson and Vogt 2012, 154), suggest the emergence of a new configuration of informality and formality in today's societies. Since we do not fully understand the implications of this configuration, there is a clear need to re-open a debate on the role and consequences of the new relationships between informality and formality in the contemporary setting. To develop a new analytical approach, we should first scrutinize the processes that are behind this configuration, especially, the impact of the growing complexity of the world and our increasing reliance on digital technologies, on ways in which we interact and make decisions. These two trends, as they alter the boundaries between private and public life and conditions of cooperation, influence the scope and function of informality in this new context. Thus, the aim of this article is to re-think these issues in the context of the new century's developments. After discussing trends that increase our hopes for informality's capacity to enhance cooperation and debating processes that lead to the misplacement of informality and to its use as a form of control strategy, I will revise the definition of informality developed in my book on informality more than a decade ago (Misztal 2000).

The concept of informality is rather complex, unclear and ambiguous. It tends to be used in various ways: from descriptions of face-to-face, intimate, private, less rigid, less controlled interactions, and, through references to the informal economy, to descriptions of non-hierarchal or bureaucratic exchange, nepotism, old boys' networks and avoidance of formal rules. Unsurprisingly, there are always some misunderstandings and confusion surrounding this notion. Following Goffman, I have developed an understanding of informality as referring to situations with a wider scope of choices of behaviour where, in order to make the most of the possibilities in given circumstances or to reach "a working understanding" (Goffman 1983, 9), people employ various forms of action that are not pre-made (Misztal 2000, 41). In this perspective, informality is defined as a form of interaction among partners engaging in dialogue, the rules of which are not pre-designed, and enjoying relative freedom in the interpretation of their roles' requirements (Misztal 2000, 46). Such an understanding allows us to see both informality and formality as the essential and changing aspects of many processes underlined by new modes of social control, new institutions and new means of communication. To fully comprehend the new shape of their configuration, we should also include into our consideration analyses of previous forms of their mutual interdependence.

It is not surprising that the significance of informality has been always recognized. We know many historical accounts of the role of informality on the world stage. For example, Mann (2012) in his discussion of sources of power argues that WWI destroyed the regime of informal international cooperation that existed prior to that war. Furthermore, he argues that one of the consequences of WWI was America's imperial dominance, which unlike its European predecessors, was informal and which, without much formality, was able to bend the course of events in the direction of its interests. Although political scientists in general tend to put central emphasis on the status of formal institutions, for the last couple of decades they have also been raising questions about the importance of informal institutions in the process of political transformation and their relevance for democracy. Noticing the varieties of informal institutions, and their different impacts on the transparency of political processes and public communication, they propose that in order to evaluate informality's relevance for democracy we need to develop the typology of informal arrangements (Lauth 2000; Cormack 2013).

On the bases of many empirical social studies, we also know that any significant social changes require face-to-face informal efforts of strategically positioned actors within any field, organisation or system (Walker 2012). Informality is the universal element of relationships in every society, although its importance and the intensity of its application differ from country to country. The shift in emphasis reflects broader social transformations, having moved from being predominantly directed toward the intensive reliance on informality for access to resources through a focus on informality as a means of control, to a total concentration solely on formal structures. Many of the relaxations of restrictions on relations and conduct constitute instances of a "controlled decontrolling of emotional control" (Wouters 1986, 3) and lead to "a shift from relational and emotional management through command to a management through negotiation" (de Swann 1990, 270). As the process of the relaxation of restrictions on ways we behave in public has spread to increasing numbers of people, this informalisation has simultaneously been accompanied by the process of formalisation.

The interplay of both processes (formality and informality) has always been visible in many spheres of life, although strong informality, which thrives locally and is used as a means for control in familiar communities, seems to belong to the past. For example, social order in 1950s Britain, in the "era of trust", "self-restraint" and "carefully calibrated politeness", was helped by the informal control of public spaces "by bus conductors, by park keepers, by lavatory attendance and by a police force that was largely admired" (Kynaston 2009, 542). The reliance on informality not only reflects local communities' cohesion, but also can reflect the limits of state regulation or legal devices (Farrell 2004). A weak central government provides the ground for the flourishing of various types of informal deals, exchanges and bargaining, which – although not necessarily illegal – are often outside of the law (Hart 1988). However, in reality the picture is even more complex as much evidence suggests that also under strong, centralized governments, informality can play a significant role in modifying the rigid and direct state control in the economy (Misztal 2000). More generally, it can be said that although there is a tendency to perceive informality as some form of favouritism and nepotism and as associated with corruption, bribery and malpractice, in fact impropriety is not an inherent characteristic of informality. This can happen under any type of government when there is no adequate system of laws and regulations. Where there are "tight rules and regulations, and their strict enforcement", like in Singapore, they prevent "widespread corrupt practices" (Chan and Ng 2006, 56).

Although discussions of informality often focus on its role in business, where informality plays a silent role in many agreements and contracts, the role of informality extends to all areas of socio-economic life. Informality is used as an effective strategy for many social purposes, for instance to sells products, as illustrated by the trend of "personalized" products, from coffee shops to airlines. Furthermore, its impact has always been acknowledged in the financial sector where, since its foundation in 1801, the London Stock Exchange's motto has been "My word, my bond", suggesting that bargaining and deals can be made in an informal way, with no contracts or documents. It has continually been present in the legal system where, even during the period in which legal formalism was dominant, the role of informality was recognized. "We are under a Constitution, but the Constitution is what the judges say it is" (Charles Evans Hughes Justice of the US Supreme Court quoted in Unah 2009, 154).

The growing informalisation of many spheres of public life, which started in the 1960s, also saw the development of "informal" or "popular" justice (van Krieken 2001). This new wave of informal justice is defined as:

> encompassing legal institutions which are non-bureaucratic in structure and relatively undifferentiated from the larger society, minimize the use of professionals, and eschew official law in favor of substantive and procedural norms that are vague, unwritten, commonsensical, flexible, ad hoc, and particularistic (Richard Abel quoted in van Krieken 2001, 7)

spread from the 1970s onwards. The foundations for the shift to legal informalism were laid down by the growing critiques of legal formalism and efforts "to bridge and link the realm of formalized legal ideas and procedures with extralegal forms of social ordering, often within a framework of attempting to modify the workings of power relations" (van Krieken 2001, 6). Behind legal informalism in a range of fields including family, criminal, administrative, commercial, discrimination and equal opportunity law was both the development of the welfare state and critical attitudes towards the state in the 1960s (van Krieken 2001, 7).

To sum up so far, each society permits some space for informality that is socially, culturally, and economically determined. All societal sub-systems strive to find their own mixture of rule-bound formality and rule-independent informality. In every case, informality of conduct and formality of rules are joined together notwithstanding their opposition and tensions. Their relationship is far from immutable and their dynamism results in the evolution of styles of interaction. Thus, our question is: What are the nature and dynamics of the relations between informality and formality today? How stable is such a configuration and what role does it play in shaping the quality of social life? In order to answer these questions, we need to take a closer look at our capacities for interpersonal concordance and our opportunities for informality in today's circumstances. Since the private-public shift and the erosion of conditions facilitating cooperation are seen as affecting the scope and role of informality, in the following section we will analyse the impact of these processes on the configuration of informality and formality.

Informality and the private-public shift

The relationship between informality and formality, and its perception as orderly – and thus acceptable – are historically and culturally contingent. If the sixties were about the processes of the informalisation and liberalisation of society, the twenty-first century is about the digitalisation and the growing complexity of the globalized world. These two processes have been making a profound impact on the transformation of the relationship between public and private life. Thus, the demarcation between private and public, which has been continuously eroded for several decades "as the two realms indeed constantly flow into each other" (Arendt 1958, 33), is now further undermined and complicated. The shift in the relationship between the public and the private that has accompanied and shaped the development of today's societies alters the relationship between formality and informality. In other words, the new media and the complex nature of the globalised world, by blurring the boundaries between the private and public, further increase the scope for a more informal, not role-bound and role-obedient, conduct.

The productive and effective functioning of intricate societies within the complexity of the global system, defined by networks of dense, non-linear interactions that change over time, requires finding solutions to many new uncertainties and unpredictabilities.

> We do not live in a governed world so much as a world traversed by the "will to govern", fuelled by the constant registration of "failure", the discrepancy between ambition and outcome, and the constant injunction to do better next time (Rose and Miller 1992, 191).

The modern labyrinthine international order, with its sets of possibilities and constraints, together with the system's open boundaries and plays of force inside of the system, means that today's main challenge is to maintain the coherence of the system itself (Human and Cilliers 2013). Responses to this challenge in the context of unpredictability and "the permanent suspicion of the authority of authorities" (Rose and Miller 1992, 191), are not only "a matter for government regulations but a concern of individuals as the key decision makers" (Chandler 2013, 4). The problems of complexity mean that leaders need to be able to respond in a more reflexive way, optimize their chances and "chart their way through the choppy waters of a globalised economy" (Grist 2009, 16). In other words, with the increase in the complexity of the socio-economic system, we tend to rely on informality as a less rigid, more flexible means of overcoming the sheer size and density of global institutional arrangements and regulations, be they economic, financial or political in nature. However, there are some unintended consequences of such developments.

The first unintended consequence of the reliance on informality in transactions or dealings with others, which normally is prompted by increases in uncertainty, risk and the absence of pertinent knowledge, refers to the use of informality as an example of a governance mechanism, which results in collusions, breaching rules and contributes to the malfunctioning of the whole system. This can be illustrated by the recent failures of the UK financial system to govern its own complexity, which was brought about by deregulation and globalisation. This complexity has been addressed by the growing reliance on informal managerial estimates and valuations (Lanchester 2013a, 3–9), as it was assumed that the reliance on informality tends to lower transaction costs and reduce uncertainty in the market environment as well as helping to deal with the contingencies of maladaptation or failure. Thus, this inflated financial sector, entrenched in its cartel-like culture, is more than ever amenable to illicit quid pro quo deals, insider trading, informal bargaining and scandals such as the Libor (the London Interbank Offered Rate) fixings scandal (Luyendijk 2013). As "the big banks have simply become too complex and too big to manage" and their market-based information becomes more managerialized and informalised, even their employees do not believe that "their top people know what's going on" (Luyendijk 2013). Although informality has always been part of legitimate international financial and commercial markets, the extent to which today's global complex economy increases opportunities for informal trading, false contracts, bribery and the manipulation of prizes leads some to argue that the globalisation and complexity of the global economic system have vastly increased the scope of informal actions that are not necessary illegal, but nonetheless are not in the best interest of the public (Cockcroft 2013).

The second unintended consequence of the reliance on informality in the context of the multifaceted system refers to the reliance on informality as a strategy of control.

As the boundaries between markets and organisations and between external and internal reporting are being blurred, there are new initiatives to control a new interplay of formality and informality. Presently many organisations, in their attempt to adjust to these new conditions, are adopting a broad scale of measures to informalise their structure and practices. For example, they opt for flatter, more flexible, organisational structures, decentralisation of decision making, less formal relations between superiors and subordinates, informal, more colloquial speech, shifting boundaries between work and private time and they cultivate "informal Fridays" within their workplaces (Misztal 2000, 63).

> But this apparent "relaxing" of controls, this opening of emotional exchange to greater variety, individual nuance, and the growth of emotional alternatives, also involves at one level an *intensification* of demands on affect economy (Hughes 2010, 44 – emphasis in original).

In the relative absence of explicit and formal rules governing behaviour, as Archer (2010) argued, people are expected to be more reflexive, responsible, self-controlled and to practice self-regulation and self-monitoring in order to negotiate changing networks of loosely coupled social relationships. For example, although "informal Fridays", when employees can *disregard formal dress codes*, rather than offering "a simple relaxation of pressures on how to dress", presented people "with another set of demands, and these might be even more intensely felt than those arising from the company dress code" as people need to reflect and decide what it means to dress appropriately for such non-scripted occasions (Hughes 2010, 45 – emphasis in original). This new expectation to dress informally but correctly, "according to a blend and balance of unstated »internalized« and explicit «external» standards and concerns" (Hughes 2010, 45) can be seen as a shift in the character of social constraints towards the informalisation of standards of socially sanctioned behaviour. Yet at the same time this push towards employees' self-constraint and self-control also illustrates the use of informality as a strategy of control on the part of companies. Under the guise of the workers' liberation from formal control, we observe the usage of informalisation tactics to ensure or even intensify control. In short, the use of formalized informality as the strategy of control reflects the institutionalised power relationships.

Today's relationship between rule-bound formality and rule-independent informality is more than ever vulnerable to shifting not only because of the increase in the complexity of the socio-economic system but also because of effects of the digital revolution, which moves traditional boundaries between public and private domains while also promising to set us free.

> Today, when an individual sits in the space of his or her home or bedroom and goes online, disclosing information about himself or herself to thousands or millions of others, in what sense is this individual situated in a private sphere? (Thompson 2011, 63).

The changing relations between the public and private arenas, as a result of the internet and new social media, affect the visibility and functionality of informality. As the spread of new digital media alters the public-private relationship, it creates a new scope for informality and increases its "mediated publicness" as it sets the stage for "the flourishing of a new kind of intimacy in the public sphere" (Thompson 2011, 57). Thus, the changing role and visibility of informality is symptomatic of the development of new means of communication. With novel forms of social media giving rise to new virtual informal relations that are free from the constraints of co-presence, the public sphere becomes a complex space of information streams over which the individual does not necessarily exercise control. As the hectic informality of modern life and digital technology joined forces, the shifting boundaries between public and private life have become "a new battleground in modern societies, a contested terrain where established relations of power can be challenged and disrupted, lives damaged and reputations sometimes lost" (Thompson 2011, 49).

The access to and reliance on digital social networks can increase people's power and social experience. New social media, such as Twitter, are increasing the capacity to make people's voices immediately heard and turn "ordinary" people into broadcasters (Gitlin 2013). Some major recent events, such as the Occupy movement and the Arab Spring, were possible because of such digital social networks. Moreover, smaller scale events, which nonetheless increase people's input into the functioning of social systems, are results of the new technology. Although not every online campaign has an impact, digitally-rooted activism cannot be dismissed, as we are "moving from a vision of civics that's party-based and partisan to one that's personal and pointillist" (Zuckerman 2013, 9). Political digital activism, which shifts the boundaries between private and public, is "civics in flux", and changes with the people who practise it (Zuckerman 2013, 9). As "digital natives" participate in civic life often by "personalizing issues", they also reconfigure the relation between informality and formality (Zuckerman 2013, 9). For example, in Latvia, a country with one of the lowest levels of political engagement and trust in governmental institutions in the EU, a website called "My Voice", created by its citizens, offers an informal forum for debates and petitions, and thus contributes to rebuilding trust between people and the government (McGrane 2013, 5). With many other countries, such as Finland and Iceland as well as the European Commission, also developing online platforms for citizens' initiatives (McGrane 2013, 5), these types of informal actions have been slowly achieving legitimacy throughout the states, indicating that here again we are observing the process of the formalisation of informality. This process of formalisation is also enhanced by Google, Facebook and Twitter's solutions to problems caused by free speech. They all struggle to rule on what is permissible, as illustrated by cases including those of a man being prosecuted for his "twitter joke" about blowing up one of the British airports (Bowcott 2012, 3) as well as the tweetstorms of aggression directed against women. These and other cases have led to demands for the companies to regulate what passes through their servers, which, together with the spread of "netiquette", can be seen as the beginning of the process of the formalisation of informality in social media (McVeigh 2013).

With misplaced informality and abuses of freedom of speech leading to more and more emphasis on developing guidelines of conduct on the Internet, or "netiquette", and the introduction of "abuse" or "report" buttons, questions arise regarding not only if these developments signal the process of formalisation, but also with regard to what role this new formalized informality is taking. Although it is possible to use electronic media, as people in many countries do, as a source of informal power to increase citizens' input to the political process, we also need to be aware that, on the other hand, the social networking services, which offer real-time tracking of the public, exercise an enormous power to direct and influence public opinion. For instance, the Russian media have created Putin's image as that of a man of action and allowed him, through the personalisation and informalisation of his relations with the public, to increase his control and power (Knight 2013, 54–57). Yet this strategy of informalisation at the top is confronted with informalisation at the bottom. Thus, despite the sophisticated use of media to increase Putin's popularity, his ratings are actually declining due to the stories of the corruption, incompetence and irresponsibility of the government now circulating widely on the Web and reaching the growing number of Russians who are using the Internet as a source of news (Knight 2013, 54–57).

A new type of debate about the scope of the state surveillance of its citizens' private lives has been initiated with the publication of Snowden's files, the secrets documents about a US National Security Agency program, PRISM, which is said to tap into the customer data accumulated by corporations such as Google, Apple and Microsoft (Greenwald and MacAskill 2013, 1). Snowden's revelation seems to suggest that we are moving towards a society without privacy: "where the people with accesses to our secrets, hear, intercept and monitor everything" (Lanchester 2013b). The disclosed files show that the Internet could expand the reach of the state and the state's permanent vigilance, activity and intervention. Now not only the public but also the Web's giant corporate entities realize that the Internet is under a vast surveillance plan invisible to those being observed. During this present period of digital revolution that promised to set us free, the expanding electronic mass surveillance has raised major questions about the control of the Internet and the balance between the private and public in modern democracy.

The growing reliance on social media for information and communication is also seen as having a negative impact on various social skills directly or indirectly connected with informality. Research confirms that the preoccupation with technology hurts the social skills of young employees, affecting their ability to build relationships with others, and narrowing people's vision, empathy and understanding (McVeigh 2013). Living on the Web could not only result in inflated expectations, but it also leads us to living "within our own, restricted cocoon of information and experiences, with insufficient knowledge of the experiences of others" (Samuel Becker in Billig 2013, 29). While relying on new social media we "must make a special effort to encounter variety" as these media allow us to "have what we want and we can avoid what we don't want" (Billig 2013, 29). The reliance on digital means of communication can erode our capacity for interpersonal concordance, which is an essential element for cooperation. As such a critical component, it "comes directly not so much from a propensity to identify with others as from an ability and readiness to assume their point of view and interpret their intentions" (Burns 1992, 74). Thus, one of the most destructive consequences of the fact that we are targets of selective information gathering, together with the increasing power of social networks to regulate the expression of the individuals' views and the question of privacy, is the inflation of our skills for cooperation, seen as striving on informality.

Informality and cooperation

Sennett (2012) makes a similar point by suggesting that in order to secure cooperation, we need to respond to others on their own terms and cultivate informality regardless of the uncertainties associated with it. Stressing that contemporary societies deskill people from many of the competences they need to cooperate, Sennett (2012, 5) brings to our attention that today's more complicated world requires a new, more difficult cooperation, which he defines as "an exchange in which the participants benefit from the encounter". Arguing that the new capitalism erodes our capacity to live together, Sennett points to other than only the digitalisation causes for the recent decline of cooperation. He further stipulates that cooperation is weakened by inequality, changes in modern labour and by cultural homogenisation which produces a new type of person: "This is the sort of person bent on reducing the anxieties which differences can inspire, whether these be political, racial, religious, ethnic or erotic in character" (Sennett 2012, 8). All of these processes erode people's skills for cooperation as they undermine the scope of informality in social relations.

Sennett takes his idea of informality from Montaigne who observed that:

> in whatever position they are placed, men pile up and arrange themselves by moving and shuffling about, just as group of objects thrown into a bag find their way to join and fit together, often better than they could have been arranged deliberately (Montaigne quoted in Sennett 2012, 277).

Such a definition of informality leads Sennett to view cooperation as open, non-scripted, free exchange between "people who have separate or conflicting interests, who do not feel good about each other, who are unequal, or who simply do not understand one another" (Sennett 2012, 6). This "win-win exchange" is quite demanding in contrast to destructive cooperation, which is "cooperation of the us-against-you sort, or about cooperation degraded into collusion" (Sennett 2012, 6). Sennett believes that such responsiveness to others on their terms "emerges from practical activity" and that the most important fact about hard cooperation is that it requires "dialogic skills" (Sennett 2012, 6). In the course of the skilled dialogue practices both informal and empathic people gain a measure of self-respect and autonomy without becoming either winners or losers. In short, since cooperation means engaging in dialogue without pre-designed rules, it requires informality, which contemporary economic forms and cultural influences and forces continuously undermine.

Sennett attributes the decline in the space for and role of informality and thus, the erosion of our skills for cooperation, mainly to the nature of today's work place, which enhances momentary transitions, the ethos of transaction and fosters a culture that explicitly devaluates informality. In the past "manual labourers forged strong informal bonds at work which took people out of their niches. These informal relations consisted of three elements composing a social triangle" (Sennett 2012, 148). They are "earned authority, mutual respect and operation during a crisis" (Sennett 2012, 148). These three sides of the social triangle were essential for any organisation that wanted to encourage informal bonds of these sorts to cohere socially. However, in the new global capitalism, power diverges from authority, distrust increases and the elites live detached from responsibilities, and hence, informality declines.

The contemporary workplace is in total contrast to its past, in which cooperation rewarded all participants since each gained a measure of self-respect and autonomy. Nowadays, we are faced in the workplace with shallow, distrustful, superficial, short term relationships with colleagues, where informality is repressed by the duties and rules of the formal contract. The domination of "superficial relations and short institutional bonds together function to recycle »superficial« informality", which is often used as a tool by management to enhance their control (Sennett 2012, 8). These short-term projects, with their feigned solidarity, the superficial familiarity with others and forced informality represent the very opposite of cooperation. At the end of the day, such arrangements "reinforce the silo effect: people keep to themselves" (Sennett 2012, 8). The more cooperation declines, the more people are exhorted to perform their roles as "team players", which illustrates how the managerial imperative empties informality of any substance. While developing the argument that without institutional stability and long-term prospects, there will be no opportunity for mutual obligations, respect and trust to flourish, Sennett (2012) brings the idea of informality closer to the Chinese idea of *Guanxi*, which is seen an important element of cooperative relationships. Defining *Guanxi* as informal interdependence networks in which trust is achieved, Sennett (2012, 136) stresses that informal Guanxi networks are meant to be sustainable, and that people in "*Guanxi* networks are not ashamed of dependency". However, today's short-termism and rhetoric of self-responsibility distort our capacity to cooperate by undermining the conditions in which it flourishes and draining informality of any meaning through its formalisation.

The importance of cooperation is also recognized by other scholars, although they do not necessarily conceptualize its links with informality in the same way as Sennett does. Generally, it is assumed that social order is fundamentally dependent on cooperative relationships and that the efficiency of society is improved by cooperation. Thus, several approaches search for what sustains such cooperative relationships. Among them, the most prominent are middle-range theories about trust and closely related topics, such as social capital. Since the 1990s there has been an impressive proliferation of theories of trust and social capital that stress that features of social organisation such as norms and networks can facilitate coordinated action (Elias 1978, Gambetta 1988; Coleman 1990; Putnam 1993; Misztal 1996; Sztompka 1996; Edwards and Foley 1998; Woolcock 1998).

Similarities between these two concepts – trust and social capital – have led to the indirect incorporation of these notions into the stream of more general sociological discussions about the connection between cooperation and quality of life. While there are many differences between various studies addressing these issues, they all seem to adopt ad hoc claims about the capacity of informal interaction to bridge the gap between individuals and, thus, to facilitate cooperative behaviour. The majority of writers focus their attention primarily on how one may go about creating and fostering trust in order to increase social capital and thus cooperative relations. Usually in such an approach social capital is defined as a form of trust based on commonly shared norms, therefore, social cohesion is explained in terms of people's capacity to create networks of informal, reciprocal relationships. Hence, both social capital and trust are seen as linked to, and interchangeably used with, the concept of informality. To trust others is to accept the risks associated with the type and depth of the interdependence inherent in a given relationship (Shepard and Sherman 1998, 423). This is accepted by approaches seeing social capital as a public good produced by civic associations (Putnam 1993), as moral resources such as trust (Fukuyama 1995) and as the effective norm that ensures that people work together for common purposes in groups and organisations (Coleman 1990). The representatives of these perspectives argue that the nature of social ties, or "mediating structures", is essential for cooperation through which a higher quality of life is achieved. In other words, social trust, which mutually reinforces expectations about reciprocity, is seen as a lubricant for cooperation (Misztal 1996).

Fukuyama (1995, 27), who sees trust as the key to cooperation and the fixed ingredient in economic success, argues that trust is critical to economic performance since "people who do not trust one another will end up cooperating only under the system of formal rules and regulations". In high-trust countries (the USA, Japan and Germany) the existence of a supportive culture of "spontaneous sociability", that is, a readiness to cooperate with others in an economically productive way, results in the flourishing of numerous institutions and associations, seen as a good in and of themselves. Informality and sociability, viewed as constituting "a subset of social capital" or a useful kind of social capital with "the capacity to form new associations and to cooperate within the terms of reference they establish" (Fukuyama 1995, 27), sustain trust, which offers a cheaper solution than depending on extensive regulations to prevent others from cheating.

Putnam, who believes that trust produced by secondary associations facilitates democratic efficiency and cooperation, also takes trust to be virtually defining proof of social capital. A high level of social capital, or trust, within a given community is the basis of cooperation and, by the same token, for a more efficient functioning of democracy and a more innovative economy. Putnam (1993) argues that the basic problem of a democratic society is the creation of voluntary associations because only their dense networks of interpersonal trust and cooperation can overcome the free-rider dilemma. However, this approach leaves out some of the most important questions to be asked. Since a decent good society depends on trust as well as on distrust (Misztal 1996), we always need to look beyond trust and check the accountability, transparency and goals of reciprocal networks. Furthermore, the main difference in the amounts of social capital may reflect the different levels of the centralisation of networks, not necessarily a higher level of social integration, since centralisation can overcome the free-rider problem.

Assuming that the main problem faced by the USA today is a deficit of social capital, Putnam (1993, 171) promotes "norms of reciprocity and networks of civic engagement" as two related sources of trust in complex modern settings. Democratic systems can be animated by civic virtues rooted in an old tradition of civic culture, which teaches people to regard "the public domain as more than a battleground «for pursuing personal interest»" (Putnam 1993, 88). Such a reformulation of the question of civic culture assumes that a society is indifferent to government action, whereas Fukuyama's faith in inherited cultural dispositions confuses trust with familiarity. The former theorist overlooks the role of the state in generating trust relations, while the latter theorist's identification of trust with backwards looking confidence cannot be assumed to be the foundation for modern democratic interaction. Hence, both Putnam and Fukuyama, while arguing that trust is linked with stable democracy and cooperation, fail to answer more specific questions about the nature and reasons for this linkage, as well as how trust relations can be generated under today's conditions.

With recent empirical research showing the decline of trust in many modern democracies, come the realisations that not only do today's societies not provide a natural environment for trust, they also substitute trust or social capital with formal rules. Cook, Hardin and Levi in their edited book, *Cooperation without Trust* (2005), argue that the actual role of trust relations has relatively declined and that trust, although important in many interpersonal contexts, cannot any longer "carry the weight of making complex societies function productively and effectively" (Cook, Hardin and Levi 2005, 1). According to these scholars, modern democracies implemented their solution to the problem of cooperation by setting their foundations in formal procedural democracy and rational universal administration. At present, we witness a continuous shift from "a customary regulation of daily life to the growing resort to codes – explicit sets of rules" (Harre 1999, 262). The growing process of formalisation, that is, the increasing reliance on the formal rules and rights to regulate interaction, means that modern institutions, like law, provide mechanisms that mediate exchange between people. The expansion of formalism, bureaucratisation, institutionalisation and legalism replaces trust developed under the conditions of familiarity with formal rules that provide a formal source of information as to how much an individual can be trusted. Under such conditions cooperation is mediated by formal rules, expert knowledge and legal systems. Cook, Hardin and Levi (2005) seem to follow Weber's idea that increasing formality and impersonality, standardisation and regulation are justified as a means expanding the general welfare.

> Bureaucracy develops the more perfectly, the more it is "dehumanised", the more completely it succeeds in eliminating from official business love, hatred, and all purely personal, irrational and emotional elements which escape calculation (Weber 1968, 975).

According to Cook, Hardin and Levy, in modern societies we do not rely on informal social organisational mechanisms to give potentially useful partners the incentive to be cooperative. Hence, some organisations and institutions serve us well just because they are substitutes for trust relations. Their main assumption is that we cooperate not because "we have come to trust each other, but because of the incentives in place that make cooperation safe and productive for us" (Cook, Hardin and Levi 2005, 15). Worried that informalisation could lead to clientelism and corruption, that is, exchange relationships in which extrinsic or instrumental benefits or motivation dominates, they claim that some kinds of trust and informal relations, because of their exclusionary – thus undemocratic – character, should rather be avoided. They investigate what kind of informality or trust is desirable in social, economic and political life and reject informal relations that violate ethical and legal norms. However, they overlook the fact that cases of political and administrative malpractice can also occur in the bureaucratic type of exchange where impersonal control is the most significant factor. They also do not pay attention to the shift to more flexible, less formal, network-based type of organisations in the modern world.

From Cook, Hardin and Levy's (2005) perspective, formal rules and institutions are supposed to be the embodiment of the rationality and efficiency of bureaucratic exchange, which is a practise in compliance with the formalized norms of rationality, specialisation and conformity. Cooperation is not a matter of personal relations but depends upon the observation of abstract impersonal rules and routines. This approach reminds us of Weber's idea of the "spirit" of rational formalism, which bureaucracy embraces since otherwise "the door would be open to arbitrariness" (Weber 1968, 985). Such bureaucracy is "the most rational known means of exercising authority over human beings" (Weber 1968, 223). It stresses the depersonalisation of relationships, impersonal power and the detailed rigidity of some prescribed behaviour, all of which make the initiation of change very difficult. The ideal of bureaucracy eliminates the need to trust relationship and personal dependencies and is "a world where people are bound by impersonal rules and not by personal influence and arbitrary command" (Crozier 1967, 107). All members of organisations are dependent and controlled by formal rules and this lowers personal dependency and alleviates the tensions created by subordination. Human behaviour is made predictable, conformist, disciplined, rigid and oriented towards formal groups designed to perform according to abstract, universal criteria.

Yet, the bureaucratic system can never be so closely conforming to its idealised model and in real life the bureaucratic system has many dysfunctions and unintended consequences. Much research shows the routine and oppressive aspects of bureaucracy as well as its "vicious circle" and the role of human relations (Crozier 1967, 177). The informalisation of formal organisation is seen as a normal response to bureaucratisation, which testifies to the "limits of rationalization" of this type of institution (Stark 1989, 644). Not only does the standardisation of behaviour often result in a displacement of goals, but bureaucratic universal abstract rules also tend to produce conflict because the peculiarities of individual cases are frequently ignored. Therefore, the functioning of bureaucracy can never be totally explained by the combination of impersonality, expertness and the hierarchy of the ideal type. Furthermore, the claimed universalism of bureaucratic impersonal rules is often a rhetorical tool used by bureaucrats when they want to ignore particularistic claims that they do not wish to acknowledge. Finally, with the development of the network society, as argued by Castells (1996), we observe the transformation of bureaucratic structures into dynamic open-ended, flexible, expansive, transnational networks.

According to Castells (1996), the development of dense organisational relationships that cut across various inter- and intra-organisational networks, in contrast to traditional networks which were formal, hierarchical and based on central control, is based on the transmission of information, digitalised communication and technology and is more complex, less hierarchical, less formal, multi-dimensional and without a core centre of power. Their open structures are able to expand beyond pre-existing limits as long as new modes can share the same communication codes (values or performance goals).

> The convergence of social evolution and information technologies has created a new material basis for the performance of activities throughout the social structure. This material basis, built in networks, earmarks dominant social processes, thus shaping social structure itself (Castells 1996, 471).

The emergence of more fluid and flexible boundaries means that the industrial era's institutions and organisations have become "empty shells, decreasingly able to relate to people's lives and values" (Castells 1997, 355). Arguing that a network-based social structure is a highly dynamic, open system, suiting the capitalist economy based on innovation, globalisation, the mobility of capital and the de-aggregation of labour, Castells (1996, 278) notes that what is new about the network society is that there are "few rules about how to win and how to lose". The erosion of the social contract between capital, labour and the state, "sends everyone home to fight for their individual interests, counting exclusively on their own force" (Castells 1997, 367–368). Thus, as labour becomes more dependent on individual bargaining conditions in an unpredictable labour market, neither formal rules nor trust are bases for cooperative relations between the partners. In other words, this shift to more instrumental dealings between major groups of the global economy brings to our attention a new figuration of formal and informal, with the new vulnerability, uncertainty, instability and unpredictability of such relations.

So far, we established that there are multiple sources of cooperation, with some arguing that cooperation relies on trust tied to ascriptive characteristics or familiarity (Fukuyama 1995) or on legal regulations (Cook, Hardin and Levy 2005) or that it can only be animated by civic virtues (Putnam 1993). In contrast to these approaches, Castells' (1998) notion of network society focuses on the shift to more instrumental dealings between the major groups in the global economic system and allows one to view a new type of social relation, rooted in formalised and instrumentalised informality, as responsible for the production of cooperation in all spheres of modern economic life. The chance of sustaining cooperation in such a way is overlooked by Sennett, who, while embracing the notion of *Guanxi*, does not notice that in today's world *Guanxi* is not confined to moral obligations and emotional attachments. While Sennett rightly emphasises the fact that *Guanxi* remains an important element of relationships in modern societies, he ignores the trend that in the last decades there has been a significant increase in the importance of the instrumental, or rent-seeking, type of *Guanxi*. This type of *Guanxi*, in contrast to expressive or favour-seeking *Guanxi*, refers to "a strategy for forming advantageous relations" (Qi 2013, 310). This interdependent, privileged network, despite the instrumentalisation of informality, functions as an informal institution of assurance, which allows it to fulfil its role in sustaining cooperation in modern world, and therefore its role needs to be acknowledged.

All of the above approaches have failed to notice the shift in the dynamics of formal-informal relationships, which reflects changes in broader social transformations. Fukuyama and Putnam, like Sennett, neglect the contemporary role of instrumental *Guanxi* in gaining and maintaining trust, and thus also fail to grasp its role in providing transaction cost advantages and offering mutual support to those who share reciprocal relations. Cook, Hardin and Levy (2005) tend to be more ready to forget the positive sides of informality and they rather easily associate informality with corruption or bribery, seeing it as providing particular access to resources through personal relations rather than operating through formal structures. Although it is true that in order to avoid favouritism and nepotism there is a need for an adequate system of law and regulation that is able to ensure control, it does not mean that informality itself is a cause of any kind of malpractice. Yet, even though informality is not itself a cause of corruption, if corruption does occur, informality is likely to be one of its mechanisms. Thus, in order to avoid reaching the point when such actions become socially harmful and illegal, we need to ask what the difference is "between a dinner and a bribe" (Cormack 2013, 25), or what kind of configuration of informality and formality can produce public good and which can serve only particularistic interests.

Paraphrasing Michel Foucault's argument that power should be seen as productive as well as repressive, we can emphasise that informality can be constructive and that it enhances cooperation and achievements of public goods as well as being destructive in that it can facilitate access to material gain through inappropriate means. Furthermore, without going into a detailed discussion of Foucault's (1982) notion of power, and referring only to his arguments that power extends beyond the state and that power is exercised as much through what is permitted as through what is forbidden, we can say that both informality and formality are used to control and construct conditions for cooperation.

> A stupid despot may constrain his slaves with iron chains; but a true politician binds them even more with the chain of their own ideas [which is] all the stronger if we do not know of what it is made and we believe it to be our own work (Foucault, quoted in Hughes 2010, 47).

In short, cooperation can be reinforced through the synchronisation of formal and informal interactional practices – a situation that reflects broader socio-economic circumstances.

Concluding remarks

Following Goffman (1983), I view both informality and formality as the crucial and dynamic aspects of many processes that depend upon the piecing-together of new modes of social control, new institutions and new means of communication. To comprehend ongoing informal and formal contributions made by people to the constitution of today's social institutions, including new configurations of informality and formality, it is useful to employ Elias' (1978; 1992) notion of figuration. The reliance on Elias' concept of figuration and "figurational change" means that our objects of analysis are the changing figurations, or continuously re-patterned configurations of informality and formality. Such an approach allows us to explore a "continuum of changes" (Elias 1992, 46), or changes in informality's relations with formality, and to grasp the complexities, dynamics, interdependencies and heterogeneity of both informality and formality.

While arguing that formality and informality should not be conceptualized as dichotomous or contradictory concepts, the paper asserts that the interdependent and dualist structure of formal/informal, defined by the notion of configuration, allows for an understanding of how changes in their relations translate into the penetration of social life and the widening net of social control. By focusing on their dualistic span, we have demonstrated that the institutional conditions typically involve both informal and formal constraints. In the first part of this paper, we established that the shift in the private and public arenas often expands the scope for informality, changes its role and leads to the informalisation as well as to the formalisation of informality. In the second part, we discussed how the instrumentalisation of informality becomes a new source of cooperation.

The formalisation and instrumentalisation of informality, in the context of when trust is produced neither by generalized morality nor by institutional arrangements, are important strategies for sustaining cooperation and exercising control. The instrumental type of informality, on the one hand, can uphold the stability of transactions between individuals of known reputations, while on the other hand, the formalized informality can be used to as a strategy to control and to guard against troubles.

When taking into account that in these new socio-economic conditions, the usage of informality as the controlling device and as the new co-figuration of formal and informal processes, we come to realize that the notion of informality needs to be grounded in a theory of power. While my original approach was rooted in Goffman's and Elias' theories, in order to better comprehend today's features and functions of informality, there is also a need to reach for Foucault's notion of power and his concept of the technology of domination. While the process of informalisation refers to Elias' (1978; 1992) ideas about the shifting character of social restraint towards self-restraint, adopting Foucault's (1982) theories reduces our hope that practices of informality can operate outside power relations. Following Foucault's ideas, we can say that relations of power characterise all practices of informality and we can view the process of the formalisation of informality as a technology of domination.

Taken together then, the analytical possibilities presented in the work of Goffman, Elias and Foucault provide us with complementary insights into the changing configuration of informality and formality in the contemporary world. Since the nature of the configuration of formal and informal is detrimental to the levels of cooperation, and thus to quality of life, our understanding of the relationships between formal and informal is of enormous social importance.

References

Archer, Margaret S. 2010. "Routine, Reflexivity and Realism." *Sociological Theory* 28: 272–303.

Arendt, Hannah. 1958. *The Human Condition*. Chicago: Chicago University Press.

Billig, Michael. 2013. *Learn to Write Badly. How to Succeed in the Social Sciences*. Cambridge: Cambridge University Press.

Bowcott, Owen. 2012. "Twitter Joke Trial." *The Guardian*, July 27.

Burns, Tom R. 1992. *Erving Goffman*, London: Routledge.

Cockcroft, Laurence. 2013. *Global Corruption. Money, Power, and Ethics in the Modern World*. London: Tauris.

Coleman, James S. 1990. *Foundation of Social Theory*. Cambridge (MA): Belknap Press.

Castells, Manuel. 1996. *The Rise of the Network Society*. Oxford: Blackwell.

Castells, Manuel. 1997. *The Power of Identity*. Cambridge: Polity Press.

Castells, Manuel. 1998. *End of Millennium*. Cambridge: Polity Press.

Chan, Kwok Bun, and Beoy Kui Ng. 2006. "Singapore." In: *Chinese Business in Southeast Asia*, edited by Edmund Terence Gomez and Huang Hsiao, 38–61. London: Routledge.

Chandler, David. 2013. "Democracy Unbound? Non-linear Politics and the Politicization of Everyday Life." *European Journal of Social Theory* 17: 1–18.

Cook, Karen S., Russell Hardin, and Margaret Levi. 2005. *Cooperation without Trust*. New York: Russell Sage Foundation.

Cormack, Zoe. 2013. "Everyone's at it." *Time Literary Supplement*, September 27.

Crozier, Michel. 1967. *The Bureaucratic Phenomenon*. Chicago: University of Chicago.

de Swann, Abram. 1990. *The Management of Normality*. London: Routledge.

Edwards, Bob, and Michael F. Foley. 1998. "Civil Society and Social Capital: Beyond Putnam." *American Behavioral Scientist* 42: 124–140.

Elias, Norbert. 1978. *The Civilising Process*. Oxford: Blackwell.

Elias, Norbert. 1992. *Time: An Essay*. Oxford: Blackwell.

Eriksson, Kai, and Henri Vogt. 2012. "On Self-service Democracy." *European Journal of Social Theory* 16: 153–173.

Farrell, Henry. 2004. "Trust, Distrust and Power." In *Distrust*, edited by Russell Hardin, 27–45. New York: Russell Sage Foundation.

Fukuyama, Francis. 1995. *Trust. The Social Virtues and the Creation of Prosperity*. New York: Free Press.

Foucault, Michel. 1982. "The Subject and Power." In: *Michel Foucault: Beyond Structuralism and Hermeneutics,* edited by Hubert. L. Dreyfus and Paul Rabinow, 208–226. Chicago: University of Chicago Press.

Gambetta, Diego, ed. 1988. *Trust. Making and Breaking Cooperative Relations*. Oxford: Basil Blackwell.

Gitlin, Todd. 2013. "Occupy's Predicament: The Movement and the Prospects for the Movement." *The British Journal of Sociology* 64: 3–25.

Goffman, Erving 1983. "The Interaction Order." *American Journal of Sociology* 48: 1–17.

Greenwald, Glenn, and Ewen MacAskill. 2013. "NSA Prism Program Taps in to User Data of Google, Apple and Others." *The Guardian*, June 7.

Grist, Matt. 2009. *Changing the Subject: How New Ways of Thinking about Human Behaviour Might Change Politics*. London: RSA.

Hart, Keith. 1988. "Kinship, Contract, and Trust." In: *Trust: Making and Breaking Cooperative Relations*, edited by Diego Gambetta, 176–193. Oxford: Blackwell.

Harre, Rom. 1999. "Trust and Its Surrogates." In *Democracy and Trust*, edited by Mark E. Warren, 249–272. Cambridge: Cambridge University Press.

Hughes, Jason. 2010. "Emotional Intelligence: Elias, Foucault, and the Reflective Emotional Self." *Foucault Studies* 8: 28–52.

Human, Oliver, and Paul Cilliers. 2013. "Towards an Economy of Complexity." *Theory, Culture and Society* 30: 24–44.

Knight, Amy. 2013. "Putin's Downhill Race." *New York Review of Books*, September 26.

Kynaston, David. 2009. *Family Britain*. London: Bloomsbury.

Lanchester, John. 2013a. "The Banks' Barely Believably Behaviour." *London Time Review*, July 18.

Lanchester, John. 2013b. "The Snowden Files: Why the British Public Should Be Worried about GCHQ." *The Guardian*, October 3. Accessed October 6, 2013. http://www.theguardian.com/world/2013/oct/03/edward-snowden-files-john-lanchester.

Lauth, Hans-Joachim. 2000. "Informal Institutions and Democracy." *Democratization* 7: 21–50.

Luyendijk, Joris. 2013. "Our Banks Are Not Merely Out of Control. They're Beyond Control." *The Guardian*, June 19. Accessed October 6, 2013. http://www.theguardian.com/commentisfree/joris-luyendijk-banking-blog/2013/jun/19/banking-britain-beyond-control.

Mann, Michael. 2012. *The Sources of Power*, vol. 2. Cambridge: Cambridge University Press.

McGrane, Sally. 2013. "In Latvia New Laws via the Web." *The Observer*, April 12.

McVeigh, Tracy. 2013. "Short on Social Skills?" *The Observer*, September 15. Accessed October 6, 2013. http://www.theguardian.com/politics/lord-mcalpine.

Misztal, Barbara A. 1996. *Trust In Modern Societies*. Cambridge: Polity Press.

Misztal, Barbara A. 2000. *Informality. Social Theory and Contemporary Practice*. London: Routledge.

Putnam, Robert. 1993. *Making Democracy Work: Civic Tradition in Modern Italy*. Princeton: Princeton University Press.

Putnam, Robert. 2000. *Bowling Alone*. New York: Simon and Schuster.

Qi, Xiaoying. 2013. "Guanxi, Social Capital Theory and Beyond." *British Journal of Sociology* 64: 308–324.

Rose, Nikolas, and David Miller. 1992. "Political Power beyond the State: Problematics of Government." *British Journal of Sociology* 43: 173–205.

Sennett, Richard. 2012. *Together. The Rituals, Pleasure and Politics of Cooperation*. London: Allen Lane.

Shepard, Blair H., and Dana M. Sherman. 1998. "The Grammars of Trust and General Implications." *Academy of Management Review* 23: 422–438.

Stark, David. 1989. "Bending the Bar of the Iron Cage." *Sociological Forum*, 4: 637–664.

Sztompka, Piotr. 1996. "Trust in Emerging Democracy." *International Sociology* 11: 37–62.

Thompson, John B. 2011. "Shifting Boundaries of Public and Private Life." *Theory, Culture and Society*, 28: 49–70.

Unah, Isaac. 2009. *The Supreme Court in American Politics*. London: Palgrave.

van Krieken, Robert. 2001. "Legal Informalism and Liberal Governance." *Social and Legal Studies* 10: 5–22.

Walker, Edward T. 2012. "Social Movements, Organizations and Field." *Contemporary Sociology* 41: 576–587.

Weber, Max. 1968. *Economy and Society*, edited by Guenther Roth and Claus Wittich. Berkeley: University of California.

Woolcock, Michael 1998. "Social Capital and Economic Development." *Theory and Society* 27: 151–207.

Wouters, Cas. 1986. "Formalization and Informalization." *Theory, Culture, and Society* 3: 1–18.

Zuckerman, Ethan. 2013. "Political Activism Is as Strong as Ever, But Now It's Digital and Passionate." *The Observer*, October 13.

Mikko Lagerspetz

When Formal and Informal Rules Meet: The Four Sets of Rules of the Estonian Language and Minority Regime

Introduction

For the social sciences, the post-communist countries of Central and Eastern Europe (CEE) and the former Soviet Union (FSU) are of interest for two principal reasons. First, they are interesting as a region, the development of which will in many senses determine the future face of greater Europe. And second, the rapid change makes post-communist countries a testing ground for more general theories and models on social development.

One of the specificities of that development lies in what Elster, Offe and Preuss (1998) describe as "re-building a ship at sea", i.e., a need during a rapid change of the whole society to create and re-create the very social institutions that are supposed to manage that change. Research on post-communist development began in the early 1990s with a focus on "transition", or the way in which the institutions of liberal democracy and capitalism could be established along the models known from the West. The later paradigm of "consolidation" lays more stress on the environment in which they function (Beyme 1999). At the same time, there of course exists no consensus about which aspects of the environment are crucial for successful consolidation; likewise, there is also no consensus about the criteria for success. As for democracy, for example, one can stress either stability or representation and either the legitimacy of output or input (Risse 2006, 185). Among the factors influencing whether consolidation will be successful is not only the design of the new institutions, but also their relationship with other institutions and the rest of society, geography more generally as well as long-term historical processes.

Within sociology and political science, New Institutionalism pays attention to the structural context within which individual interests and group norms emerge, and their role in institutional change (Nee and Brinton 1998, xv). In this usage of the term, "an institution" is a collection of rules and organized practices that creates continuity and stability in society, relatively immune to changes in external circumstances and in the participators' individual preferences. They are embedded in a functioning environment that provides them with resources, expectations and frameworks within which rules and goals are interpreted (March and Olsen 2005).

Whereas formal rules are explicit and rely on formal mechanisms of reinforcement and control, informal ones may or may not be explicitly stated and will rely on informal mechanisms of monitoring (Nee and Ingram 1998, 19). Differences between the two different types easily lead to a decoupling between official norms and actual practices, when rules remain unimplemented or routinely violated (Bromley and Powell 2012, 7). At the same time, decoupling does not necessarily need to be seen only as impairing the institution's efficiency. In some situations, disregarding the formal framework of rules can indeed better serve its normatively desired goals (Helmke and Levitsky 2003, 15). Because of New Institutionalism's effort to integrate the analysis of institutions with that of their practical functioning environment, the broadening of the approach it calls for seems like the right step to take in post-communist studies also. I will apply these ideas for an analysis of the language and minority regime of Estonia, one of the countries that restored their independence in 1991, during the fall of the Soviet Union and the Real Socialist political systems in CEE.

The relevance of informal rules and practices for the analysis of countries that have undergone democratic transition has been noted by several scholars (Helmke and Levitsky 2003, 4, 15; Tilcsik 2010; Đurić 2011). In such countries, new norms have been introduced, often from outside, in an environment where the legacy of the former authoritarian rule is incompatible with the new norms' underlying principles of, e.g., civil society, rule of law and protection of minorities. This easily leads to decoupling – the new norms are not implemented or become modified in ways that contradict their original intention. To these insights one should add, that the very decoupling between norms and practices is one of the powerful legacies of the Real Socialist regimes. This phenomenon, called *doublethink* in George Orwell's famous novel, *1984*, meant that officially-declared values and norms did not necessarily guide the day-to-day activities of individuals and institutions. It characterised the highest political leadership and the government and Communist party bureaucracy, as well as the population at large. A common accusation by the "dissidents" was that the state leadership in fact did not "*uphold its own constitution or the laws* which guarantee freedom of thought, speech, the press, and political activity, and the right to public trial" (*Lisandusi mõtete ja uudiste vabale levikule Eestis. I köide. Kogud I–VII.* 1984, 169, cited in Lagerspetz 1996, 113). At the same time, an important skill for a local administrator, a party boss, or the manager of a production unit was the ability to find protection and additional resources for his or her workers or constituencies. This often required disregarding or modifying official regulations and officially-stated goals. Activities verging on the criminal, such as stealing from the employer (a state-owned production unit or a collective farm) were tacitly accepted as economic survival strategies, and this continues to affect patterns of norm compliance and norm enforcement in post-socialist societies also (see Allaste and Lagerspetz 2005).

The possibilities for monitoring and assessing different elements of the new political order vary greatly. As for elements of language and minority regimes, for instance, some of them are easily accessible for outside observers, while others can be difficult to grasp even by those thoroughly cognisant of the overall context (see *table 1*). The accessibility of different types of information of course varies between countries and between policy fields, but in general, it correlates with the degree of formality or informality of the various policy elements. As a result, observers will be faced with discrepancies between policy measures and outcomes that are difficult to explain without knowledge about the informal rules at stake. So, for instance, all new CEE member states of the EU have adopted constitutions that include guarantees for the protection of and respect for national minorities (Agarin and Regelmann 2011, 82), while at the same time, minorities are more often than not shown to be disadvantaged as to their socioeconomic position or access to political power. On the other hand, there are examples of legislation that verge on ethnic discrimination (such as the Estonian and Latvian citizenship laws), but have not led either to radical marginalization or to large-scale permanent political protest and mobilization among the minorities.

Table 1: Elements of language and minority regimes, and possibilities for monitoring them

Type of element	**Example**	**Ease of monitoring**	**Sources of information**
Constitutional guarantees for minorities	Guarantees for minority representation	Easy	Constitutions
Legislation on language and minority protection	Right for education in minority language	Easy	Laws
Institutions for minority representation and protection	Minority ombudsman	Easy	Laws, government decrees, state programs
Resources used for minority protection	Financing of an agency	Moderate	State and municipal budgets
Enforcement of language and minority protection laws	Provision of public services in minority language	Moderate to difficult	Evaluative reports, administrative and court decisions
Actual functioning of agencies for minority representation and protection	Activity and influence of an agency	Moderate to difficult	Evaluative reports
Values and attitudes	Existence of hidden discrimination	Difficult	Population surveys, media analyses, case studies
Policy outcomes	Labour market position of different ethnic groups	Easy	Statistics, population surveys

In this chapter, I will discuss the Estonian language and minority regime as it has emerged as a result of the development that began shortly before Estonia restored her independence in August, 1991. The country's policies have been met with mixed responses, both from international organizations and from researchers. In the 1990s, the high number of stateless persons among its ethnic minorities caused some observers to draw parallels with the "ethnic democracy" of Israel (Järve 2000; Pettai and Hallik 2002). Estonian social scientists are concerned about widening socio-economic differences between the country's Estonian and Russian speakers (Vetik and Helemäe 2011) and about the latter's diminishing trust in governmental institutions (Kivirähk and Lauristin 2013, 96 f.). On the other hand, Estonia's secession from the Soviet Union took place without ethnic violence; even later, ethnic protest has been virtually (even if not completely – see below) absent. Organizations such as the Council of Europe (CoE) and the EU have regarded Estonia's minority legislation and policies as consistent with their own requirements, and the prevailing view is that Estonia's democracy is among the most successfully consolidated in the CEE region and the FSU (e.g., Pettai and Mölder 2013).

The chapter's aim is to analyse the policy outcomes as a result of the interplay between formal and informal rules and practices. In other words, I will not be looking for explanatory factors in the minority nationals' individual resources and attitudes, nor in the dynamics of ethnic identity and mobilization. First, I will identify four key sets of rules underlying the Estonian language and minority regime, both formal and informal; and second, I will illustrate their way of functioning with the help of a specific case – the government-induced change of the tuition language of those upper-secondary schools that previously utilised Russian. The analysis will show how the outcome is produced and also highlight some of the potential shortcomings of the resulting regime.

Four key sets of rules of the Estonian language and minority regime

Throughout post-socialist and post-Soviet Europe, new or thoroughly re-written constitutions were adopted in the early 1990s. Not unlike those of other countries in the region, the Estonian Constitution of 1992 defined the state as the protector and expression of one "nation" with a distinct language and culture (Agarin and Regelmann 2011, 82). From the outset, the state administration and all the mainstream Estonian political parties have strongly committed themselves ideologically to the idea of a monolingual nation state. This, however, has been a factor in the country's continually strained relationship with the Russian Federation.

On several occasions, Russia has accused Estonia of discriminating against its Russian-speaking minority, a view with which both the Estonian government and many international organizations disagree (see, e.g., Smith 2002, 11–14). This means, however, that all issues concerned with language or with minorities carry a heavy political and symbolical meaning. Despite Estonia's official monolingualism, the actual composition of the population is ethnically diverse, with some 30% of all inhabitants (and close to 20% of the citizens) speaking Russian as their first language. Estonia's language and minority policies of the past twenty years can be read as efforts for coping with this discrepancy.

It should be kept in mind that Estonia's (and neighbouring Latvia's) Russian-speaking population is not just any ordinary minority; it is one that in scholarly language could be called post-colonial. Until Estonia's annexation by the Soviet Union in 1940, the share of Russian speakers in the population was around 8%. The Soviet industrialization policies brought to Estonia large numbers of mainly Russian-speaking workers and administrative personnel from the other Soviet republics, especially in the 1960s and 1970s. Most of Estonia's present Russian-speakers were either themselves settlers, or their parents or grandparents were among the settlers. In the eyes of ethnic Estonians, the very presence of Russian speakers in Estonia is a reminder of the country's five-decade history as a Soviet Republic. "Civil Occupation" was the term for this presence, coined in 1994 by a politician of the then Prime Minister's party, *Pro Patria* (*Isamaa;* literally, "Fatherland") (Riigikogu 1994). Even if references to the Russian-speakers as former occupiers cannot be found in any legal texts or policy documents, and even if such language nowadays would be avoided by all politicians except a handful of nationalist hardliners, a tacit assumption still held by many Estonians is that the Russian speakers' loyalty to the Estonian state is in one way or another compromised by their loyalty to their ethnic kin.

To make a long story short, I suggest that Estonia's minority regime forms a pattern with four distinct sets of rules, with varying degrees of formality. One of them is based on the idea of re-nationalising the society, i.e., securing the Estonian language and ethnicity's leading position in the state. Another consists of the different constitutional and legal provisions designed to guard the minorities and to balance the dominance of the Estonian language with the privileges given to minority languages. The implementation of such protections is, however, mostly dependent on the political will of the government. A third rule set that is crucial for the regime's practical functioning is the actual pragmatism shown by the state and local governments in language-related issues; and the fourth is the absence of institutionalised channels for minority representation. Obviously, this combination is compatible with the Estonians' prevailing view of the Russians, as discussed above: they are othered from the core nationality, both as an ethnic group and as citizens. But it also acknowledges the practical need for a *de facto* multicultural society, and the lack of the resources that would be needed for any strict implementation of the language laws (e.g., for replacing large segments of health care personnel or the police forces). Together, the four sets of rules form the foundation on which the Estonian society has so far been able to combine official monolingualism with only a minimal degree of ethnic unrest – with the obvious exception of the "Bronze Soldier" riots of April, 2007. On 26–28 April, 2007, street rioting and looting mainly by Russian-speaking youth in Tallinn was triggered by the Government's decision to remove a Soviet-era war monument from the city centre. Around 1,300 people were detained by the police during the two nights. The events have received considerable scholarly attention as well (see, e.g., several articles in 2008's last issue of *Journal of Baltic Studies*; Petersoo and Tamm 2008; Berg and Ehin 2009).

(Re-)nationalising policies

In the wake of the events of the 1990's, Estonia and most other countries of CEE and the FSU opted for what Brubaker (1996) has termed "nationalising policies". They thus consist of laws, regulations and practices securing the dominance of the ethnic majority's language and cultural heritage in public life. Maybe the example with the most profound influence on further developments was the *Citizenship Act* (enacted in September 1992 and enforced beginning in July, 1993), which gave the automatic right of citizenship only to people who were citizens of Estonia in 1940 and their descendants. As a result, a majority of the country's more than 400,000 Russian speakers were transformed into aliens (Jurado 2003, 399). (At present, some of the Russian speakers have left Estonia, some have gone through the naturalization procedure or opted instead for Russian citizenship; the number of stateless persons in Estonia is still around 85,000. For details, see Lagerspetz 2014b). Another example is the *Language Act*, the present version of which was adopted in 2011. Among other things, it specifies the degree of proficiency in the state language that is required for different categories of employees (in the relevant Government Decree specified for both public and private employees), and the punishments for not complying with the regulations (Riigi Teataja I 1995, 23, 334; Riigi Teataja I, 18.03.2011, 1; Riigi Teataja I, 29.12.2011, 169). These laws are also important as symbolic gestures and are seldom publicly challenged by mainstream politicians (even in cases when they might not correspond to an individual politician's private opinion). Throughout the period of renewed independence the government coalitions have virtually been dominated by rightist-conservative and rightist-neoliberal parties, who in their programs and campaigning routinely appeal to nationalist ideas and sentiments.

Guarantees for minorities

Another set of rules with sometimes symbolic, sometimes real substance consists of several legal provisions safeguarding minority rights. The Constitution states, among other things, the minority nationals' right to create institutions of non-territorial cultural self-government, "according to the conditions and regulations stated in the law on the cultural autonomy of national minorities" (§ 50); other paragraphs guarantee the right of national minorities' educational institutions to decide upon their language of tuition (§ 37), and everybody's right to preserve his or her ethnic belonging (§ 49) (*Eesti Vabariigi Põhiseadus* 1992). However, efforts at establishing bodies for the purpose of implementing cultural autonomy have in practice come to be blocked by bureaucratic obstacles (to be discussed further on). Likewise the language of tuition in upper-secondary education (also to be discussed below) has, during the past few years, become the hottest issue of controversy between the government and the Russian speakers. The *Law on Local Government* allows resident non-citizens to vote in local elections, which has endowed Russian speakers with a channel of political influence in Tallinn and a few other cities. A paragraph (§ 11) in the Language Act indeed also allows municipalities with a non-Estonian linguistic majority to use the other language in its administration – but only if the Government of the Republic grants the relevant permission. Up to now, this provision has never been implemented, despite reoccurring requests by municipalities with large Russophone majorities. Thus, a characteristic element of this and many other laws, and of the relevant paragraphs of the Constitution, is that their actual implementation is dependent on the Government of the Republic, which has hitherto not shown the political will necessary.

Unofficial bilingualism

However, neither has the implementation of nationalistic legislation been stringent. Reflecting this, the Council of Europe's Advisory Committee monitoring the Framework Convention for the Protection of National Minorities was pleased to note in its opinion of 2001 that the use of Russian was widely accepted in contacts with authorities, despite a lack of legislation that would guarantee that right (Council of Europe 2002, 10). As a solution to practical problems in a city, where the great majority of both the population and the members of the municipal council are Russian speakers, the municipal counsellors of the city of Narva in north-eastern Estonia have been reported to organise preliminary meetings, in which the agenda is first discussed in Russian, followed by short regular meetings conducted in Estonian, in which the decisions are officially made (a practice that by some is considered legally suspect – see Riigikogu 2005). In child care and education, which are administered by the local governments, systems working in Estonian and Russian exist in parallel. Basic health and social services are easily available in Russian, and private businesses serve their clients in both Estonian and Russian as a matter of course. These are just some examples of the actual pragmatism and flexibility with which language issues are treated in everyday life. They form the third crucial element of the language regime.

While the society in general seems to have adopted unofficial bilingualism, this tolerance has no formally codified basis and does not imply any easy recognition of formal language rights or bodies of minority representation. The creation of such protections dates back to the early 1990s, but they have by now become increasingly imperceptible in politics and publicity.

Blocking interest representation

An early example of a semi-official consulting body on minority issues was the *President of the Republic's Roundtable of Non-Citizens and National Minorities,* set up by President Lennart Meri in the midst of a political crisis on 25 June 1993, at the suggestion by CSCE's (now OSCE) High Commissioner on National Minorities (HCNM), Max van der Stoel (Kemp 2001, 135). However, after years of declining visibility, the Roundtable was finally in 2010 transformed into a commission within a state-financed private foundation (*Vabariigi President Eesti Koostöö Kogu rahvuste ümarlaua kokkukutsumise puhul,* 26.05.2010). Another agency, originally influential in bringing minority issues to the political agenda, was the *Minister Without Portfolio of Population Issues* and his or her bureau, that existed first in 1990–1992 (as the *Minister of Ethnic Relations*), then in 1997–2005 and again in 2007–2009. The Bureau was in charge of integration, citizenship, refugee and population policies and initiated Estonia's first policy documents on minority integration strategies in 1998, but was finally disbanded and its tasks divided between different ministries.

The Constitution's provision for the establishment of bodies of non-territorial cultural autonomy found expression in the enactment of the relevant law on the *Cultural Autonomy of National Minorities* on 26 October 1993, by a solid parliamentary majority. Following the example of a similar law from 1925 (see, e.g., Alenius 2007), it grants groups of Estonian citizens with a distinct cultural heritage and reaching a minimum number of 3,000 registered members, the right of establishing through an election procedure Cultural Councils that are given the task of coordinating the minority's cultural and educational activities and of forming relevant public bodies (National Minorities Cultural Autonomy Act 1993). However, the reality has turned out rather differently from that which was originally promised. The law requires that in order to establish cultural autonomy, a roll of citizens belonging to the national minority should be created. The government decree regulating the procedure was issued in October, 1996 (Riigi Teataja I 1996, 72, 1272). Further, the decree that specified the election procedure of the Cultural Councils was issued only in 2003, ten years after the adoption of the law itself (Riigi Teataja I 2003, 40, 275). As of today, Cultural Councils have only been elected by two tiny minorities – the Ingrian Finns in 2004 and the Estonian Swedes in 2007. They have not received the status of legal persons and both groups have had to establish parallel NGOs in order to run their activities. Applications on behalf of the far more numerous Russians have been filed four times – in 1996, 2006, 2009 and 2011, but have not led to any positive decision by the Ministry of Culture despite years of administrative processing, and thus the applying organizations have questioned the legality of their treatment (see Lagerspetz 2014a). By the time the first application was filed, none of the necessary by-laws had yet been decreed. As to the application filed in 2006, the Ministry of Culture started processing it only in 2009, after an administrative court decision obliged it to do so, and finally dismissed it with the justification that the initiating organization was not sufficiently representative of the whole Russian minority community (a requirement not to be found in the law or its acts of implementation). The Ministry also stated that the establishment of Russian cultural autonomy might complicate the pending change of tuition language in hitherto Russian secondary education (see below). The Ministry has not, as of January 2015, made any decision on the two later applications either, with the rationale that the existence of several parallel applications makes the decision process legally complicated.

The legislation on political parties includes no obstacles for the formation of parties representing ethnic minorities; some do exist, but none of them managed to win a seat either in the 2003, 2007 or 2011 parliamentary elections (in 1999, the Russian-speakers' United People's Party won 6 seats out of 101). However, the Centre Party in particular – now the largest opposition party – has been successful in appealing to Russian-speaking voters.

In the next section, I will illustrate the four sets of rules of the Estonian language and minority regime with the help of a topical controversy: the debate over the tuition language of upper secondary education. The plan for introducing Estonian as the main language of tuition in the hitherto Russian Gymnasiums has received surprisingly little scholarly attention. It has been met with cautious criticism by Estonian educational scientists, who base their opinions on surveys and interviews with Russian speaking teachers, students, their parents and education experts (Kello, Masso and Jakobson 2009; Masso, Kello and Jakobson 2013; Masso and Soll 2014). The related political processes during the 1990s and early 2000s have been discussed by Jurado (2003) and Galbreath (2005, 170–172).

The issue of the Russian gymnasiums

The present controversy over the language of tuition in upper-secondary schools is a case in point, showing the working of all the sets of rules discussed above: the ideologically motivated decisions and the difficulties in actually carrying them out; legal, political and pragmatic ways of overcoming the difficulties; the existence of formal guarantees for minorities and the dependence of their implementation on political will; and finally, the limits that exist for the participation of the minority citizens themselves. Even if the Constitution (§ 37) gives the "national minorities' educational institutions" the right to decide upon the language of tuition themselves, and the Law on the Cultural Autonomy of National Minorities (§ 4) gives persons belonging to a national minority the right to "form and support" such institutions, a legal definition for them is nonetheless nowhere else to be found. As long as the latter law is not fully implemented in practice, there cannot exist any such educational institutions that would be entitled to use this constitutionally guaranteed right; the state or municipal schools are not legally recognised to belong to the category of "national minorities' educational institutions", whatever the language of tuition or the mother tongue of the students.

The *Law on Gymnasiums and Primary Schools* was originally passed on 16 July 1993 and slightly amended on 15 September of the same year after an intervention by President Lennart Meri. The law defined Estonian as the sole language of tuition in all public upper-secondary schools (gymnasiums) and ordered a shift to Estonian to be carried out in Russian gymnasiums by the year 2000 (Riigi Teataja I 1993, 63, 892). In the parliamentary discussion of the law, Estonia's Russian-speaking population was referred to as representing the former occupiers, and the law as essential for securing the Estonian language's future survival (Riigikogu 1993a; Jurado 2003, 412f.). It was finally signed by the President, despite quick and alerted responses from both the Council of Europe and the OSCE and its HCNM (Jurado 2003, 411f.). The President's remarks to the *Riigikogu* justifying his initial intervention were not at all about issues of minority protection, but about how the general compulsoriness of secondary education was formulated in the law (Riigikogu 1993b). However, already during the parliamentary discussion, the

Minister of Culture and Education, Paul-Eerik Rummo, expressed serious doubts about the time frame:

> What we need to accomplish is that a pupil graduating from a Russian-language basic school will in fact be capable of continuing his or her studies in an Estonian-language gymnasium or vocational school. To reach that by the year 2000 is virtually impossible; in order for that, the whole process should have been started already two years ago. To speed up the process now would first of all require much larger resources and secondly, that a part of the students of Russian schools would be prepared to enrol in the new, much more demanding Estonian-language curricula. That would in turn require talent above the average, very strong study motivation and large numbers of very well-qualified teachers. All these components we are lacking now (Riigikogu 1993a).

According to Rummo, the hurried timetable did at first sight seem "beautiful, nice, ideologically correct and patriotic", but would in practice prove to be a short-sighted decision that would never bring about "the noble results intended" (Riigikogu 1993a). The necessary preparations in fact did not start until several years later. A Language Strategy Centre designed to train Russian-speaking teachers in Estonian and to provide teaching materials was officially set up in 1995, but it worked with limited finances and did not produce an action plan to assist the transition sooner than 1998 (Jurado 2003, 414). In 1997, the law's implementation was indeed postponed from 2000 to 2007 (Jurado 2003, 415), and even that deadline later came to be postponed again. A new version of the Law from 9 June, 2010 (Riigi Teataja I 2010, 41, 240) states the academic year 2012/2013 as the time of implementation.

However, a more principal change to the law was made by an amendment in June 2000. It defined "the language of tuition" as the language in which at least 60% of the teaching of the curriculum is performed. This allows schools to continue teaching some subjects in Russian. Jurado (2003) attributes this new approach to attitude changes among the politicians, who had been increasingly socialised into the more "ethical" and minority-friendly way of thinking represented by the Council of Europe. This may be one part of the story; however, the preservation of 40% of the tuition in Russian also certainly makes the transition cheaper and more realistic. Moreover, it made explicit that a merger of the then Russian gymnasiums with the Estonian ones was not what was intended – despite frequent talk about integration. The resulting pedagogical problems will be experienced by the Russian students and their teachers, but remain irrelevant for the Estonian schools and for most of the Estonian-speaking electorate.

Despite the amendment, the reform still seems difficult to carry out without considerable losses in teaching quality, which is a major reason for the opposition from the Russian-speaking teachers as well as students and their parents

(Kello, Masso and Jakobson 2009, 5). A survey of Russian Gymnasium teachers showed that in 2009, a mere 25% had received any supplementary training in the methods of teaching their Russian-speaking pupils in another language (including those teachers who were already doing so) (Masso and Kello 2010, 24). The Estonian Language Inspectorate's regular assessments of 2012 revealed that 49.5% of the assessed teachers in Russian-speaking schools did not fulfil the requirements for fluency in the state language (*Keeleinspektsioon* 2012). The Ministry of Education has given no clear instructions about how much Russian may be used in a supporting role in a lesson supposedly taught in Estonian (Kello, Masso and Jakobson 2011, 6), and, according to some newspaper reports, the teachers seem to be developing their own *ad hoc* pedagogies that include a rather liberal use of Russian, while the Ministry of Education is turning a blind eye. At least until very recently (see Afterword), the formerly Russian gymnasiums' pedagogical problems have had a low priority within Estonia's education policies.

According to the letter of the law, the change of tuition language is in fact not obligatory – something that the Estonian government did not forget to mention in its 2010 report to the Council of Europe on the protection of national minorities (Council of Europe: Secretariat of the Framework Convention for the Protection of National Minorities 2010, 44). In the law's present, 2010 version, one paragraph (§21 (3)) indeed gives the option of using a language of tuition other than Estonian. The applications for doing such are to be forwarded to the Government of the Republic by the local government upon the request of the gymnasiums' school councils. However, the Government's Decree of 6 January, 2011 on gymnasium curricula ignored this option; only after an intervention by the Chancellor of Justice did the Government amend its decree so as to correspond with the law (Chancellor of Justice 2011). The Government has hitherto – January 2015 – rejected all such applications made by municipalities on behalf of Russian schools (out of a total of 18 applications that the Government has made a decision on, the only exceptions were made in August 2011 for two adults gymnasiums in Tallinn and Narva, and none have been made for ordinary secondary education). The Minister of Education and Science of the previous cabinet (until March 2014) did also state publicly that he intended to continue doing so. In September 2010, the NGO *Vene kool Eestis* [Russian School in Estonia] was founded in order to represent and inform the parents of Russian school children, but the Minister of Education, Mr. Tõnis Lukas, immediately declared that the organization was involved in politicking in a manner hostile to the Estonian state (Raiste 2010; Lobov 2011). The 2011 *Annual Review* of the Estonian Security Police [*Kaitsepolitsei*] suggested that the NGO in question was in fact an example of the Russian Foreign Ministry's efforts of using NGOs as means to "influence the sovereign decisions of other countries and to divert attention away from its own problems through the manipulation of these groups [of Russian speakers]" (Security Police of the Republic of Estonia 2012, 10). Because of their activities within the organization, two Russian-speaking politicians were branded as serving the interests of Russia. As the Security Police summarises its stance in the matter,

> [t]he Russian-language educational system and the special status of the Russian language were established as part of the Soviet Union's Russification policy. Preserving them is a priority of Russian influence operations. The Russian Embassy in Estonia supports these activities [...]. It is regrettable that the Russian Federation attempts to use young people as instruments in its influence operations as the future of young Russians in Estonia and Europe depends, above all, on them receiving a competitive education (Security Police of the Republic of Estonia 2012, 11).

Obviously, the Security Police's opinion is that "competitive education" equals tuition in Estonian, irrespective of the teachers' and students' prior fluency in that language. The NGO's and the two politicians' activities that the Security Police condemned were of course fully legal as such: informing the school councils of Russian gymnasiums about the possibility of applying for the continuous use of Russian as the language of tuition – information that the Ministry of Education itself is not announcing. The two politicians sued the Security Police for libel, and one of them has already won her court case. Clearly, much of the future of the Russian upper-secondary education in Estonia will be decided in courtrooms.

Conclusions

Of the four sets of rules underlying Estonia's language and minority regime, two belong to the sphere of legislation: (1) the laws intended to secure the Estonian language and ethnicity's leading position in the state and (2) the legal provisions designed to guard the minorities, which are usually made dependent on the Government's consent. These two are balanced by the informal practices of (3) less than stringent implementation of the language laws, often resulting in solutions that are essentially bilingual; and of (4) blocking and discouraging potential channels for the Russian speakers' collective interest representation. In one sense, these rules do indeed counteract each other; however, they can also be seen as functionally complementary. In the way suggested by Nee and Ingram (1998, 35), the formal rules serve to "satisfy external constituents that provide the organization with legitimacy", while the informal ones "guide [its] day-to-day business". It was argued earlier that taken as a whole, the resulting regime is in harmony with the views, or the "institutionalised myths", about Russians prevailing in the public opinion. But taken individually, each set of rules also corresponds to the expectations of important external constituencies: the part of the electorate guided by nationalist sentiments, the international and European organizations concerned with minority issues (cf. Smith 2002), and the municipalities and state agencies implementing the policies.

The expectations of these external constituencies are different. Expressing pride for the newly won independence and nostalgia for the more ethnically homogeneous country of the 1920s and 1930s have shown to be important for a large segment of voters. At the same time, Estonia's "Return to the Western World", i.e., the country's accession to the EU and NATO in 2004 was dependent on its reputation as a democratic country that respects its minorities and human rights. It was, however, clear from the outset that one of those two constituencies – the voters – would always be there, while the attention from the other one – the international and European organizations – was likely to disappear after successful accession negotiations. The EU accession also meant the end of external pressure for minority rights. The lack of minority organization and representation minimises domestic pressure; at the same time, the need to run everyday business in a society with a large linguistic minority makes it impossible to implement nationalising policies without exceptions and modifications. In sum, the answer to one external constituency was promises without strict legal commitment, and to the other, commitment without strict implementation. A question is how much strain between formal and informal rules can be tolerated without endangering the precarious balance between them; how sustainable is the settlement thus reached?

In the process of changing the Russian gymnasiums' language of tuition, all the four sets of rules discussed previously are present. The original law of 1993 was adopted much as a symbolic gesture, ignoring the doubts that were expressed already at that time about its possibilities for practical implementation. The first experiences after the language change suggest that the actual use of Russian and Estonian as tuition languages is not being strictly monitored. Instead of allowing the use of existing legal and constitutional guarantees for education in minority languages, the legislators and the government have opted for a softening of both the legal and the unofficial language requirements. On the other hand, both the school boards and the Russian-speaking parents' organization have been denied the possibility of influencing this aspect of their children's education. Also, the present practice leaves the Russian teachers and school directors in a potentially vulnerable position when trying to cope with the resulting pedagogical problems, with them thus possibly operating in a legal grey zone.

Together, the formal and informal practices now prevailing in Estonia contribute to a minority incorporation regime that has usually shown itself able to guarantee a smooth functioning of society and to prevent the surfacing of ethnic strife. However, the regime's heavy reliance on informal practices also makes it dependent on political goodwill and on a balance of power between different actors. That balance means not only a relative equity between more nationalist and more liberal-minded parties but also a balance between the more ideologically committed central authorities and the municipal governments favouring a more pragmatic, down-to-earth approach (and with constituencies consisting of both Estonian citizens and non-citizens). During all political election campaigns, an effort to change that power balance is likely to be made. The lack of acknowledged interest representation means that the Russian-speaking minority itself has little say on the future continuation or discontinuation of these practices; the Estonian minority and language regime remains a strictly top-down enterprise controlled by the majority. It also weakens the possibilities for moderating between the opinions of Estonian nationalists and the more militant among the minority activists. The riots of 2007 showed the dangers inherent in such a situation.

The events that led to the "Bronze Nights'" riots gained momentum during the campaign before the March 2007 parliamentary elections, when Prime Minister Andrus Ansip's neoliberal *Reform Party* vied for the nationalist vote with its coalition partner, the conservative nationalist *Union of Pro Patria and Res Publica* (see Lagerspetz and Vogt 2013, 57). As reported by correspondence from the U.S. Embassy in Tallinn, Res Publica's general secretary said "off the record" in September 2006 that an ethnically motivated incident "would be extremely good for Res Publica's campaign and they would make good use of the publicity" (PLUS D 2006). It is not difficult, either, to see a connection between the March 2011 elections and the Government's sudden determinacy in January 2011 to enforce the change of the Russian gymnasiums' tuition language. Many of Estonia's political parties will be tempted in the future to mobilise Estonian nationalist sentiments not only because the number of Russian-speaking voters is smaller, but also because of the recent economic slowdown. As Offe (1996, 63) notes, during economic hardship, appeals to (national) "pride" often are more effective in gaining votes than appeals to (economic) "hope".

The rule of law is an important precondition for democratic consolidation. However, both the élites and the populations of the formerly Real Socialist societies have profound experience with a political system in which the distance between officially stated norms and the rules guiding everyday life was larger than usual. A decoupling between norms and practices was essential both for the survival of the system and for individuals to accommodate to it. Finding ways to disregard regulations coming from central authorities could even be seen as a patriotic act of passive resistance. To apply a common Soviet phrase, it is "not incidental" that minority issues have become the field of policy where decoupling now is most easily visible (Đurić 2011). They have, at least formally, ranked highly on the agendas of the EU, the OSCE and the CoE. Now, as before, a perceived situation has emerged in which formal rules are being set by a power centre outside the control of local constituencies. Legislators and the people supposed to implement the laws respond with a mixture of both formal and informal modes of compliance and resistance. In Estonia, the resulting minority and language regime has not been without success, but from the point of view of the minorities, it includes a great deal of arbitrariness and insecurity. This is not to say that all aspects of a society's functioning should be regulated by formal law; on the other hand, if formal and informal practices stand in obvious contradiction to each other, the resulting balance will remain fragile – an easy target for both legal and political criticism.

Afterword

Soon after this chapter was first finished in March 2014, important changes took place in Estonian politics. The Prime Minister Andrus Ansip resigned his position in order to become a member of the EU Commission. The new prime minister's Cabinet that was installed on 26 March was formed on the basis of a new coalition between the neo-liberal Reform Party and the Social Democrats, who thus replaced the nationalist-conservative Union of Pro Patria and Res Publica. Importantly, the position of Minister of Education and Science was given to a Social Democrat, the 28-year-old Jevgeni Ossinovski with an education in Political Science from the London School of Economics and Political Science. He was the first ethnic Russian ever to make it to the Cabinet after the reestablishment of Estonian independence. In June 2014, he appointed a commission in order to assess the process by which the tuition language was changed in formerly Russian Gymnasiums. The Commission's report, delivered in September, confirmed many of the findings of previous studies by educational researchers, effectively stating that there was a lack of suitable teaching material, and that the language fluency of both the teachers and the students was in many cases insufficient in order for the reform to be successful. Among other recommendations, the report called for more flexibility with regard to the requirement that a minimum of 60% of tuition be in Estonian (Haridus- ja Teadusministeerium 2014). No political consequences have been drawn from these conclusions yet, nor has any secondary school yet been granted the permission to continue using Russian as the main medium of instruction. However, the developments hitherto show that there exist chances for overcoming the ideological obstacles rooted in nationalist policies and thus, for bringing the framework of formal rules closer to the demands of actual day-to-day practices.

References

Agarin, Timofey, and Ada-Charlotte Regelmann. 2011. "Status Quo Multiculturalism: The Crux of Minority Policies in Central Eastern Europe's New EU Member States." *Journal of Minority Studies* 5: 69–98.

Alenius, Kari. 2007. "The Birth of Cultural Autonomy in Estonia: How, Why, and for Whom?" *Journal of Baltic Studies* 38: 445–462.

Allaste, Airi-Alina, and Mikko Lagerspetz. 2005. "Drugs and Doublethink in a Marginalized Community." *Critical Criminology* 1: 267–285.

Berg, Eiki, and Piret Ehin, eds. 2009. *Identity and Foreign Policy: Baltic-Russian Relations and European Integration.* Farnham (UK) and Burlington (USA): Ashgate.

Beyme, Klaus von. 1999. "Osteuropaforschung nach dem Systemwechsel. Der Paradigmawandel der »Transitologie«." *Osteuropa* 3: 285–304.

Bromley, Patricia, and Walter W. Powell. 2012. "From Smoke and Mirrors to Walking the Talk: Decoupling in the Contemporary World." *The Academy of Management Annals* 6: 1–48.

Brubaker, Rogers. 1996. *Nationalism Reframed: Nationhood and the National Question in the New Europe.* Cambridge: Cambridge University Press.

Chancellor of Justice. 2011. *Märgukiri* [Letter of Notification to the Ministry of Education and Science, July 11]. Accessed August 26, 2013. http://oiguskantsler.ee/sites/default/files/field_document2/6iguskantsleri_margukiri_gumnaasiumi_oppekeel.pdf.

Council of Europe. 2002. *Advisory Committee on the Framework Convention for the Protection of National Minorities. Opinion on Estonia,* ACFC/INF/OP/I (2002)005, September 14, 2001.

Council of Europe: Secretariat of the Framework Convention for the Protection of National Minorities. 2010. *Third Report Submitted by Estonia Pursuant to Article 25, Paragraph 2 of the Framework Convention for the Protection of National Minorities,* ACFC/SR/III(2010)006, April 13, 2010. Accessed March 5, 2014. http://www.refworld.org/docid/4bd7ed652.html.

Đurić, Ivana. 2011. "The New Institutionalism(s): A Framework for the Study of Public Policy in Post-conflict and Post-communist Countries." *Politička misao* 48: 85–105.

Eesti Vabariigi Põhiseadus [The Constitution of the Republic of Estonia]. 1992. *Riigi Teataja 1992 26, 349*. Accessed February 20, 2013. https://www.riigiteataja.ee/akt/633949.

Elster, Jon, Claus Offe, and Ulrich K. Preuss. 1998. *Institutional Design in Post-Communist Societies: Rebuilding the Ship at Sea*. Cambridge: Cambridge University Press.

Galbreath, David J. 2005. *Nation-Building and Minority Policies in Post-Socialist States. Interests, Influences and Identities in Estonia and Latvia*. Stuttgart: *ibidem*-Verlag.

Haridus- ja Teadusministeerium. 2014. Raport vene õppekeelt kasutavate gümnaasiumide õppe tulemuslikkusest eestikeelsele (vähemalt 60% kohustuslikest kursustest) gümnaasiumiõppele üleminekul. [Report on the Study Results of Russian-medium Gymnasiums Changing Their Language of Instruction (to a Minimum of 60% of Obligatory Courses Taught in Estonian)]. Tallinn: Haridus- ja Teadusministeerium. Accessed January 27, 2015. https://www.hm.ee/sites/default/files/raport_uleminek_viimane_0.pdf.

Helmke, Gretchen, and Steven Levitsky. 2003. *Informal Institutions and Comparative Politics: A Research Agenda*. The Helen Kellogg Institute for International Studies, Working Paper #307.

Järve, Priit. 2000. *Ethnic Democracy and Estonia*. Flensburg: European Centre for Minority Issues, ECMI, Working Paper #13. Accessed May 22, 2013. http://edoc.vifapol.de/opus/volltexte/2009/1887/pdf/working_paper_7.pdf.

Jurado, Elena. 2003. "Complying with European Standards of Minority Education: Estonia's Relations with the European Union, OSCE and Council of Europe." *Journal of Baltic Studies* 34: 399–431.

Keeleinspektsioon. 2012. *Keeleinspektsiooni 2012. aasta tegevuse aruanne.* [Report on the Language Inspectorate's Activities in 2012]. Accessed August 27, 2013. http://www.keeleinsp.ee/?menu=30&news=931.

Kello, Katrin, Anu Masso, and Valeria Jakobson. 2009. "Eestikeelsele õppele üleminek venekeelses gümnaasiumis" [*Transition to Estonian as the Language of Tuition in Russian Gymnasiums*]. *Haridus* 11–12: 1–7.

Kello, Katrin, Anu Masso, and Valeria Jakobson. 2011. *Uurimisprojekti "Vene laps venekeelse üldhariduskooli eestikeelses õppes" koondaruanne* [Composite Report on the Project "Russian Children within Estonian-language General

Education"]. Tartu: Tartu Ülikool, Haridusuuringute ja õppekavaarenduse keskus.

Kemp, Walter A. 2001. *Quiet Diplomacy in Action: The OSCE High Commissioner on National Minorities.* The Hague: Kluwer. Accessed February 22, 2013. www.osce.org/hcnm/78633.

Kivirähk, Juhan, and Marju Lauristin. 2013. "Public Assessments of the State's Development and Trust in Government Institutions." In *Estonian Human Development Report 2012/2013: Estonia in the World*, edited by Mati Heidmets, 95–99. Tallinn: Eesti Koostöö Kogu. Accessed January 27, 2014. http://www.kogu.ee/wp-content/uploads/2013/05/EIA2013_eng.pdf.

Lagerspetz, Mikko. 1996. *Constructing Post-Communism. A Study in the Estonian Social Problems Discourse.* Turku: Annales Universitatis Turkuensis.

Lagerspetz, Mikko. 2014a. "Cultural Autonomy of National Minorities in Estonia: The Erosion of a Promise." *Journal of Baltic Studies* 45: 457–475.

Lagerspetz, Mikko. 2014b. "Estonia." In *European Immigration: A Sourcebook*, edited by Anna Triandafyllidou and Ruby Gropas, 87–98. Aldershot: Ashgate.

Lagerspetz, Mikko, and Henri Vogt. 2013. "Estonia." In *The Handbook of Political Change in Eastern Europe*, edited by Sten Berglund, Joakim Ekman, Kevin Deegan-Krause and Terje Knutsen (3rd edition), 51–84. Cheltenham (UK) and Northampton (MA): Edward Elgar.

Lisandusi mõtete ja uudiste vabale levikule Eestis. I köide. Kogud I–VII. 1984. [Contributions to the Free Distribution of Ideas and News in Estonia, Volume I, Issues I–VII]. Stockholm: Eesti Vangistatud Vabadusvõitlejate Abistamiskeskus.

Lobov, Andrei. 2011. *Interview* by the author with Andrei Lobov, Council Member of the NGO Russian School in Estonia, Tampere 25 March, 2011.

March, James G., and Johan P. Olsen. 2005. *Elaborating the "New Institutionalism"*, Working Paper no. 11. Oslo: Arena - Centre for European Studies, University of Oslo.

Masso, Anu, and Katrin Kello. 2010. *Vene õppekeelega koolide valmisolek eestikeelsele gümnaasiumiõppele üleminekuks* [The Readiness of Russian-language Schools to Take Over Estonian as the Language of Tuition]. Tartu: Tartu Ülikool, Haridusuuringute ja õppekavaarenduse keskus.

Masso, Anu, Katrin Kello, and Valeria Jakobson. 2013. "Üleminek riigikeelsele gümnaasiumiõppele: vilistlaste seisukohtade Q-metodoloogiline diskursusanalüüs" [*With Summary in English: Transition to State-language Instruction in Upper-Secondary Schools According to Alumni: Q-methodological Discourse*

Analysis of Individual Positions]. *Eesti rakenduslingvistika aastaraamat* 9: 161–179.

Masso, Anu, and Maie Soll. 2014. "Change in Language of Instruction in Russian Medium Schools: Multilevel Analysis of Attitudes and Language Proficiency." *Journal of Baltic Studies* 45: 517–544.

National Minorities Cultural Autonomy Act. 1993. Accessed September 23, 2013. http://www.legaltext.ee/et/andmebaas/tekst.asp?loc=text&dok=XX00038K1&keel=en&pg=1&ptyyp=RT&tyyp=X&query=kultuuriautonoomia.

Nee, Victor, and Mary C. Brinton. 1998. "Introduction." In *The New Institutionalism in Sociology*, edited by Mary C. Brinton and Victor Nee, xv–xix. Stanford: Stanford University Press.

Nee, Victor, and Paul Ingram. 1998. "Embeddedness and Beyond: Institutions, Exchange, and Social Structure." In *The New Institutionalism in Sociology*, edited by Mary C. Brinton and Victor Nee, 19–45. Stanford: Stanford University Press.

Offe, Claus. 1996. *Varieties of Transition. The East European and East German Experience*. Cambridge: Polity.

Pettai, Vello, and Klara Hallik. 2002. "Understanding Processes of Ethnic Control: Segmentation, Dependency and Co-optation in Post-communist Estonia." *Nations and Nationalism* 8: 505–529.

Pettai, Vello, and Martin Mölder. 2013. "Estonia." In *Nations in Transit 2012. Democratization from Central Europe to Eurasia*, edited by Sylvana Habdank-Kolaczkowska, Katherin Machalek and Christopher T. Walker, 201–217. New York: Freedom House. Accessed 23 May, 2013. http://www.freedomhouse.org/sites/default/files/Estonia_final.pdf.

Petersoo, Pille, and Marek Tamm, eds. 2008. *Monumentaalne konflikt: Mälu, poliitika JA identiteet tänapäeva Eestis* [The Monumental Conflict: Memory, Politics and Identity in Contemporary Estonia]. Tallinn: Varrak.

PLUS D. 2006. *Estonain* [sic] – *Russian Relations: No Longer Hot but Still Potatoes*. [Memorandum from the U.S. Embassy in Tallinn to the Secretary of State, Washington, on September 15, 08:35]. Accessed January 24, 2014. http://www.wikileaks.org/plusd/cables/06TALLINN845_a.html.

Raiste, Anne. 2010. *Lobov: vene kooli eestistamine peab lõppema* [Lobov: The Estonification of Russian Schools Must Come to an End]. Accessed May 30, 2013. http://uudised.err.ee/index.php?06215828.

Riigi Teataja I. 1993, 63, 892 [The Official Estonian Bulletin Publishing Laws and Government Decrees]. *Põhikooli- ja gümnaasiumiseadus* [The Law on Basic

School and Gymnasium]. Accessed March 5, 2014. https://www.riigiteataja.ee/akt/28542.

Riigi Teataja I. 1995, 23, 334 [The Official Estonian Bulletin Publishing Laws and Government Decrees]. *Keeleseadus* [The Language Act]. Accessed March 5, 2014. https://www.riigiteataja.ee/akt/28746.

Riigi Teataja I. 1996, 72, 1272 [The Official Estonian Bulletin Publishing Laws and Government Decrees]. *Vähemusrahvuse kultuuriautonoomia seadusest tuleneva õigusakti kinnitamine* [Decree Pursuant to the National Minorities Cultural Autonomy Act]. Accessed March 6, 2014. https://www.riigiteataja.ee/akt/25169.

Riigi Teataja I. 2003, 40, 275. [The Official Estonian Bulletin Publishing Laws and Government Decrees]. *Vähemusrahvuse kultuurinõukogu valimise eeskiri* [Regulations for the Election of a National Minority Cultural Council]. Accessed March 5, 2014. https://www.riigiteataja.ee/akt/578630.

Riigi Teataja I. 2010, 41, 240 [The Official Estonian Bulletin Publishing Laws and Government Decrees]. *Põhikooli- ja gümnaasiumiseadus (lühend - PGS)* [The Law on Basic School and Gymnasium]. Accessed March 5, 2014. https://www.riigiteataja.ee/akt/13332410.

Riigi Teataja I. 18.03.2011, 1 [The Official Estonian Bulletin Publishing Laws and Government Decrees]. *Keeleseadus (lühend - KeeleS)* [The Language Act]. Accessed March 5, 2014. https://www.riigiteataja.ee/akt/118032011001.

Riigi Teataja I. 29.12.2011, 169 [The Official Estonian Bulletin Publishing Laws and Government Decrees]. *Avaliku teenistuja, töötaja ning füüsilisest isikust ettevõtja eesti keele oskuse ja kasutamise nõuded* [Requirements for Fluency and Use of Estonian Language for Civil Servants, Employees and the Self-Employed]. Accessed March 5, 2014. https://www.riigiteataja.ee/akt/129122011169.

Riigikogu. 1993a. *Minutes of the Session on June 1, 1993.*

Riigikogu. 1993b. *Minutes of the Session on September 15, 1993.*

Riigikogu. 1994. *Minutes of the Session on May 9, 1994.*

Riigikogu. 2005. *Minutes of the Session on April 11, 2005.*

Risse, Thomas. 2006. "Transnational Governance and Legitimacy." In *Governance and Democracy. Comparing National, European and International Experiences*, edited by Arthur Benz and Yannis Papadopoulos, 179–199. London and New York: Routledge.

Security Police of the Republic of Estonia. 2012. *Annual Review 2011*. Tallinn: Kaitsepolitsei: 10. Accessed March 8, 2013. http://www.kapo.ee/cms-data/_text/138/124/files/kapo-aastaraamat-2011-eng.pdf.

Smith, David J. 2002. "Framing the National Question in Central and Eastern Europe: A Quadratic Nexus?" *The Global Review of Ethnopolitics* 2: 3–16.

Tilcsik, András. 2010. "From Ritual to Reality: Demography, Ideology, and Decoupling in a Post-Communist Government Agency." *Academy of Management Journal* 53: 1474–1498.

Vabariigi President Eesti Koostöö Kogu rahvuste ümarlaua kokkukutsumise puhul (26.05.2010) [The President of the Republic on the Occasion of Establishing the Roundtable of Nationalities within the Estonian Foundation for Cooperation]. Accessed March 8, 2013. http://www.president.ee/et/ametitegevus/koned/2557-vabariigi-president-eesti-koostoeoe-kogu-rahvuste-uemarlaua-kokkukutsumise-puhul/index.html.

Vetik, Raivo, and Jelena Helemäe, eds. 2011. *The Russian Second Generation in Tallinn and Kohtla-Järve: The TIES Study in Estonia*. Amsterdam: Amsterdam University Press.

Hans-Joachim Lauth

Rule of Law and Informal Institutions

Introduction

Over the course of the widespread third wave of democratization, a key observation has been made frequently: The introduction of elections alone does not guarantee a functioning democracy. The main indicators used in this determination are weak rule of law and a lack of checks and balances due to an unsatisfactory level of institutionalized, horizontal accountability (O'Donnell 1998; Schedler, Diamond and Plattner 1999). Additional criteria, which are present in modern, Western constitutional states, must be met. Unsurprisingly, O'Donnell (1999; 2004) explicitly urges that the existing foundations of democracy in the West and the implicit requirements of democracy should be identified and analyzed. He considers the concepts of *Rechtsstaat*, rule of law, and the constitutional state (or constitutionalism) to be essential to the analysis. For this reason, the fundamental constitutional order, rather than system of government, is addressed. This applies to the legal form or the judicial system, the basis for governance, in which state activity manifests itself. The analysis of systems of law is relevant not only to democracies, but also to authoritarian regimes, because it sheds light on their dynamics and stability. Investigating both kinds of regimes leads us to the following questions: Do legal systems always have characteristics of the rule *of* law, or in some regimes, do they possess only an instrumental nature ("rule *by* law")?

Passing references to missing or limited rule of law are not sufficient to gather differentiated empirical findings. In order to appropriately integrate the legal level into the analysis of young democracies, as well as authoritarian regimes, this article employs a dual perspective. First and foremost, it allows different characteristics of formal legal structures to be examined. Moreover, by employing the neo-institutionalist understanding of informal institutions (Lauth 2000; Merkel and Croissant 2000), strong emphasis can be placed not only on legally codified institutions, but, at the same time, existing informal structures can be taken into account. Both are relevant to the analysis and have to be considered, as not all judicial systems are regarded in the analysis of official legal systems. Informal law and systems of rules are not systematically included in the analysis of rule of law up to today. The connection between formal and informal legal systems has not been appropriately addressed in the current literature on the topic of democratization; even comparative research on the development of rule of law is rare (Fukuyama 2010). There are numerous case studies, however, which investigate special forms of informal systems (such as indigenous and religious law or corruption and violence in the research area of rule systems; Brinks 2012).

The analysis of formal and informal law structures allows the empirical findings to be differentiated and assessed. In using this approach, the following questions emerge and allow this article to be structured into corresponding chapters:

- What is the nature of law and legal systems? What distinguishes them from rules and informal judicial systems?
- What are the core principles of *Rechtsstaat* and of related concepts (rule of law and constitutional state, or constitutionalism)?
- In the empirical analysis, attention will subsequently be given to legal systems that exist outside of the outlined forms of constitutional order. The discussion focuses on the parallel existence of *Rechtsstaat* and informal legal systems and includes thoughts about a connection between *Rechtsstaat* and alternative systems of rules based on a factual and a normative level. In order to make these considerations conducive to empirical use, the findings about rule of law will be integrated into a proposal for the creation of a typological differentiation of legal systems.

Law and legal systems

Firstly, the term "law" has to be distinguished from the term "rights." Rights, in the sense of fundamental, human rights, draw their validity in the traditional jurisprudential thinking from their jusnaturalistic status. During the Age of the Enlightenment, this justification strategy became based on rationality. Thus, the jusnaturalistic argument was replaced by a rational law justification (Kant 1982). Accordingly, rights have validity, even if they are not preceded by laws, because they are generally tied to the dignity and freedom of the individual. Rights can become positive law. However, does the law gain its legal status only through the inclusion of rights?

A look at the evolution of law in different cultural backgrounds shows that rights are incorporated in varying degrees. First of all, this implies that the nature of law does not necessarily require the inclusion of rights into the official legal system. However, a common feature of all empirical legal systems is the authoritatively binding nature of rules. In this way, legal systems are binding systems of rules; law is, therefore, always set through government power and authority, whose effectiveness can be enforced through coercion if needed (Alexy 2011). This understanding of law finds its expression in laws that are understood in the tradition of legal philosopher John Austin as "the generalized commands of a sovereign" (Campbell 1993, 186).

The separation of law and rights or, in other words, of law and morality, is a hallmark of the traditions of the positivist theory of state law and the positivist philosophy of law (Kelsen 1960 [1934]; Hart 1961). However, this position has not gone unchallenged (Dworkin 1977); even in Hart's (1961) writings, the question arises whether his "formal" acceptance criteria for "valid law" already set legal standards that are difficult to reconcile with positive law in totalitarian regimes. Due to the comprehensive political subordination of law in these kinds of regimes, it seems less convincing that this idea of law – in terms of Hart's understanding – is tenable. Here, the discussion should shift to a perversion of the law. This interpretation would be emphasized if relevant considerations of legal theory were followed according to an understanding of law as a defense against political despotism.

In using the term legal or judicial system, however, we will refer back to a basic concept of the term, which includes any (positive) law written by the state, regardless of how fair it is; this definition, however, does not include perversions of the law. This decision is based simply on the empirical facts; regardless of a regime's status, a legal system is attributed to almost all states. Were the legal term tied to certain normative standards, this classification would have to be revoked from some states, and another term would have to be used. Indeed, it is preferable to consider the legal system, which is characterized by certain normative standards, as a *Rechtsstaat* or analogous as the rule of law[1].

In considering different constitutional traditions, it becomes obvious that binding the law to rights has largely prevailed in the wake of the emergence of rule of law in its different forms. In liberal and democratic societies, the law virtually draws its legitimacy from this fact. For this reason, it is appropriate to use the term "rule of law" or "*Rechtsstaat*" in order to make it clear that legal systems are not inherently bound to rights, although legal systems are barely viable without a minimal reference to them. Without this connection, the possibility of legal systems being legitimized is lower, and they must be increasingly based on coercion. It is assumed, however, that legal systems in other cultures also include rights. The understanding of these rights has to be examined, however, to determine to what extent these interpretations differ from those of constitutional traditions in the West.

1 It is possible, of course, to discuss the legal nature of specific autocratic regimes. Doing so, however, entails a cultural bias, which could tempt one to deny legal status to legal systems of other cultures if they do not meet the normative standards of a preferred legal sphere. Therefore, it is appropriate to employ a basic concept of this legal term.

The concept of rule of law (*Rechtsstaat*)

If the discussion of *Rechtsstaat* or rule of law from a continental European and Anglo-Saxon point of view is condensed, the following core idea can be recognized (Raz 1979; Lauth 2004; Becker and Zimmerling 2006; Waldron 2008; Schulze-Fielitz 2011; Enzmann 2012; Shapiro 2012). A *Rechtsstaat* or rule of law is based on a functioning state and the commonality of law, which prohibits any law specific to one single individual, as well as retroactive laws. The principle of the rule of law requires equality before the law and the general application of the law, regardless of the social status of the people involved (i.e., fairness). This includes state institutions specifically. The legal bond of the state refers to the conformity of the constitution and legislation. Actions by the government and the administration ("legality of administration") have to comply with the laws. In addition, state intervention is limited by the proportionality principle.

Thus, there is a connection between formal justiciable guarantees (legal procedures) and individual citizens, who can exercise their constitutionally-granted rights against the government (court protection). For this purpose, laws must be transparent, well-defined, and consistent. At the same time, the legal guarantees require the public to be aware of them. A certain stability in the laws is also necessary to gain a familiarity with them and allow rational calculations (legal certainty). An essential prerequisite for litigation is a qualified procedural law, in addition to many other features, including the existence of an independent and professional judiciary that is accessible to all citizens and has ultimate control over the actions of the executive. The various criteria culminate in a realization of legal thought, which includes a prohibition on state despotism and can be understood as a fundamental contribution to justice.

The separation of powers among the judiciary and the other branches of government is a central criterion for determining the validity of rule of law (Böckenförde 1976; Grimm 1994). In understanding the separation of powers, the precedence of democratic legislation in comparison to other forces is assumed. Neither the judiciary, nor the executive, nor the administration, can create its own law. Administrative decrees are subject to the law. By providing institutions, standards, and procedures, the rule of law demonstrates a most striking expression of horizontal accountability, which can be differentiated into various institutional forms (Lauth 2007). The legally-secured design of public space and the political sphere means protection not only from state arbitrariness, but also from social actors who either disregard laws or try to manipulate them unconstitutionally (e.g., by means of corruption). The quality of the rule of law is restricted to the extent where it fails to curb these actors. In this regard, the rule of law outlines rights and responsibilities for the state and citizens while limiting both of them.

All previously introduced characteristics of the rule of law apply to the formal rule of law. The assertion of fundamental rights, however, appears logically necessary to ensure that the institution of legal due process has meaning. Such an assertion is also imperative if the goal of limiting state action – an idea intrinsically linked to the legal process – is to be taken seriously. Relating this limitation solely with binding government actions on laws would ultimately mean accepting only a low threshold for future actions taken by the majority, because law-making could change accordingly.

For these reasons, it is quite plausible to conceive of fundamental rights as a material component that is – in addition to other formal procedural guarantees of the rule of law – a constituent component of rule of law (Zippelius 1991, 281); nevertheless, there is considerable room for interpretation when it comes to defining and applying abstract rights. With the essential aspects of the *Rechtsstaat* substantiated, the questions remain: What fundamental rights should be included in the understanding of rule of law, and how should they be interpreted?

The outlined understanding of a *Rechtsstaat* should not simply be separated from the thoughts about a constitutional state, because the latter has various meanings. There are positions which consider the constitutional state to be nearly identical to the substantive concept of a *Rechtsstaat*, while others understand it in the context of a positivist perspective – a position which is closely connected with the principle of "absolute" parliamentary sovereignty in Great Britain. However, key features of the *Rechtsstaat* (fundamental rights and procedural rights) are also often found in the constitutional state (constitutionalism). With this in mind, a constitutional state is not necessarily identical to a state that, while possessing a written constitution, does not demonstrate the required normative conditions. Nevertheless, compared to a *Rechtsstaat*, a constitutional state could have other additional features, particularly with regard to the political order, which given its institutional rigidity expressed by oversized majorities, is difficult to change. In this sense, Article 20 of the German *Grundgesetz* (constitution – *Grundgesetz für die Bundesrepublik Deutschland* 1949) crosses into the realm of the constitutional state, because it not only establishes the substantive aspects of the *Rechtsstaat*, but also the federal political order.

Table 1: Principles of the (formal) Rechtsstaat

(1)	The universality of the law (framing laws while being unaware of the specific cases in which they will be applied, not *ad personam*).
(2)	The knowledge of the law among those concerned.
(3)	The prohibition of retroactive laws.
(4)	The clear and comprehensible formulation of laws.
(5)	The absence of contradictory laws (in and of themselves, with regard to other laws, and with regard to the constitutional norms).
(6)	The absence of behavioral requirements which are impossible to meet (unfair laws).
(7)	Relative stability of the laws (changes not made too often – legal certainty).
(8)	The prohibition of excesses (proportionality of ends and means).
(9)	Equality before the law, general application of the law, i.e., applied independently of the social status of those concerned (fairness imperative, impartiality of the law).
(10)	The application of the law to the state and all its institutions (legal liability of the government, all are subject to the law, an explanation of the areas of legal basis for action, primacy of the law, caveats).
(11)	Independence and effective controlling ability of the courts (effective legal protection from the state, protection of the courts).
(12)	Adequate procedure and due process of law (no sentencing or imprisonment without a trial, time limits for processes, accessibility for all, legal counsel, professional judges, penalties that fit the crime, the chance to appeal, fairness, transparency and public nature of the process, equal treatment of equal cases).
(13)	Right to payment for damages to the extent applicable; government liability.
(14)	Realization of the principle of justice (relinquishing of arbitrariness and contributing to justice).

Informal institutions

If we understand law in the sense of norms and binding systems of rules, we are referring to institutions. This understanding of institutions, which is common in classical institutional theory, will be explicitly focused on in the neo-institutionalist debate. There will be frequent references to the definition by North (1990, 3), who regards an "institution as a norm or set of norms that have a significant impact on the behavior of individuals" (concerned by or included in the institution). Thus, institutions constrain the actions of individuals. Although North did not emphasize the role of sanctions, in the neo-institutional debate, one can find different interpretations of constraints that are linked with them.

General agreement exists that institutions restrict individual behavior to some extent (Peters 1999). The extent and the mechanisms through which this occurs vary. Some authors (March and Olsen 1989) highlight the internalization of norms during processes of primary or secondary socialization (family, kinship – school, military, companies, etc.). In this case, those who do not follow the rules have a guilty conscience, and deviations from the rules are sanctioned by an internal mechanism. External sanctioning mechanisms also exist (social discrimination or exclusion, loss of status, arrest, etc.). Rational choice perspectives include the latter, as rational choice approaches imply the possibility of suffering from disadvantages when rules are not followed. In this case, actors violating the institutions will not benefit from incentives linked to the institution.

In all types of enforcement mechanisms, defecting from the rules set by informal institutions implies losses for rule-breaking individuals. To avoid a catch-all category, which includes all sorts of inconveniences (caused by a particular sanctioning mechanism), it seems appropriate to consider institutions to be institutions only when they maintain (their own) external sanction mechanisms (which can be introduced by third parties). This obviously applies to formal institutions, such as legal systems[2].

Even if sanctions are a defining feature of institutions, they are not the only reason why actors comply with institutions. Actors conform to institutions, because they regard them as given or "natural". Actors also respect institutions, because they display a legal character or because they regard them as legitimate. In accordance with North, these reflections on sanctions and the reasons why actors follow rules relate to the main purpose of institutions: "Within an institutional perspective, a core assumption is that institutions create elements of order and predictability" (March and Olsen 2006, 4).

To summarize these reflections, institutions are defined as follows: Institutions constitute a set of rules, which implies rights and responsibilities. A set of rules also creates and shapes a social order in such a way that the behavior of all actors involved in that social order is predictable. Institutions affect an individual's performance by his/her voluntarily following the rules or being motivated by the threat of sanctions.

By definition, systems of law can be compulsorily enforced by state actors. For this reason, they will be labeled as formal institutions. This term already indicates the existence of informal institutions. Indeed, a diversified and widespread set of informal rules exists, which partially has a considerable influence on the workings of rule of law.

---

2 This does not mean that internal sanctions have to be absent. They can also exist in the case of formal institutions (not obeying the rule of law can create such internal mechanisms). The meaning here is simple: Internal sanctions create no defining characteristic of an informal institution.

To differentiate between formal and informal institutions, the following serves as a useful starting point: *Informal institutions* are institutions that are *not formally codified* in official documents (in constitutions or laws). *Formal institutions* are officially codified in written documents. Thus, not only are regulations that have the status of constitutional clauses and laws included, private contracts and norms that have legal consequences also fit in this category. According to this line of thought, all private contracts or rules of association that are protected by the state are formal institutions.

Formal institutions are guaranteed by state agencies, and deviations from these institutions are sanctioned by the state. In contrast, the existence of informal institutions is the result of the emergence of social or political practices and the effectiveness of these practices. Informal institutions are known and recognized publicly; however, they are often not codified. Informal institutions also have sanctions in place. These sanctions either include mechanisms of social exclusion or mechanisms that restrict access to much needed goods and services. Under the special conditions of communist rule, we can speak of "tertiary social control", as Podgórecki and Łoś illustrate:

> If behind the given legal system (which is rejected by the population at large as unjust, undemocratic, etc.) there operates a complicated infrastructure of mutually interdependent interests, then this legal system may become accepted, not on the basis of its own merits, but because it creates a convenient cover-system for the flourishing phenomenon of "dirty togetherness" (Podgórecki and Łoś 1979, 203).

The authority of informal institutions stems from various sources. Firstly, informal institutions are socially accepted, which provides them with a basic degree of legitimacy. The fact that these informal institutions are socially acceptable also serves as a major source of motivation for actors when they follow these patterns of social conduct prescribed by these informal institutions. Actors pursue different purposes when they enter these patterns of conduct; purposes can be defined either narrowly or broadly. These purposes can be linked to results, as well as to certain patterns of behavior. Institutions facilitate interaction between individuals and groups. They foster stability by creating known and accepted behavioral structures that cannot be changed by individual people. Even if actors disagree with these structures, they obey them because, in accordance with rational calculation, the costs involved in rejecting them can only be offset when behavioral alternatives are available. These considerations correspond with the definition proposed by Helmke and Levitsky (2004, 727): "We define informal institutions as *socially shared rules, usually unwritten, that are created, communicated, and enforced outside of officially sanctioned channels*."

Informal systems of law and informal systems of rules

If law systems are conceived of as a set of formal institutions, it becomes necessary to clarify in which sense we speak of informal law systems or informal legal systems. Does it make sense to speak of "alternative" or "informal" systems of law inside a country if the status of legal systems is linked to formal institutions? If we take up this line of argumentation, we would always have to speak not only of competing judicial systems, but also of competing states inside a national territory. This is quite conceivable if we use the monopoly on force as the main indicator to identify a state. Thus, areas occupied by armed units (e.g., as is the case in guerrilla warfare) could be regarded as states and, likewise, the sphere of influence of organizations which are able to enforce their rules and assert their authority through violence (e.g., mafia)[3].

However, the sharp contrast in the area of stateness can be mitigated if we take a nuanced look at the functionality and the motives behind legal compliance. Thus, it is possible that legally analogous systems of rules or informal judicial systems – the systems of rules which make legal claims without being codified by the state – exist and are only limited by coercion. Compliance with these legal systems can occur "voluntarily" if they are based on accepted social traditions or arrangements[4]. In this case, the systems establish themselves on the basis of internal, not external, ties. In the case that "voluntary" acceptance fails, however, social sanctions can be put in place. The strength of "living law" can be observed not only in traditional or transformational countries, but also in Western countries which have experienced massive migration, for instance, the legal behavior of Kurds in London (Tas 2014).

From a functional viewpoint, informal legal systems can be distinguished from each other. Legal spheres, such family law, property law, or criminal law, which can work according to their own rules, are worthy of consideration. These kinds of rules can be based on traditions of indigenous systems of rules (common law, clan law, and tribal law)[5] or come from larger legal systems, as mentioned, for

3 Almond (1960) uses the term "political system" in the same way.

4 That does not mean that formal systems of law are based only on the threat of violence. Also, in constitutional legal systems, there is a high level of voluntary observance of the law.

5 Custom law does include all non-codified rules and behavioral patterns that prove to be enforceable in state or private tribunals. An important area has developed in the field of economic relations – especially at the international level – where private bodies conduct conflict resolution according to the self-created right of the economy. Customary law also includes folk traditions, which cannot be completely (but only partially, as in they are actually permitted) brought before state tribunals.

instance, in the legal thought of certain legal traditions that are regionally bound (Bryde and Luchterhandt 1997). Examples include Islamic law, Hindu law, and Far Eastern law, among others. We can speak of informal law systems or legally analogous systems of rules if, and only if, they develop social effectiveness, and the typical characteristics of the law can be observed. This includes a recognizable, coherent system of legal norms, which are fixed and known.

In addition, these norms are associated with a procedural law, in which the state jurisdiction is regulated by the analogous enforceability of the law. Furthermore, there must be authoritative bodies that adjudicate the application of the law and whose decisions are normally followed. An understanding of norm systems such as law can be justified by comparing it to the analogous functionality of international law vis-à-vis state law. This kind of law does not necessarily have to be established and authorized (and enforced) by the state. Even international law cannot be enforced by any one state and is still referred to as law. A key feature of the validity of international law corresponds to custom law or, more generally, informal law[6]. In addition to the element of behavioral development, the subjective belief in a legally binding relationship for the parties involved – the group of people that fit into the law system – must be present.

Informal constitutional rules form a specific variant of informal law. Schulze-Fielitz (1984, 20) defines it in the following manner: "informal constitutional rules constitute the totality of those unwritten rules for the behavior of leading officials in the highest governing bodies of the state, but also for the political parties and publically significant social groups, whose compliance is considered an essential prerequisite for long-term orderly constitutional life, according to the prevailing beliefs." Examples of these kinds of informal constitutional rules are proportional representation rules (gender, region, factions), which can be observed in the composition of committees, as well as coalition agreements, that establish the principles of intergovernmental cooperation. Significantly, these rules cannot be brought before and examined by general jurisdiction courts. They create customs, however, that cannot be changed without protests or political sanctions.

This definition sends a clear message: Informal constitutional rules are compatible with the existing constitutional norms and are closely associated with them. They are considered to be a necessary condition for an orderly constitutional life,

6 In international law, customary law denotes a continuous practice of behavioral patterns, and the states believe that they have a legal obligation to this behavior. Accordingly, the custom-law standard is composed of an objective (the practice) and a subjective (the recognition) element (Raustiala and Slaughter 2002); without the latter, the behavior is still considered a custom.

because they help formal rules function or even enable their use. In this sense, they form part of the constitutional culture, which refers to the totality of individual attitudes towards the constitution and the law, which, in turn, is correlated to corresponding action by individuals and corporate actors.

Furthermore, there are *informal rule systems* that can be distinguished from informal legal systems. These informal rule systems can also be effective institutions and are connected to further sanctions[7]. Examples of such informal institutions are corruption and clientelism or specific forms of networks. These systems, however, are not legal in nature and exhibit little or no analogous legal functions (Lauth 2000; Nuijten and Anders 2007). Unlike formal systems of rules in the official legal system, these informal institutions are not classified into a clearly defined category of informal legal systems. An essential part of a legal system (i.e., court proceedings) is missing. Even so, relationships exist. For example, the rule system of "clientelism" and legal system of "tribal law" are closely connected with one another. In addition to "clientelism," corruption – in its different forms – can be associated with clan structure. As shown by the remarks by Waldmann (2001), the boundaries between an informal system of rules and an informal legal system can be fluid. Likewise, together they can create complex informal patterns, which are condensed nearly into a second "proper" informal constitution (Ledeneva 2001, 9f) or "delegative code" (O'Donnell 1994, 55ff)[8]. In this case, they have a significant impact on the workings of the official legal system (Meyer 2006).

Relationship between rule of law and informal institutions: Hybrid legal systems and deficient rule of law

Rule of law has been found to be lacking not only in authoritarian regimes, but also in many young democracies (Zakaria 1997; O'Donnell 1999). Significant deficiencies have been observed, such as incoherent, non-transparent judicial systems that are not accessible to all citizens, inadequate respect for laws – also by the state actors, who act without a sufficient legal basis for action, little presence of court protection, and unfair litigation practices and procedural law. Although

7 If there are no sanctions in these systems, they are considered to be practices or conventions.

8 The scholarly concept of the neo-patrimonial state (Erdmann and Engel 2007) is relevant to this issue, although here, there tends to be only a slight predominance of informal rules. The formal institutions are not only manipulated, but also have intrinsic value.

more criteria could be added to the list, the more interesting question is: What factors lead to these kinds of findings? Do their causes lie in an incompetent application of the law, and would better instruction of judicial representatives lead to stricter, better enforcement of the rule of law?

There is reason to suspect that the state of affairs is often organized in a more complicated fashion. The weakness of the rule of law is not only limited to the fact that it was not fully implemented or not used to control lawless areas. The problem is also that, in the same country, preexisting, informal systems of laws or rules exist and compete and come into conflict with the rule of law. This can occur on a functional and/or a territorial level.

It has already been pointed out that, in different states, some areas of law (family law, property law, or criminal law) can work according to rules that are based on traditions of indigenous systems (clan and tribal law) or originate from larger legal systems (Islamic law, Hindu Law and Far Eastern law)[9]. In these kinds of cases, the official law could be neglected or merely considered a potential competing alternative. Nevertheless, it makes sense to assume that competing legal systems produce a problem if, and only if, the informal legal status exists and the different legal systems are not compatible with constitutional principles. While the first criterion can be verified via its empirical characteristics and its impact, clarifying the compatibility proves to be more difficult, because it requires an additional comparison to the formal and material principles of the rule of law. Besides clear contradictions, the findings can reveal partial inconsistencies, which do not correspond to complete incompatibility. It is also possible to identify functional equivalents, as seen with customary law in an Anglo-Saxon context. Moreover, private law, which develops through international trade relations, can be compatible with principles of the rule of law; the same is true for the kind of law that is created by contractual arrangements in self-help organizations (Eckert 2004). These examples refer to other legal sources, whose constitutional codification is fundamentally possible. In the above examples, informal law forms a functional equivalent to constitutional practices and can, theoretically, be transformed into formal law. This, of course, also applies to the aforementioned informal rules of the constitution.

Although it would make sense at this point, having a normative discussion about the character of different sources of law is not appropriate, on account of the complexity of the subject matter. Contradictory findings exist in the current state of research within this comprehensive and complex field. Such a discussion

9 Compare Glenn (2008) to comparative law families and comparative legal traditions.

should also focus on the empirical findings in individual cases, because informal legal traditions often exist in specific forms and thus, are not totally conducive to a deductive approach (Zips and Weilenmann 2011).

However, at first glance, present findings (KAF 2006) give evidence that human rights and civil liberties are not protected analogously to the rule of law in all informal legal systems. Doubts about the compatibility with the rule of law grow significantly when other forcibly established legal systems, ones not based on established legal traditions, are considered. Informal legal institutions, in which regulation based on private power occurs, are addressed. Local political bosses (*caciques*) – be they in the country or in urban slums, or gentry or warlords on the regional level – who regulate, monitor, and enforce their own rules, should be considered. Such systems also include mafia organizations in their different forms, or guerrilla organizations, which govern their conquered territory. These examples allow the territorial component to be addressed. Many of these phenomena can be bundled into the concept of "brown areas", which focuses on areas in which state control is mostly absent (O'Donnell 1993, 1359f.). The explosive nature of informal legal systems becomes apparent if – as illustrated – competing incompatible informal systems of laws and rules exist. The following remarks are based on this issue.

If the competing and contradictory systems of laws or rules, which are not compatible with rule of law traditions, are dominant, the rule of law is, in fact, undermined or nullified. This rare constellation, however, will not be closely examined in what follows. A closer look will be taken at cases where the interference due to incompatible systems of laws or rules is weaker even though serious consequences are possible. Two possibilities can be distinguished: (a) equilibrium, in which the rule of law and competing legal and rule systems are in relative balance and (b) domination, where the rule of law dominates the competing law and rule systems, although it cannot completely eliminate them. The first case can be seen as a *hybrid legal system*; the second as a *deficient rule of law* (Lauth and Sehring 2009)[10].

Competing legal systems can exist and persist in different ways. (1) On the one hand, they can exist in the shadow of formal law. Classic examples are indigenous traditions, which have survived the introduction of modern legal systems. These traditional systems persist due to their social acceptance. They are partially

10 A hybrid legal system is not totally the same as "legal hybridity" as proposed by Myint (2014). In that proposal, the rule by law – as a formal construction which is *individually* and informally manipulated by the rulers – is combined with elements of rule of law (especially, an almost completely independent judiciary).

compatible with rule of law. Serious tensions arise, however, when we are confronted with legal systems that are enforced by social actors. Several examples, which will be addressed later, come to mind: oligarchies, which use private "security forces" to protect their privileges; militant groups, which reserve the right to make illegal interventions when they see their interests – i.e., their understanding of the law – threatened; mafia cartels, which act analogously; and guerrilla organizations, which enforce their own rules in the territory they control.

(2) On the other hand, formal and informal legal systems can be interwoven in various forms. The possibility exists that competing areas of the law can be completely or partially adopted in the official legal system and be authorized to regulate certain functional areas (e.g., as in the Bolivian Constitution, which regulates the areas of family and criminal law). In this way, the competing legal system loses its informal status, but still remains in conflict with the principles of rule of law. The disparity is simply codified and incorporated into the legal system. From the perspective of the rule of law, the presence of these kinds of solutions does not seem very likely; even so, there is sufficient empirical evidence that substantiates their existence under certain conditions (Benda-Beckmann 2002; Beyer 2006). For example, the rules of land distribution and usage in several African countries are guided by tribal law, which is difficult to reconcile with the guarantee to property which is provided by the existing constitutions. Another example concerns the inclusion of religious traditions in family law, which curtails civil rights and liberties, often those of women[11]. With respect to the rule of law, such a practice is unacceptable. Another pertinent example would be the incorporation of Sharia in criminal law, which does not comply with all of the principles of the rule of law (Possamai, Richardson and Turner 2013). These kinds of legal adoptions can be in force at the national level (as in Egypt) or apply only to certain states in a country (as in Nigeria).

While in the first case, two legal systems – the rule of law and an incompatible informal one – separately oppose each other, in the second case, they are intertwined, and the competing legal system has an official character. The latter case has the advantage that the state's monopoly on force remains intact; the disadvantage is that the resulting system of law is incoherent. The situation is complicated when multiple systems of law compete with the rule of law and, in doing so, overlap with each other. This variant can be empirically confirmed simply by looking at African legal systems (Ruppel and Winter 2011). In parts of

11 Such a constellation could be observed in India regarding personal law (Rudolph and Rudolph 2001).

Africa, remnants of colonial law, indigenous tribal law, and religious courts exist parallel to the official, constitutionally-based law system.

With these thoughts in mind we have to decide how to classify the findings. When, for example, is a legal system considered a hybrid one, and when does a deficient rule of law exist? According to the above definition, the classification depends on the strength of the competing legal system. Despite their unwieldy character, incorporated legal practices hardly ever challenge the dominance of the rule of law unless important functional areas of law, like the criminal justice system, are substantially affected; thus, a hybrid legal system is denoted by the separation of the two systems. However, a deficient rule of law can be present in both cases if the precedence of the rule of law persists. Although a hybrid legal system is no longer considered a *Rechtsstaat*, it still possesses features that are common to rule of law. The remaining constitutional element, however, is only one part of the total legal system. In order to be identified as a hybrid legal system, empirical evidence of a competing legal system, which is largely incompatible with the rule of law, has to be provided. The rival system has to be stronger than the deficiencies observed in "brown areas". This constellation is true in only some cases. Despite the existence of one or more competing legal systems, and provided that the precedence of the rule of law is not called into question, a deficient rule of law is far more likely.

When the rule of law is deficient, it is assumed that the legal system in question is predominantly a *Rechtsstaat* and that the state's ability to function is impaired only in smaller areas, which, in turn, significantly exceed the level of deviations in functioning constitutional states. In the case of the existence of separate and competing legal systems, a deficiency in the rule of law is present if these deviations generate only a low level of activity ("enclave right"), or the validity of constitutional decision making is only slightly affected.

The rule of law can also deteriorate when confronted with *systems of rules* that alter constitutional logic. This change can happen through the persistence of clientelistic structures and/or corruption. Clientelistic structures, in their various forms, can infringe on equality before the law in different ways; in general, corruption undermines the law[12]. These types of deficiencies can also be found in functioning constitutional states. However, they appear there rather sporadically and do not indicate any established patterns of action. In states with a deficient

12 In the empirical research, these kinds of systems of rules can appear in different combinations and create specific patterns. They acquire special effectiveness in connection with informal legal systems.

rule of law, the systems of rules, however, have acquired an institutional status, which leads to a permanent connection with the formal legal system (as "parasitic" institutions).

It is exactly this argument of the institutional status of deficiencies that separates a functioning *Rechtsstaat* from a deficient one. A far-reaching impact is associated with the respective functional logic of both types. For example, the elimination of defects in a functioning *Rechtsstaat* is easier – as it just involves the correction of individual acts – than in the case of deficit rule of law, which requires institutional changes.

Thus, the existence of competing rule systems is a sufficient condition for a deficient rule of law. It is evident that a massive accumulation of these kinds of informal interventions could ultimately undermine the rule of law. This particular case would no longer be considered deficient rule of law, but a state of lawlessness. Unlike hybrid legal systems, a deficient rule of law does not necessarily exist in tandem with competing and incompatible legal systems.

In addition to the threat to the rule of law by competing rule systems and informal parasitic institutions, a further risk arises vis-à-vis poor handling of the rule of law itself. This potential risk stems from limited sensitivity to problems on the basis of a traditional perception filter and/or the inadequate suitability and competence of institutional actors (Garzón Valdés 1999). The social "blindness" that is found in the realm of legal protection is also worth mentioning. Socially marginalized groups usually lack sufficient access to the legal system. Because they lack access, these groups cannot appropriately make use of the law and, in administration of justice, frequently find themselves disadvantaged in comparison to socially privileged actors (O'Donnell 1999). In this regard, the judicial system is often in a precarious condition, which is characterized by its poorly-staffed personnel facilities, drawn-out legal proceedings, and poor prison conditions. Moreover, there is also a lack of transparency in the body of law itself, which arises from excessive, unchecked legislation. This legislation is fraught with inconsistencies, which makes it difficult even for "judicial staff" to work with it[13]. This discrepancy, however, is not identical with the addressed incompatibility of different legal systems, as it is, in principle, immanently revocable.

13 These discrepancies or tensions can occur within the same area of law or among different areas of law (e.g., fundamental rights, criminal law, and civil law). This non-transparency can be fostered by the lack of coherent bodies of law. In this case – which is not unlikely – legal knowledge becomes exclusive expert knowledge, which can gravely undermine the exercise of rights.

As a result of our considerations we can adhere the following. A lack of rule of law leads to the differentiation between two legal systems: a *hybrid legal system* and a *deficient rule of law* or *Rechtsstaat*. While a hybrid legal system is characterized by the existence of competing and largely incompatible legal systems – and can no longer be understood as rule of law – in a deficient rule of law, central principles of the rule of law remain intact, despite significant defects. These principles can be endangered in three ways: by competing legal systems, by incompatible informal systems of rules, and by the actors inside the legal arena. The deficient character of the rule of law is largely reflected in the institutional status of the threats, which, in turn, are not allowed to exceed a certain amount. Once this point is reached, a deficient rule of law becomes a hybrid legal system.

The relevance of regime types

The relationship between rule of law and informal legal systems is similar as between the latter and democracy because of the inherent connection between democracy and rule of law. The relationship between rule of law and autocracies is more complex. Because each application of rule of law limits political power and incorporates this limitation constitutionally, there are very few examples (such as Singapore) of authoritarian regimes with respect for rule of law. Typically, authoritarian regimes use the formal judicial system for their interests or limit it via informal systems of law[14]. Therefore, the concepts of a "deficient rule of law" and a "hybrid legal system" are compatible with autocracies. However, the formal legal systems in these kinds of regimes could be undermined in two ways: by informal systems of laws, as well as by a system of rules. Either these informal legal systems exhibit legally analogous characteristics to rule of law (e.g., as with customary law) and contradict formal authoritarian law (as could be observed in the former USSR), or they arise out of conflict between two informal systems, like clan structures vs. Sharia. The latter phenomenon can be observed in some North African countries (Kemper and Reinkowski 2005).

Scholars examining the relationship between formal and informal institutions should note that these relationships differ in regards to the regime type. In democracies, informal institutions such as clientelism and corruption constitute the conflictive type. In autocracies, they can be classified as complementary, as the study of neopatrimonial and sultanistic regimes demonstrates. In totalitarian regimes, however, clientelism and corruption oppose formal institutions, because

14 One example of this kind of unfair behavior is the establishment of neo-patrimonial rule with appropriate "rent-seeking" mechanisms.

they challenge the control of the ruling class over the political system. The same reasoning applies once we create subtypes of clientelism and corruption, and we relate them to different subtypes of democracy and autocracy. It is important to note that the relationship between formal and informal institutions is not always the same, but depends on the regime type in place. The same informal institution can lead to very different outcomes in different government systems. These considerations also shed light on the change of formal institutions in transformation processes. These kinds of informal institutions, which were practiced and received a high acceptance in authoritarian times, also shape the emergence and application of the new formal institutions to a significant degree. Grzymala-Busse (2010) demonstrates this idea by emphasizing the importance of democratic competition in the change of informal institutions.

To better understand the stability and dynamics of authoritarian regimes, it is appropriate to more closely analyze their legal condition. In conducting empirical research, there are two integral perspectives on the rule of law, which can be underscored by two questions: (1) To what extent have the characteristics of the rule of law developed? (2) To what extent do competing systems of laws and rules, which affect the functioning of the substantive rule of law – in a positive or negative way – exist?

If we look at the empirical findings (such as the assessment of civil liberties in Freedom House surveys) with regard to the rule of law, not only in autocracies, but also in many democracies, it seems that various forms of deficiencies are the Achilles' heel of the regimes in which such deficiencies exist. The question of how to fix this shortcoming certainly has the same relevance as the question concerning possibilities to improve economic development. For this reason, it would be appropriate to discuss what perspectives appear to be viable for further research.

Conclusion and reform perspectives

In the discussion of law and the rule of law, the concept of a formal and a substantial rule of law was introduced and specifically defined. Because rule of law has not been fully implemented in autocracies or many young democracies yet, its relation with competing informal legal systems and systems of rules was more closely examined. Two concepts – a *deficient rule of law* and a *hybrid legal system* – were introduced and explained in the typological discussion, which should improve the classification and analysis of the empirical findings. While a functioning rule of law is compatible with democracy and is considered to be a central basis for it, this applies only limitedly to a deficient rule of law and cannot be said of a hybrid legal system. In fact, the latter undermines central elements of democratic rule.

The limited results of promoting rule of law through external actors are known through empirical studies (Carothers 2006). What reform strategies then have potential? Reform efforts that seek to strengthen the rule of law should be related to the causes of defects in the rule of law. Thus a priority is preventing disturbances in the rule of law, which are based on private violence (mafia, etc.). Likewise, problematic systems of rules (e.g., corruption) have to be confronted head-on. The situation becomes more ambiguous if competing legal systems are based on autochthonous and socially-entrenched traditions. Because complete incompatibility with traditions of the rule of law cannot be assumed in these cases, a gradual integration of both legal systems – not an outright merger – should be considered as a solution strategy. In the process of association, certain elements would be emphasized, elements which could be successfully connected with the rule of law (Kaneko 2008; van Rooij 2009). The advantage of this kind of gradual integration is that it does extend access to judicial systems and does not entail damaging the sense of justice of the parties involved, and, if the system is transformed over long enough periods of time, the changes remain manageable in practice, as well as in the minds of the people who are affected. In this way, both the cognitive and the affective dimension of human actions can be respected. Nevertheless, in this development, social conditions also have to be created, conditions which ensure the effectiveness of the rule of law procedures for those who, up to that point, have had only an uncertain guarantee of the rule of law.

This goal can be realized through two concrete strategies: (1) Competing legal systems are allowed in specific, clearly defined functional areas, while, at the same time, the official legal channels are still available. In this way, conflicts can be dealt with and solved on a voluntary basis and through traditional means[15]. (2) Competing legal systems are connected with each other, but are functionally or territorially integrated into the existing official legal system. Here, it should be noted that the disparity between the two cannot be too great; in addition, an approximation of laws and a streamlining of legislation should follow in due

15 These kinds of possibilities can be considered as analogous to the rule of law if the state has expressly authorized these entities and delegated competencies to them (compare to courts of arbitration) or accepts them (e.g., mediation procedures), because it considers them to be functional equivalents. This arrangement has the advantage of having functioning legal structures in place. Due to insufficient development, the official channels of the rule of law may not always be accessible in the same way that these structures are.

time[16]. This strategy is not possible, however, if fundamental principles or rights blatantly contradict each other. These comments highlight only initial thoughts vis-à-vis overcoming verified deficiencies in the rule of law, whose strengthening requires greater research efforts in the field of comparative research on the rule of law.

References

Alexy, Robert. 2011. *Begriff und Geltung des Rechts* (3rd edition). Freiburg and München: Alber Studienausgabe.

Almond, Gabriel A. 1960. "Introduction: A Functional Approach to Comparative Politics." In *The Politics of Developing Areas*, edited by James S. Coleman, 3–64. Princeton (NJ): Princeton University Press.

Becker, Michael, and Ruth Zimmerling, eds. 2006. *Politik und Recht*. PVS-Sonderband 36. Wiesbaden: VS Verlag für Sozialwissenschaften.

Benda-Beckmann, Franz von. 2002. "Who's Afraid of Legal Pluralism?" *Journal of Legal Pluralism* 47: 37–82.

Beyer, Judith. 2006. "Revitalisation, Invention and Continued Existence of the Kyrgyz Aksakal Courts: Listening to Pluralistic Accounts of History." *Journal of Legal Pluralism* 53–54: 141–176.

Böckenförde, Ernst-Wolfgang. 1976. *Staat, Gesellschaft, Freiheit: Studien zur Staatstheorie und zum Verfassungsrecht*. Frankfurt: Suhrkamp.

Brinks, Daniel M. 2012. *The Judicial Response to Police Killings in Latin America: Inequality and the Rule of Law*. New York: Cambridge University Press.

Bryde, Brun O., and Otto Luchterhandt. 1997. "Rechtssysteme." In *Lexikon der Politik*, vol. 4, *Die östlichen und südlichen Länder*, edited by Dieter Nohlen, 450–466. München: Beck.

Campbell, Tom D. 1993. "Legal Studies." In *A Companion to Contemporary Political Philosophy*, edited by Robert E. Goodin and Philip Pettit, 183–211. Oxford: Blackwell Publishers.

Carothers, Thomas, ed., 2006. *Promoting the Rule of Law Abroad: In Search of Knowledge*. Washington: Carnegie Endowment for International Peace.

Dworkin, Ronald. 1977. *Taking Rights Seriously*. London: Duckworth.

16 It is possible, however, that the developments do not follow legal universalism, but legal pluralism (Rudolph and Rudolph 2001, 55f).

Eckert, Julia. 2004. "Urban Governance and Emergent Forms of Legal Pluralism in Mumbai." *Journal of Legal Pluralism* 50: 29–60.

Enzmann, Birgit. 2012. *Der Demokratische Verfassungsstaat: Entstehung, Elemente, Herausforderungen*. Wiesbaden: VS Verlag für Sozialwissenschaften.

Erdmann, Gero, and Ulf Engel. 2007. "Neopatrimonialism Reconsidered: Critical Review and Elaboration of an Elusive Concept." *Journal of Commonwealth and Comparative Studies* 45: 95–119.

Fukuyama, Francis. 2010. "Transitions to the Rule of Law." *Journal of Democracy* 21: 33–44.

Garzón Valdés, Ernesto. 1999. "Rechtsphilosophische Überlegungen über Verfassungsreformen in Lateinamerika." In *Rechtsreformen und Demokratieentwicklung in Lateinamerika*, edited by Helen Ahrens and Detlef Nolte, 110–132. Frankfurt am Main: Vervuert.

Grimm, Dieter. 1994. "Entstehungs- und Wirkungsbedingungen des modernen Konstitutionalismus." In *Die Zukunft der Verfassung*, edited by Dieter Grimm (2nd edition), 31–66. Frankfurt: Suhrkamp.

Hart, Herbert L.A. 1961. *The Concept of Law*, Oxford, Oxford University Press.

Glenn, Patrick. 2008. "Comparative Legal Families and Comparative Legal Traditions." In *The Oxford Handbook of Comparative Law*, edited by Mathias Reimann and Reinhard Zimmermann, 422–440. Oxford: Oxford University Press.

Grundgesetz für die Bundesrepublik Deutschland. 1949. Accessed January 28, 2015. http://www.gesetze-im-internet.de/bundesrecht/gg/gesamt.pdf.

Grzymala-Busse, Anna. 2010. "The Best Laid Plans: The Impact of Informal Rules on Formal Institutions in Transitional Regimes." *Studies in Comparative International Development* 45: 311–333.

Helmke, Gretchen, and Steven Levitsky. 2004. "Informal Institutions and Comparative Politics: A Research Agenda." *Perspectives on Politics*: 725–740.

KAF (Konrad Adenauer Foundation), ed. 2006. *Rule of Law. The KAF Democracy Report 2006*. Berlin: Konrad Adenauer Foundation.

Kaneko, Yuka. 2008. "Catalistic Role of Legal Assistance between Formal Law and Social Norms: Hints from Japanese Assistance." *Journal of International Cooperation Studies* 15: 49–86.

Kant, Immanuel. 1982. "Die Metaphysik der Sitten" [1797] In *Werkausgabe*, edited by Wilhelm Weischedel (5th edition), Bd. VIII, 309–634. Frankfurt (aM): Suhrkamp.

Kelsen, Hans. 1960 [1934]. *Reine Rechtslehre* (extended 2nd edition). Wien: Deuticke.

Kemper, Michael, and Maurus Reinkowski. 2005. *Rechtspluralismus in der Islamischen Welt. Gewohnheitsrecht zwischen Staat und Gesellschaft.* Studien zur Geschichte und Kultur des Islamischen Orients. Berlin: De Gruyter.

Lauth, Hans-Joachim. 2000. "Informal Institutions and Democracy." *Democratization* 7: 21–50.

Lauth, Hans-Joachim. 2004. *Demokratie und Demokratiemessung.* Wiesbaden: VS-Verlag.

Lauth, Hans-Joachim. 2007. "»Horizontal accountability«: Aktuelle Aspekte der Gewaltenteilung: Ein Vorschlag zur Systematisierung der Kontrollfunktion der Gewaltenteilung." In *Gewaltenteilung und Demokratie*, edited by Sabine Kropp and Hans-Joachim Lauth, 45–71. Baden-Baden: Nomos-Verlag.

Lauth, Hans-Joachim, and Jenniver Sehring. 2009. "Putting Deficient *Rechtsstaat* on the Research Agenda: Reflections on Diminished Subtypes." *Comparative Sociology* 8: 165–201.

Ledeneva, Alena. 2001. *Unwritten Rules: How Russia Really Works.* London: CER Publications.

March, James G., and Johan P. Olsen. 1989. *Rediscovering Institutions. The Organizational Basis of Politics.* New York: Free Press.

March, James G., and Johan P. Olsen. 2006. "Elaborating the «New Institutionalism»." In *The Oxford Handbook of Political Institutions*, edited by Sarah A. Binder, R. A. W. Rhodes and Bert A. Rockman, 3–20. Oxford: Oxford University Press.

Meyer, Gerd, ed. 2006: *Formal Institutions and Informal Politics in Central and Eastern Europe: Hungary, Poland, Russia and Ukraine.* Opladen: Barbara Budrich Publishers.

Merkel, Wolfgang, and Aurel Croissant. 2000. "Formal Institutions and Informal Rules of Defective Democracies." *Central European Political Science Review* 1: 31–47.

Myint, Aung Phone. 2014. "Legal Hybridity: Rule of Law under Authoritarianism." *Honors Theses.* Paper 96.

North, Douglass C. 1990. *Institutions, Institutional Change and Economic Performance.* Cambridge: Cambridge University Press.

Nuijten, Monique, and Gerhard Anders, eds. 2007. *Corruption and the Secret of Law: A Legal Anthropological Perspective.* Aldershot: Ashgate.

O'Donnell, Guillermo. 1993. "On the State, Democratization and Some Conceptual Problems: A Latin American View with Glances at Some Post-communist Countries." *World Development* 21: 1355–1370.

O'Donnell, Guillermo. 1994. "Delegative Democracy." *Journal of Democracy* 5: 55–69.

O'Donnell, Guillermo. 1998. "Horizontal Accountability in New Democracies." *Journal of Democracy* 9: 112–126.

O'Donnell, Guillermo. 1999. "Polyarchies and the (Un)Rule of Law in Latin America: A Partial Conclusion." In *The (Un)Rule of Law and the Underprivileged in Latin America*, edited by Juan E. Méndez, Guillermo O'Donnell and Paulo Sergio Pinheiro, 303–337. Notre Dame: University of Notre Dame Press.

O'Donnell, Guillermo. 2004. "Why the Rule of Law Matters." *Journal of Democracy* 15: 32–46.

Peters, B. Guy 1999. *Institutional Theory in Political Science: The New Institutionalism.* London and New York: Pinter.

Podgórecki, Adam, and Maria Łoś, eds. 1979. *Multi-dimensional Sociology.* London, Boston and Henley: Routledge and Kegan Paul.

Possamai, Adam, James T. Richardson, and Bryan S. Turner, eds. 2013. *Legal Pluralism and "Shari'a" Law.* London [etc.]: Routledge.

Raustiala, Kal, and Anne-Marie Slaughter. 2002. "International Law, International Relations and Compliance." In *Handbook of International Relations*, edited by Walter Carlsnaes, Thomas Risse and Beth A. Simmons, 538–558. London: SAGE Publications.

Raz, Joseph. 1979. *The Authority of Law.* Oxford: Clarendon.

Rudolph, Susanne, and Lloyd I. Rudolph. 2001: "Living with Difference in India. Legal Pluralism and Legal Universalism in Historical Context." In *Religion and Personal Law in Secular India: A Call to Judgment*, edited by Gerald James Larson, 36–65. Bloomington and Indianapolis: Indiana University Press.

Ruppel, Oliver C., and Gerd Winter, eds. 2011. *Recht von innen: Rechtspluralismus in Afrika und anderswo* [Justice from within: Legal Pluralism in Africa and beyond]. Hamburg: Verlag Kovač.

Schedler, Andreas, Larry Diamond, and Marc F. Plattner. 1999. *The Self-Restraining State. Power and Accountability in New Democracies.* Boulder and London: Lynne Rienner.

Schulze-Fielitz, Helmuth 1984. *Der informale Verfassungsstaat. Aktuelle Beobachtungen des Verfassungslebens der Bundesrepublik Deutschland im Lichte der Verfassungstheorie*. Berlin: Duncker & Humblot.

Schulze-Fielitz, Helmuth. 2011. "Zur Geltung des Rechtsstaates: Zwischen Kulturangemessenheit und universellem Anspruch." *Zeitschrift für Vergleichende Politikwissenschaft* Jg. 5, 1: 1–23.

Shapiro, Jan, ed. 2012. *The Rule of Law*. Delhi: Universal Law Publishing Co Ltd.

Tas, Latif. 2014. *Legal Pluralism in Action: Dispute Resolution and the Kurdish Peace Committee*. Farnham (Surrey): Ashgate.

van Rooij, Benjamin. 2009. *Bringing Justice to the Poor: Bottom-Up Legal Development Cooperation*, March 25. Accessed September 5, 2014. http://papers.ssrn.com/sol3/papers.cfm?abstract_id=1368185.

Waldmann, Peter. 2001. "Alternative Normensysteme zur staatlichen Rechtsordnung in Lateinamerika." In *Rechtsstaat und Demokratie*, edited by Michael Becker, Hand-Joachim Lauth and Gert Pickel, 112–129. Wiesbaden: Westdeutscher Verlag.

Waldron, Jeremy. 2008. "The Concept and the Rule of Law." New York University School of Law, Public Law & Legal Theory Research Paper Series, Working Paper No. 08–50.

World Bank. 2003. *Legal and Judicial Reform: Strategic Directions*. Washington: The World Bank.

Zakaria, Fareed. 1997. "The Rise of Illiberal Democracy." *Foreign Affairs* 76: 22–43.

Zippelius, Reinhold. 1991. *Allgemeine Staatslehre* (11th edition). München: C.H. Beck Verlag.

Zips, Werner, and Weilenmann Markus, eds. 2011. *The Governance of Legal Pluralism: Empirical Studies from Africa and Beyond*. Münster (u.a.): LIT Berlin.

Part IV: The Sociology of the Emergence and Transformation of Formality and Informality

Timothy Eccles

Identifying a Formality Hinterland: Trans-informality and Meta-formality Within UK "Better Regulation" Discourses

Introduction

There is an obvious conceptual distinction between formal and informal social actions in both the sociology literature and in common usage. This is evident from the earliest classical sociology literature: in Tönnies' *Gemeinschaft* and *Gestellschaft*, in Durkheim's "mechanical" and "organic" solidarity and Weber's "traditional" and "legal-rational" modes of authority. Since then the interaction between the two has been a central focus of empirical work and the duality remains the dominant form of analysis.

In the business world, formalisation through standardisation, managerialism, legitimacy functions, benchmarking of best practices, codification, performance measurement and bureaucracy are pitted against the vagaries of adhocracy and deviancy. These contrasts make formality, the way in which formal meaning is generated and transmitted, and, more recently, the acceptance of informalised equivalents through heuristics, a useful conception for examining and describing issues within regulatory frameworks.

This paper falls into three distinct parts. Firstly, an overview of regulation is provided within a UK context. Secondly, an analysis of formality and informality as a means of explaining the regulatory process, especially its evolution into a system of "better regulation" is then developed. These two parts are deliberately integrated since this paper is intentionally concerned with the sociology of formality and informality rather than a technical paper on regulation. Thirdly, two case studies are then undertaken to apply this analysis and draw out the themes of meta-formality and trans-informality that are being proposed. Methodologically, the two cases are selected through personal bias, and are not intended to provide a representative sample of regulatory systems. Therefore, whilst it is not intended to argue that the conclusions are generalizable, these specific cases offer a typology of the stages of development of "smart regulation" that are presented in the concluding section as a grounded theory. The research established a systematic set of procedures to develop this inductively derived general framework with the intention that it will also be relevant outside its smart regulation setting.

What is regulation?

Regulation involves a framework that establishes the appropriate standards that should be met by a business when carrying out its activities and an enforcement regime with powers of inspection and punishment.

> In an era of globalisation, in which barriers to movement of goods, services and people are falling, citizens expect their governments to ensure their safety and welfare. Businesses expect public authorities to ensure a level playing field and boost innovation and competitiveness.
>
> Regulation is key to meeting these challenges. It serves many purposes – to protect health by ensuring food safety, to protect the environment by setting air and water quality standards, to set rules for companies competing in the marketplace to create a level playing field. Regulation is a necessary and accepted aspect of modern society (European Commission 2013).

For the purposes of sociological analysis, regulation involves, or has traditionally involved, a number of themes. Notwithstanding any potential difficulties with a precise definition of the term, regulation is a formal process. Most immediately, it is a visible process with clear guidelines that have a concrete and "measurable" reality. There is an inspection regime that can check that these standards are achieved. If not actually desired by all parties, it is universally recognised as necessary and forms part of a legal and behavioural code of business conduct. The process has also certain symbolic value.

All of this requires a single point of control, a determining authority. This is usually the state, of course, which is seen as a fair arbiter of what is and is not acceptable practice. To the point of this work, this provides a formal system of "command and control" regulation.

The starting point for this analysis follows the fall of this idealised situation. Giddens (1990; 1991; 1994) points to the inevitable decline of the state within Modernity. His description of Late Modern scepticism features a widespread lack of trust and confidence in the probity of institutions and scepticism regarding the competence of governments and state regulators to put in place effective systems of control. This generates a loss of authority, a refusal by parties to accept external regulation, and even disagreement over the technical competency of those approving procedures. Thus the preconditions of formality fall, leaving "command and control" increasingly untenable. The dialogue of regulation becomes that of deregulation, allowing individual firms to best judge how to carry out their business. This provides for informality as a creator of regulated goods and services.

Of course, Giddens also includes all institutions within his idea of the Late Modern decline in authority. Informal systems developed by self-interested companies are regarded as equally invalid, especially in the wake of the 2007–08 financial collapse. Deregulation has fared no better than state regulation in recent years (see, for example, Eccles and Pointing 2013 for a fuller analysis of this historical evolution).

Where formality and informality offer an interesting avenue of examination is in how these two extremes of regulation are being drawn together by practitioners seeking to gain the advantages of both in what is being described as "smart regulation". This paper utilises formality and informality to examine this UK regulatory evolution.

My conception of the formality hinterland

The introduction above leads to the first problem – that of defining what formality and informality mean in the context of regulation. Tönnies, Durkheim and Weber all provided idealised (if not ideal) types in their analyses, and all draw on a dual typology. Duality, through two extremes, mirrors the regulatory paradigm – command and control versus deregulation. The concept of formality provides an ideological simile for the two contending philosophies towards regulation. Formality allows for measurable benchmarking to a universal standard, whilst informality reflects the authority of expertise and knowledge embedded within individual experiences. Formality provides a clear form, and the likes of Stinchcombe (2001) can provide concrete examples to round out a definition – rules, blueprints, and citation practices, for example. Formality is perhaps partly symbolic, but offers a certainty through its visibility in regulatory action. Informality is a fuzzy, potentially deviant, formless will o' wisp.

From this contention, I generated my first simple model (*figure 1*) of how the formality of command and control interacted with the informality of deregulation. This was a competition wherein the former was trying to dominate; however, its "voice" was becoming less dominating the further from the source it went. Equally, informal responses (complaints to members of Parliament, compliance variances, breaches of regulations, etc.) resound back to the regulator, though with less force.

Figure 1: Formality vs. informality

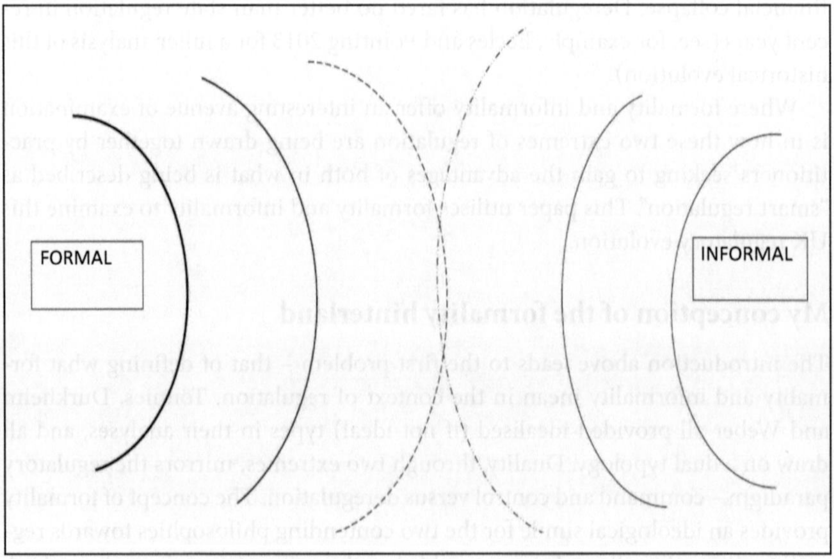

I immediately complicated this (*figure 2*) by contemplating the communicative "ripples" of *figure 1* as various forms of communication and interference. In this figure, I was considering the way that regulations might be related to distance; some businesses feel the demands upon them with more or less intensity than others. Responses would also be irregular, as befits the informal nature of ad hoc replies and reactions. The informal lacks a clear form, whilst the single determinant authority of formality provides a clear point of regulatory potency.

Figure 2: Formality, clarity and the strength of the need to respond to regulatory requirements

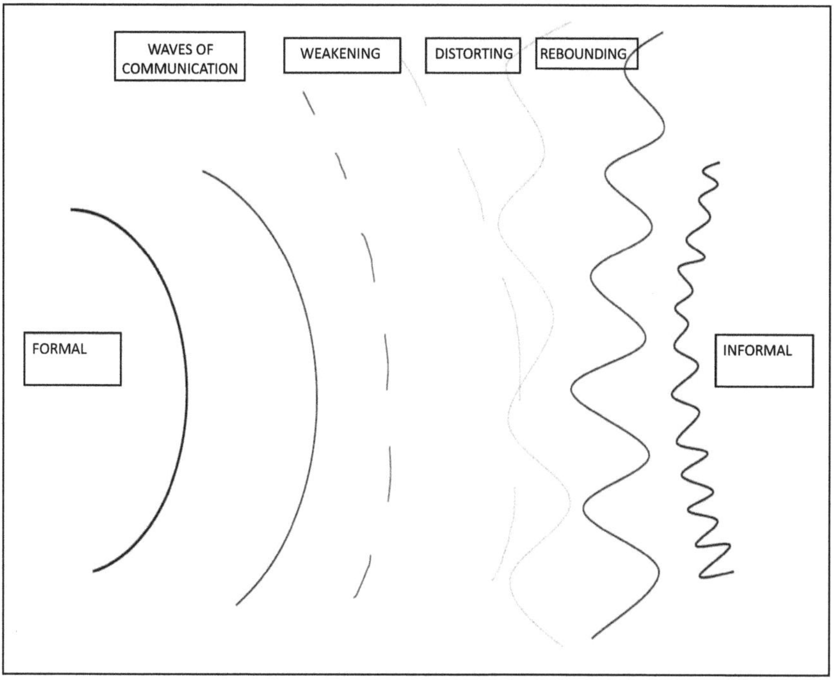

Figure 3 represents the final part of my original thinking, namely that a central premise of informal responses to regulation is that they are based in the individual business, and that there are many competing responses to a single point of formality. Hence, we find that there is a veritable orchestra of ad hoc responses of different strengths, opinions and direction with informal responses being targeted not only at the regulator, but also between businesses. At times, some of these might combine to return a unified response.

Figure 3: Formality versus multi-informality

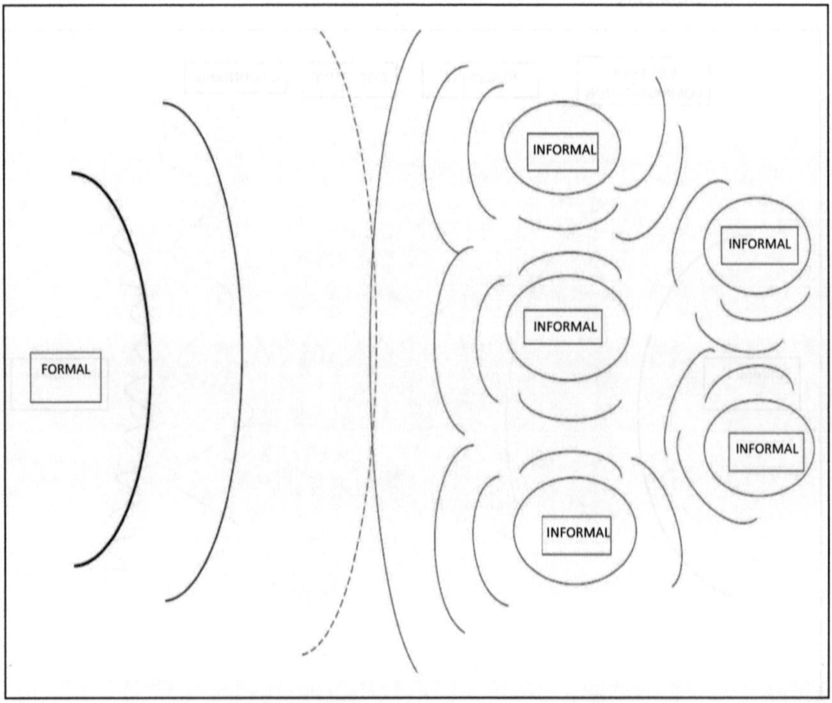

From these ideas, I was drawn to the notion of a formality "hinterland", where neither of these pure forms existed. Rather, regulators and the regulated established protocols and interacted in the real world to achieve profitable businesses operating within established legal norms.

Conceptually, my approach in this paper is drawn to that of Weber. His description of the "official" is quite evidently appropriate to that of "the regulator". Similarly, the bureaucratic requirements of "methodical provision" and of "jurisdictional competency" resonate when analysing the process of regulation. In the same way, his views on formality and informality in the working of rational-legal social systems will be seen mirrored in much of the following arguments. Finally, Weber's analysis on the role of knowledge in society and his (interchangeable) use of the terms "expert" and "specialist" remain potentially useful tools to analyse the practical implementation and supervision of such knowledge in the sphere of business.

However, this typology fails to recognise the fragmentation of the role of the regulator in this process. For me, the above shows that regulation is associated with formality because of the place of a single, determining authority. Formality derives from a determinant authority. Within the timeframe considered by this paper, which is placed within Giddens' (1990; 1991) Late Modernity, this no longer exists. He refers to competing, transient dominant authorities that have replaced the single, determinant authority. Within the field of regulation, this transformation away from a single authority has been characterised by *deregulation*. In an idealised form, this involves the removal of all authority, as part of a wider "small government" movement associated with economic liberalism. This lack of authority is associated with informality since there is no single meaning, no measurement and no approved method for a firm carrying out its business practices. Hence, informality is associated with a fragmenting of authority.

Figure 4 illustrates this fragmentation of authority. It also places the concept of authority at the centre of any study of formality in regulation. Like each separate field of study that is being called upon in this work, there is a well-respected tradition of examining the concept of authority (see, for example, Raz 1990). In regard to "command and control" regulation, however, the regulator performs exactly those functions of Hobbes' Law Giver. Further, traditional regulation worked on the principle that the lack of such a figure would render the same "nasty, brutish and short" life as Hobbes' anarchic state of nature that exists without a law giver to provide order (Hobbes 1968 [1651]; Peters 1956). Of course, such an authority raises questions of governance and therein lays a topic for further literature. Within the field of regulation, Black (2009) certainly regards problems of legitimacy as central to problems of regulation. Where there is no determining authority, then generating and maintaining legitimacy becomes a primary motive for each competing authority. Whilst legitimacy management becomes an aspect of regulation, the duality of formality – informality is offered here as a better overarching focus for study.

Figure 4: Competing dominant authorities attempt command and control

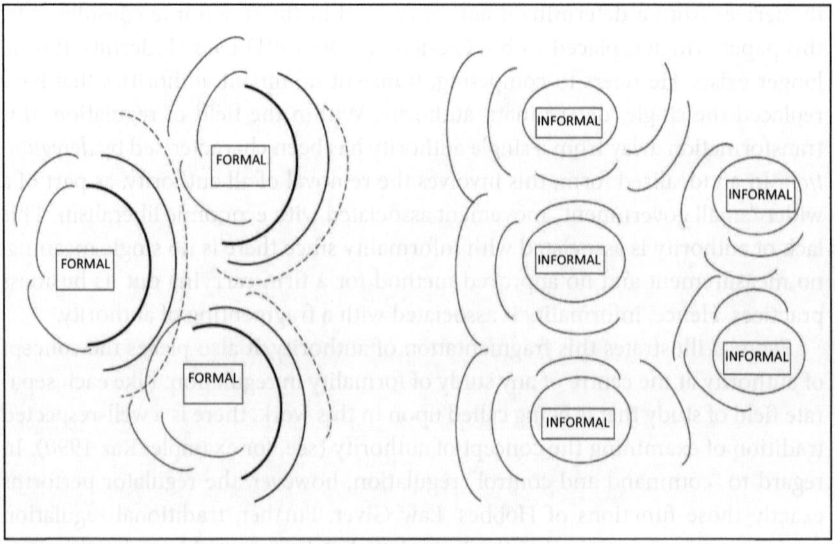

Formality in regulation: Command and control

An ideal, "pure" form of regulation is the "command and control" system of regulation. In this system, standards are set by a single, determining authority and disseminated through a formal system of statutory legislation, codes of practice and the like. This formality derives from three social preconditions, which are:

(1) A general (that is, political) will to concede authority to the regulator;
(2) A legal right to regulate, enforce and penalise;
(3) Technical competence to determine what the standard ought to be, and how to measure performance.

One of the evolving social themes within the process of regulation is the erosion of these preconditions. For example, we can derive from Giddens (1990; 1991; 1994) the following schemata:

(1) Ideological opposition to a single, determining authority is endemic.
(2) Without a political will, the legal right to regulate is increasingly constrained.
(3) Similarly, questions concerning the competence of the regulator are not only ideological, but based upon continual empirical evidence of failings in regulatory processes.

Regulation is therefore moving away from the "command and control" model, because it is seen as ideologically unsound, but also impractical and expensive. The place of the formal in this is quite interesting, because practical issues of how to manage regulation through a formalised system are central to criticisms of the command and control model. Considering Black's (2009) concept of legitimacy, the danger is that competing authorities are simply rebuilding the rejected "command and control" model in an inferior form. These dominant, but not determinant, authorities seeking to mirror a Hobbesian law giver, I will later label with the term meta-formality. There is, however, an omission in this analysis, since it overlooks regulatory evolution prior to the 2007 financial collapse, and outside financial markets. In financial markets there remains a dialogue concerning corruption against which the need for a strong authority endures. In other regulatory spheres, this is either absent or less powerful.

Under "command and control", regulators have tended to see the informal as a deviancy to be stamped down upon, favouring approved procedures, systematic inspection and adherence to pre-determined standards. Stinchcombe (2001) argues that these work, though he is open to criticisms of a selective sampling. However, businessmen have long criticised that this is not how firms operate, and it is simply not a practical method of regulation. One of the first management theorists, Chester Barnard, in analysing his own business experiences as a business executive, described the management process as informal, intuitive, non-routinized and involving primarily oral, two-way communication. Barnard (1938) refers to management as an art and not a science, emphasising the need to empathise the whole organisation rather than focus upon context-less objectives. The terms that he utilises are "feeling", "judgement", "sense", "proportion", "balance", and "appropriateness". Regulation ought to reflect the way businesses actually operate, so the argument runs.

Outside of the ideological arguments, then, "command and control" regulators face problems within three archetypes:

(1) Generating a formal system that is understandable. This involves the use of language.
(2) Actual enforcement in the real world, when many individuals respond to the same requirements in their own ways. These might be influenced culturally, ideologically and by other factors.
(3) Lack of expertise in technical competence by regulators, compared with those being regulated, leading to disagreement on the "best" way to regulate.

The question faced by society is how to best deal with these three technical problems. The first case study, which follows below, shows that the response was to determine the challenges as insurmountable, and approach regulation from another direction, that is, to place emphasis on Barnard's managers to act in a balanced and proportionate manner. The answer, therefore, was self-regulation.

Study 1: Building control

It is, of course, dangerous to ascribe movements and tendencies to an individual or a particular date. At the same time, I think it is useful in this analysis of regulation and its relationship with the formal and the informal to do precisely this. In 1979, the British world changed. Not only was a Conservative government elected under Margaret Thatcher, but with it there was an acceptance and an admission that the future could not continue as the past. Not everyone realised what this might mean at the time, and many of those who supported these changes would later be discarded or become enemies, but I do think that there was an acceptance that change had to take place.

From the perspective of this paper, the primary change in regulatory frameworks was their removal. The language was of deregulation, and the regulatory discourses concerned minimising external intervention. Formal systems of regulatory control were removed, and there was an agreement that informal systems could prove equally as effective. Regulation as a means – through formality – was replaced by regulation of ends, and this was seen as measured by a number of voices rather than just that of the regulator, to now include the market, the consumer, the professions, one's peers. Regulation had been seen as requiring formality, and this could only be generated by a single, determining, authority. The Thatcherite ideology overthrew this, and replaced it with competing formalities, with the ultimate aim of discarding any authority. This creates what I refer to as trans-informality. In essence, individual experts are allowed to resolve the questions of how best to regulate themselves, and their industries. Informal systems are given primacy as ad hoc systems, previous experience and heuristics are all accorded the authority to determine what form regulation should take. What were previously regarded as personal, informal ways of working, of solving problems, of doing a job were accorded recognition.

There is still a clear regulatory outcome required, and this demands a formal statement of expectations. There is still methodical provision and there is jurisdictional competence. However, there is no singular formality of process by firms in how they achieve this, and, ultimately, there is no uniform formality in how their achievement might be measured – or at least by whom this might be measured.

This conceptualisation is derived from a number of studies on regulatory processes. For the purpose of this paper, my practical example of this theme concerns the deregulation of building control. Building control was one of the first areas of professional work to be privatised in the 1980s. Its role is the examination and, if compliant, approval of new buildings to confirm that they have been built in accordance with the appropriate Building Regulations. Traditionally, local authorities exercised control through a monopoly of the service in order to guarantee quality. Its removal from state control to that of private businesses generated two distinct themes. First, the original intention was to allow private individuals to undertake building control work. Ideologically, the presumption was that "the market" would determine those competent to do so. Therefore, informal systems of what might be regarded as no regulation were envisaged as acceptable. It was very quickly realised that the potential dangers required a second theme, that of non-centralised regulation of competence. Formal rational-legal systems of approval were required, but under a different model of control than the "command and control" system. The result was to endorse professional self-regulation, since professional associations were identified as a priori competent to regulate building control work in the final legislation. Thus, regulation of building control was to be carried out by co-operation between a statutory authority and a professional association with the emphasis on the latter's self-regulation guarantees. This was deregulation in that it allowed competition, but restricted the market to members of professional associations.

Building control was the very first profession to face this new regulatory regime. The 1984 Building Act and the statements set out within the Building (Approved Inspectors etc.) Regulations 1985 proved to be key pieces of legislation in developing the deregulation agenda within the building industry. It removed building control from the exclusive domain of public sector Building Control Officers and replaced this with a market-orientated competition between public and private sector provision.

The primary driver within building control – and the enforcement of Building Regulations generally since the Building Act 1984 came into force – has been to minimise the need for (and expense of) inspecting building work. This led to the search for so called "competent persons", whose work is of a high quality such that it does not need checking and may be self-certified (DETR 1997). The Department of the Environment, Transport and Regions (DETR) emphasised the importance of costs, that "building control must also be efficient, to minimise cost and delay for those carrying out the building work" (DETR 1999a, 3). The result was that membership of a professional body was deemed essential to qualify in terms of expertise and to regulate the competence of building control service providers (DETR 1999a; 1999b). Of course, this returns to the notion of professions as regulators: not exactly self-regulation, but close enough at the time to appease the wider profession, whose private sector members saw business opportunities at the expense of their public sector colleagues. Similarly, the Royal Institution of Chartered Surveyors (RICS) saw fresh opportunities for itself and for its members in becoming the dominant professional authority under this new regulatory paradigm (see, for example, Finn 2007).

Building control is important primarily as it was the first example of modern privatisation, of deregulating independent, local authority inspection within the UK construction industry. A single superstructure of formality was deliberately dismantled, and replaced. Economic efficiency through market competition was the primary driver. What it achieved was the transfer of expertise from the local authority to professional associations, whose own standards of membership and regulation were deemed to be a suitable replacement.

The "ideal type" of deregulation perceived by the government was one of heuristics and informality. All that actually matters is that buildings do not fall down. Placing further restrictions on the regulatory process incurs unnecessary costs. Experts should simply be allowed to "get on with it" and there was political hyperbole supporting this extreme.

However, implementation of this ideal type was more pragmatic. Government recognised the need to have transparency in the deregulated process, and this, inevitably, required some degree of formality to allow comparison between suppliers and to protect against building control carried out ineffectively, on the cheap or not at all. In this pragmatic form, deregulation generated building control that was fractured from a formal hegemony, but competing suppliers of building control retained formal processes for dealing with their customers, the government and the wider public. The primary difference was that these were not the same, not standardised and sometimes not even comparable. This generates what I refer to as a formality hinterland, where formality remains seen as a requirement of regulation and there is a tension between the needs for transparency and comparison versus the prime motivator of the deregulation, which is a reduction in the economic cost and sweeping away inefficient "red tape".

The search for the "competent persons" described above followed a formal procedure of checking credentials. The focus, though, shifted away from a single, formalised procedure to formally recognising the importance of less formal qualifications, such as statements of competence. It sought to regularise the informal, to take working knowledge and practical experience and accord them recognition. That said, building control has drifted back towards a unified formality inasmuch as specified professional qualifications are seen as providing the proof of competence sought. Where the informal has been transformed is that professional associations have endorsed such previously unacceptable indicators of competence and given them status and a degree of formality. In Weberian terminology, traditional forms of occupational activity have been accorded legal equivalence.

Conceptually, building control provides empirical evidence of a number of phases in the deformalisation process.

(1) Removal of a single, formal "voice" on regulation; and
(2) Concentration on measuring "ends" rather than "means";
(3) Whilst accepting that different parties might have different "ends" and measure them in non-similar ways;
(4) Which requires either multiple formalities, or trans-informal interpretations, such that each can be regarded as acceptable. Previously, informal regulatory procedures would be regarded as inadequate, deviant or both. Now, these can be accepted as valid because they are based upon the experiences of those who know the work best.

Trans-informality is used where informality begins to deviate from its traditional definition, i.e. when informality is transformed towards formality. Here, informal, ad hoc, heuristic modes of an occupation are given precedence. The dialogue behind this comes in two different forms. Cost is a major driver; formal systems are expensive. However, it is also recognised that occupational control is best exercised by those who actually carry through the work. Their "rules of thumb" are empirically tested methods of carrying out tasks to a standard accepted by customers. Within this paradigm, such market-led discourses are very powerful.

All that said, there remained the question as to how to then provide proof for the efficacy of these informal regulatory systems, and whether some degree of formalisation was required to achieve that. To those like Black (2009), this is an issue of legitimacy. Outside of financial regulation, I see this as more straightforward pragmatism: how does one create a working system in which informal systems derived by many sources of occupational authority were to be the drivers of building control competency systems? There is no need to generate legitimacy, because Late Modernity is typified by a lack of legitimacy, period. To the extent that legitimacy might remain a useful concept, legitimacy is simply the result of building control work being successfully carried out.

Where the issue of trans-informality becomes especially interesting is how the regulation of building control was actually developed. As discussed above, the emphasis was placed on individual practitioners (the informal). However, it rapidly became clear that some (formal) system of generating a process for regulating those "competent persons" in whom public trust was placed would be required. Whilst this was still centred upon the individual, a practical system of organisation was required, and one that was cost effective. Government concentrated on the professional associations of whom building control officers were likely members. Later, they were required to become members since the formal system of regulation that developed was embedded in the professional regulation of these associations. Professional associations thus became dominant authorities in the field of building control. However, the reason why I develop the *trans-informal* theme is that this entire process was still driven from the bottom up, by the individual members of the association, and not by the formal authoritative processes of the associations themselves. However, as the deregulation agenda was developed, professional associations did take greater control, installing authoritative command-and-control-lite regulatory systems. This is what I describe as meta-formal, and my second case study illustrates this evolution in regulatory formality.

Professional regulation and the creation of meta-formality

When regulating the manufacture of products, it is relatively easy to set standards – assuming that regulators possess technical competence and knowledge of the product. For example, in my building control case study, when building a house, foundations should be of a certain specification of concrete, laid in trenches of a certain depth and in weather conditions that do not disrupt the natural setting of the concrete. However, when dealing with services, there is no similar degree of certainty. Services are esoteric and the finished result is nebulous, immaterial and without a concrete presence. And, of course, it is this characteristic that provides a (arguably, "the") defining characteristic of a profession within the sociology literature. In the context of this paper, it is of less relevance whether this is seen as social distance (Johnson 1972) or control through generating difference (Larson 1977) inter alia. Where it features in this analysis of formality is in its role as a professional project (Larson 1977), a system (Abbott 1988) or as an institutional isomorphism (DiMaggio and Powell 1983). The question for regulators, then, is how to regulate in these circumstances. The result has been to look at process.

Within regulation, regulators are able to regulate

(1) The production process;
(2) The finished product;
(3) Both process and product.

In terms of services, regulatinc fg the finished product is problematic precisely because the service is the commodity that professions use to exercise occupational control. Clear – external – regulation is one function that they are actively engaged in stymieing. At the same time, professional associations do seek to gain ever-greater control of their own membership. Associations themselves are transforming from peer-controlled membership clubs to dominant command-and-control authorities. It is these dual processes that generate meta-formality: command-and-control, but within a number of competing authorities, in this case professional associations.

Within the regulation of services, therefore, the focus has been on creating (the "command and control", again) a standardised system of producing this service (process). The argument goes that if the process of delivering the service is correct, then the finished result will automatically be satisfactory. This regulation of the "means" of service provision involves three stages:

(1) Generate a regulated provision mechanism;
(2) Regulate the person carrying out the process;
(3) Provide a concrete agreement on what the service actually is.

Again, the importance of formality is central. Any deviancy risks destroying the service being provided, rendering it unsatisfactory, illegal and unproven. These three parts all require a formalised system of generation and monitoring, but the responsibility for generating this formality is placed with a self-regulating authority, the professional association. The "deregulatory agenda" dialogue is maintained because there are a number of such authorities in competition with each other – competing for members and for clients to recognise the value of both their members and the association itself, and for a wider legitimacy within government and supra-government circles. Because professional associations are collections of experts in esoteric services, they have traditionally been allowed to regulate their own members and set service quality benchmarks. Ironically, deregulation has focused upon breaking up these monopolies on the basis that they, too, are anti-competitive. Meta-formality offers professions a partial return of their power, and they are keen to engage. My own work has drawn upon a number of others to ponder the question as to whether individual professionals might regret this transformation of their membership associations (Fournier 1999; Eccles 2009). Whilst not directly relevant here, it does move me to postulate that my meta-formality will always triumph over trans-informality. Within regulatory processes, the need for certainty is too ingrained to allow individual freedom to determine means without a dominant authority seizing control with the promise that formality guarantees certainty and safety. Hopefully, the second case study will flesh out these generalities with specifics.

At the practitioner level, professionalisation can be seen as a formalisation of occupational norms in order to develop a benchmark of competency. Formalisation achieves a particular end, desirous of all professional associations. To the ordinary professional, this takes the form of, for example, a Code of Practice. This Code provides the "best" approach to a job and offers some legal protection as Courts usually accept that following a Code precludes any potential case of negligence. Sociologically, a Code of Practice gives concrete evidence to the generation of "difference". It generates a particular form of "professional" service that is given a distinct social meaning that distinguishes it from "non-professional" service, from the charlatan, from the quack. The concept of formality is useful here since it illustrates how this "difference" can be given form: the measurement and validation of competence. What follows is a case study that extends across Europe, concerning the nature of the regulation of the process of providing property valuation.

Formalising meaning in a professional context – The issue of "value"

I first considered the theme of what is meant by the concept of "value" in 1993 (Eccles 1993a; 1993b) and left the issue two years later (Eccles 1995) as a discrete subject. At that time, I saw the concept as one of "culture", rather than the more focused interpretation that I am outlining here.

To the surveying profession, the central question is how one provides a commercial valuation to a non-financial asset (a building) when it has not actually been placed into a market of buyers and sellers, who, as economic actors, would be expected to determine this by agreeing upon a price for which one party is prepared to pay, and likewise for which the owner is prepared to sell. Obviously, where a sale actually takes place, there is brought into existence a clear formal value. However, assets still need to be valued for many purposes even when not actually being sold – evaluating loans, insurance, calculating the net worth of a company etc. RICS refer to these as Regulated Purpose Valuations (RPVs), those

> that are relied on by third parties and include those undertaken for the purpose of statutory financial reporting within the public accounts of listed companies. As well as financial statements, third parties rely on these valuations for takeovers, mergers, stock exchange listings and published information (RICS 2009, 5).

Because of their purpose, these are regarded by RICS as occurring "within a public interest framework" (RICS 2009, 5). These values are then accorded a formality, because they are accepted as "accurate". At the same time, they are what might be described as professional estimates of what price they would command if sold. Hence, the professional valuer follows a series of formalised procedures to achieve this value. It is the creation of such a formal value by an individual that I am going to consider as an example of the transformation of what might be seen as an informal, heuristic process into a formal, concrete value in which the professional association serves as a dominant authority, and directs expertise, rather than allow its members to each utilise their judgement. This is the meta-formal, where "command-and-control" regulation is utilised, but between different, competing, associations.

Within the case study's subject of valuation, there are two variable parameters to keep in mind when considering the valuation of assets:

(1) The need to produce a definite price that *would* be achieved if the asset were to be sold. This is considered a formal and concrete statement of reality by those utilising the value (e.g. investors, tax authorities, lenders etc.).
(2) That this value is not an actual price, since the asset has not been sold. Rather, it is an estimate, a professional opinion, of what that price *would* be.

It is self-evident that there are a series of potential differences between the reality of such a valuation and the actual market price. The concern has been how to create an accurate method to generate this estimate, and then how to monitor its function. Since valuers are experts in their field, one might expect that they be allowed to each determine the best way to carry through this work. Within a deregulatory dialogue, the trans-informalisation process described in the first case study might have been seen as apt. However, in this case, a very different approach has evolved. First, a "top-down" process has been introduced, reminiscent of the "command-and-control" regulatory system. Second, this has focussed upon regulation of the process through a very formal system, which consists of two parts: a very precise language of definition has been established; and, systems for the accreditation of the judgement of the individual have been introduced. A key feature of this case study, though, is one of fragmentary authority. Many professional associations within and between nations compete to establish the dominance of their own approach. Hence, the meta-formal, in which we see centralised systems of control, but without a single overarching determining authority. This is what makes the case of interest to sociology. Whilst there are interesting technical issues behind creating a valuation, the study transcends the particular profession ("the valuer") because all of this takes place within an international context – hence, there are international and national regulators, different languages, international and national codes of law and various cultural systems for social action in valuing assets. Thus, valuation is faced with multiple systems of formality, competing and cooperating to achieve the required valuation accuracy and consistency. Again, hopefully these generalisations will take form in the specifics to follow.

One final point that is worth noting is that the regulation of valuation does not arise out of a reaction to any particular regulatory failing. The occupational issue for practitioners concerns accuracy and how close valuations are to the market price. In practice, this can only be tested once a valued asset is actually sold, of course. However, there are continual concerns as to whether valuers might be pressured into higher valuations, for example for the purpose of security against a loan. Dishonesty aside, accuracy is a constant concern – especially at times of deflation and/or recession when asset values are likely to be in decline. Where a company suddenly finds its assets are worth less than they believed, litigation against the valuer is a frequent response. For the record, research suggests that valuation accuracy and honesty is not a particular cause, at this time, for regulatory concern (see, for example, Baum et al. 2000; RICS 2002; RICS Valuation Faculty 2003). There is no external specific driver of change that might render the case unrepresentative.

Regulating value

Regulation takes place in two different spheres. First, there is the issue of the regulating body, and, second, there is the matter of the formalised codes of procedure in carrying through the valuation.

This case study is a good example of the fracturing of controlling authority power. In terms of regulatory bodies, there is an International Valuation Standards Board (IVSB) at the EU level. There are then national standard setters, including the Royal Institution of Chartered Surveyors (RICS, in the UK) and the Polish Federation of Valuers' Associations (PFVA, in Poland). However, the UK also has a second, competing, professional body, the Institute of Revenues Rating and Valuation (IRRV). Additionally, this second body is recognised by the EU in a different manner than RICS. National professional associations are responsible for exercising occupational control over their members, unless there is a national statutory regulator. These are represented at the IVSB. The result is a series of different practice codes. *Figure 5* illustrates the hierarchy of these (competing) levels of authority. It is agreed that International Valuation Standards (IVS) are dominant internationally, European Valuation Standards within an EU framework, and that each nation state then develops its own framework within these. This is important because it emphasises the restricted nature of authority and the contested dialogue that occurs within and between these groups. It also suggests that such competition between authorities in itself is a driver of the formalisation of practices.

Figure 5: Hierarchy of valuation standards

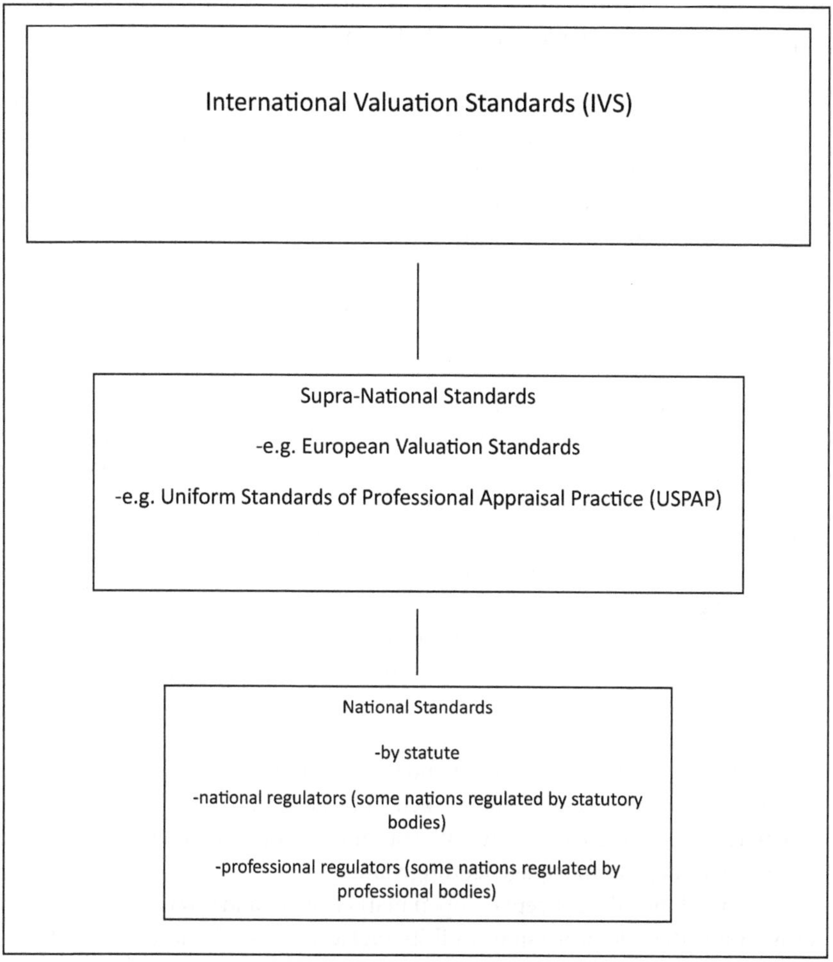

Within valuation, regulators have focussed upon the meaning of "value". Since it is an artificial construct, not an actual market price, what exactly is being measured, under what market conditions, is seen as central to determining accuracy. This is because a practice can develop from this for giving a particular asset a value as defined within that meaning. However, this is by no means simple. First, there is the problem of language. In English, the professional term adopted currently is that of "market value", but there is a second, wider, interpretation, that of "fair value" as will be discussed below. The second problem is, even if the formalisation of meaning allows for a common understanding (not guaranteed), there is the issue of how in practice this is carried through to the actual process of performing asset valuation, and how this can guarantee accuracy in the value calculated.

In both cases, the issue of the lack of a single, determinant authority generates disparate approaches. The UK provides a good example, since it has two competing systems of regulation. RICS has introduced a system of licensing professional valuers within its own membership. This introduced the term "RICS Accredited Valuer" (RICS 2009) in 2010. However, other bodies in the EU have adopted a system of "Recognised European Valuer" (REV) status. Whilst RICS has not adopted this, IRRV has. Therefore, within the UK, there are two different regulatory processes for endorsing the competence of those carrying out the same activity. Hence, we find not one, but two, competing formal structures for identifying and licensing competence. Where they are similar is in centralising the regulation of professional activity, and to illustrate this I will focus upon the differences in meaning given to the measurement of value. Whilst these might appear obtuse to the lay person, hopefully the two principles will become clear. Formality in meaning is the total focus, but there is no uniformity because there is no determining authority to enforce a universal accord. Because of this latter problem, formality is given even greater importance since there is no socialisation across occupational and national boundaries that can foster professional understanding. Individuals must fall back on the exact specificity of their practitioner definitions. Whilst Wikipedia (undated) regards the terms discussed below as "interchangeable", and, from a lay person's perspective, all these are attempting to agree to the same thing – the true value of an asset (Wikipedia undated) – this is a fundamental misreading of why these definitions exist, and in such a precise form.

The terms that the case study will examine are: open market value, market value, fair value, and mortgage lending value.

RICS developed the concept of "open market value", and it was the operational term within the UK profession until its replacement by "market value". Interestingly, the previous terminology used in the UK had also been market value, though its formal meaning differed slightly. So, between 1996 and 2003 UK valuers were faced with dealing in "open market value" in the UK and "market value" (see below) within the rest of Europe. In 2003, RICS adopted the IVS definition and abolished its use of "open market value". Market value is

> an opinion of the best price at which the sale of an interest in an asset would have been completed unconditionally for cash consideration on the date of valuation, assuming
>
> (a) a willing seller;

(b) that, prior to the date of valuation, there had been a reasonable period (having regard to the nature of the asset and state of the market) for the proper marketing of the interest, for the agreement of price and terms and for the completion of the sale;
(c) that the state of the market, level of values and other circumstances were, on any earlier assumed date of exchange of contracts, the same as on the date of valuation;
(d) that no account is taken of any additional bid by a purchaser with a special interest; and
(e) that both parties to the transaction had acted knowledgeably, prudently and without compulsion (RICS 2010, PS 3.2 2010).

This definition is found in the RICS manual, *RICS Valuation – Professional Standards* (known as the *Red Book*, RICS 2010), a mandatory practice handbook for RICS values. Only members of RICS *must* follow its guidance, because there is no statutory requirement to be a member of any particular body in order to be qualified to carry out valuations.

The IVSC (2005) first adopted the term "market value" in 1992. This is defined as

> the estimated amount for which an asset or liability should exchange on the valuation date between a willing buyer and a willing seller in an arm's length transaction after proper marketing and where the parties had each acted knowledgeably, prudently and without compulsion (RICS 2014, 30).

Note that today, whilst RICS recognises the need to work to market value, they retain their own formal meaning of the term. Two formal terms; two competing authorities.

Next we consider "fair value". IVS defines fair value as "usually" being the same as market value, but allows for one of a number of specified non-market valuations to be used if the situation requires. One might question why the need for this second definition, since market value could be subsumed within fair value. However, valuers are not the only group interested in the valuation of an asset. The point of particular interest within fair value is that a second group of competing authorities appear here – other professional groups. In this case, the accounting profession has produced International Financial Reporting Standard 13 (IFRS 13). IFRS 13 defines fair value as "the price that would be received to sell an asset or paid to transfer a liability in an orderly transaction between market participants at the measurement date" (IASB 2011, 6). Again, we have formality, but competing forms.

Mortgage lending value is also adopted by IVS as yet another precisely defined concept, and, again, the interesting point is that its rationale arises from another, competing, centralised authority. In this case, the concept originates within another generator of formality, the European Union. The term is contained within

Directive 98/32/EC of the European Parliament and of the Council of 22 June 1998. It is defined as:

> The value of the property as determined by the Valuer making a prudent assessment of the future marketability of the property by taking into account long-term sustainable aspects of the property, the normal and local market conditions, and the current use and alternative appropriate uses of the property. Speculative elements may not be taken into account in the assessment of mortgage lending value. The mortgage lending value shall be documented in a transparent and clear manner (Directive 98/32/EC of the European Parliament and of the Council of 22 June 1998).

Note the importance specifically attached to the requirement of formal practices to generate it.

The UK situation is mirrored by other countries. For example, in the case of mortgage lending value, Poland has introduced its own KSWS 3, Valuation for Loan Security Purposes. The Polish Federation of Valuers' Associations (PFVA) successfully negotiated its passage through the Ministry of Infrastructure and it became binding in 2010. It at first appears to copy the principles of market value discussed above, but does add one further twist in that the valuer is also required to take a view on predicted future changes in the market that might affect the valuation.

In restating these definitions, I am attempting to show the very narrow differences at stake and the contested authorities involved in establishing the meaning of the activity of generating a "value". What is very clear is that there is no room for informal judgement here, no trust in individual professional experience. In regulating accuracy in the calculated sum, weight is very evidently placed upon the formality of the very specific. It is generated by an authority that commands and controls. But each authority is only powerful within a small region, occupation or group of individuals. It is regulated formality, but it is meta-formality since it is a weaker authoritative form.

It is also worth repeating that there has never been what one might call a failure of the valuation profession. The profession is regarded as generally honest and the values generated have been viewed as generally accurate. RICS has instigated two independent reviews of the profession and its accuracy – the Mallinson Report (RICS 1994) and the Carsberg Report (RICS 2002) – and both gave a clean bill of health. Obviously, there have been individual failings and at times of financial distress, the spotlight has fallen onto the profession. For example, the collapse in 1993 of Queens Moat Houses, a hotel chain, involved, in part, the valuation profession since "wildly" (Stevenson 1993) different values had been placed on its assets (the hotels). This caused a lot of self-analysis and self-doubt within the profession and, undoubtedly, triggered some of the response that has left the profession in the current respected position described above. However, distressed companies in financial crises are specifically excluded from the definitions stated above, and it is unclear how the likes of Queens Moat Houses can be blamed on the valuation profession. Where values might be questioned, they are simply the visible representation of a more fundamental financial malaise – or worse. Indeed, to a lay person, the profession might seem to be highly inaccurate at times. For example, Eade (1999) offers the example of a surveyor who valued a café and maisonette at £100,000 for a loan of £80,000, which later sold for £26,000. The defendants' surveyor offered a valuation based upon comparable properties of £70,000, apparently "close enough to support the £100,000 valuation" (Eade 1999, 41) and similar properties reaching £90,000 at auction, whilst the plaintiff's surveyor calculated its value at £65,000. The defendant won, because the judge accepted the logic of his methodology in reaching the valuation.

Thus, we have:

Original surveyor values at £100,000

Defendant's surveyor values at £70,000 (claimed as close enough)

Plaintiff's surveyor values at £65,000 (the £5,000 difference apparently being enough to argue a case of professional negligence)

Ultimate selling price (i.e. market price) is £26,000 (though, I assume that this was a distressed sale from an unwilling seller)

It is the formality of *process* that saves the defendant from losing the case, and not the actual end that he came up with. His regulatory body, RICS, has developed a proven "best practice" model of carrying out a valuation. This conforms to the meta-formal process discussed above. A number of different standard-setters are engaged in establishing competing systems of formality and I find the removal of a single source to be very instructive when considering the nature of formality. In the UK, two professional associations each claim occupational jurisdiction. One relies on its own premier status and its Royal Charter, whilst the other points to its accolade as a joint member of the EU's Recognised European Valuer system of recognising competence. But, in the case of valuation, there has been no rush to accept the primacy of informal expert-led practical solutions. Whether it is informality, heuristics or valuers using their professional judgement to derive a value for an asset, none of these are seen as a legitimate option.

Smart regulation

Regulation concerns competence, and this is found in the occupational expertise of those engaged in the work. Thus, taking advantage of this expertise to draw upon empirical heuristics allows that informal procedures can be accorded greater weight in ensuring safe and competent outputs. This encourages trans-informal regulation, where the adhocracy of the individual drives forward occupational practice within some loosely coupled framework to provide the expected trappings of a regulatory regime.

It is concern about the efficacy of these trappings that drives meta-formal regulation. Here, the drive is to rebuild a "command and control" authority within some boundaries, and generate occupation-led practices. However, greater emphasis is placed on the role of the professional association (or similar authoritative body) to generate an "agree and monitor" cooperative model rather than that of a determining authority or Hobbesian "Law-giver".

This work originated in the empirical study of regulatory activities, which is published elsewhere (Eccles and Pointing 2013). However, from this analysis of a formality hinterland we can draw out a typology of the parties subject to regulation.

Smart regulation seeks to recreate what its advocates refer to as a "primary authority", which is a determinant authority within a locale or profession. However, this geographical or functional limit restricts its power. Additionally, the authority is only generated by the agreement that it recognises where expertise lies, and the regulation is a cooperative activity rather than a coercive one. This lends a great deal of emphasis to the generation of both trans-informal and meta-formal systems. Since a lot of small businesses regulation takes place where there is no obvious self-regulatory authority available, then smart regulation must work with individuals operating in disparate ways. Where associations are available, then meta-formal self-regulation allows for autonomous regulation because there is a formality in the practices adopted and approved, but it is one that flows out of occupational expertise.

Self-regulation remains the policy of the UK government, and its executive authority lies with the Better Regulation Executive (BRE), an arm of the Department for Business, Innovation and Skills (BIS), whose mission is to lead the regulatory reform agenda across the government as a whole; "better regulation" is the declared objective. Besides the aim of reducing the "regulatory burden to business and civil society organisations", the governing coalition's strategy is designed to "bring about a steady change of culture across government so that regulation is seen as the last resort and alternatives to regulation are first considered" (BIS 2011).

Smart regulation has grown out of the broader principles of deregulation, risk analysis and no-cost regulation – a process that has been exhaustively examined by a number of writers over the years (Fisher 2007). It is a somewhat broad concept:

> used to refer to an emerging form of wide-angled regulation that seeks to harness not just governments but also business and third parties to provide policy alternatives that include, but also go beyond, direct regulation (Gunningham 2009, 200).

Smart regulation in the UK is typified by the "Primary Authority scheme", which has been promoted by central government through the setting up of the Local Better Regulation Office (LBRO) and its successor: the Better Regulation Delivery Office (BRDO). The essential objectives of the Primary Authority scheme are to reduce the costs of regulation incurred by local authorities, whilst enabling businesses to operate in a simpler and less intrusive regulatory climate and bear an increasing proportion of the costs of regulation themselves. The more effective self-regulation becomes, the less the need for intrusive and expensive regulation. Consumers will be better protected because of the willingness and capability of businesses to self-regulate. As a consequence of the bringing together of these "win-win" artefacts, local authority regulators can adopt a more removed, supervisory approach to regulating businesses – "smart regulation". By adopting better regulation policies, competently run businesses having outlets in several or many locations will thus be in a position to be regulated at arm's length by a single, primary authority rather than by different local authorities depending on where each outlet is located.

Eccles and Pointing (2013) have created an "architectural triptych" model illustrating how smart regulation might be seen to work. It comprises the following three stages.

Neophyte

When businesses come into existence, they have many important considerations – finance, customers, controlling costs, obtaining premises etc. Regulation is not likely to be at the head of the list of factors. However, they will recognise the need to satisfy their legal responsibilities. They might not understand exactly what these are, and how they may be satisfied. Therefore, neophytes need encouragement rather than penalisation, and will often seek guidance, advice and help from the regulator. Neophytes may be enthusiastic, but often lack confidence and skills. The approach of the smart regulator is to provide neophytes with assistance to establish themselves, and then offer a path to the next level.

Neophytes make mistakes, but regulators will focus on preventing their continuing in error, rather than on punishing them. At the same time, the regulatory system needs to distinguish between genuine error, and those businesses that seek to evade their legal duties. Others will simply not be culturally able to engage in voluntary regulation. Small subcontractors in the construction industry, for example, have been known to prefer directive "command and control" regulation to something "smarter" (Loosemore and Adonakis 2007). Some operators will be incompetent, dishonest, or even running businesses that are actually criminal, such as commercial fly-tipping.

Regulators are thus developing both formal and informal systems of regulation here. There is an acceptance that businesses have not had the time or resources to adopt formal regulatory systems. There is also an admission that entrepreneurs' informal experience might carry through regulatory requirements, and regulators must be prepared to accept novel responses to regulatory requirements. Equally, regulation needs formality in order to establish benchmarks and inspect, penalise and prosecute those who are unable or unwilling to generate acceptable responses to regulation. Thus, trans-informal regulatory processes are condoned, but there is a desire to pursue dominant-authority centred meta-formal systems.

Self-improvers

The second stage recognises that firms will learn and develop if given appropriate support, guidance and help from the regulator. Recourse to intrusive types of enforcement – disciplinary action, serving of formal notices or prosecution – should not be necessary and smart regulators will use their powers sparingly and as a last resort. This is a transitionary stage, not an end result. Businesses are assumed to be striving towards the next stage.

I would suggest that informal-based experience and expertise are encouraged at this point to provide practical and individual responses to a required regulatory regime. The emphasis is upon the "ends" – satisfying regulatory outcomes – rather than on the means. At the same time, individualised responses offer a very fragmentary response and one that is going to be repeated many thousands of times. One would expect there to be "learning from experience" either through an informal diffusion of ideas between practitioners or through a formal system of generating "best practices". Both the Primary Authority and the professional association are organisations able to formalise many informal approaches into the "best" one. Hence, we see the informal experiences translated into more formal "best" ones. This is where I see the meta-formal regulation, because it is a dominant authority that drives this process and not the individual expert, though the latter might help create this. The exact process by which dissemination of best practice ideas is achieved has been well discussed within the professionalisation literature. DiMaggio and Powell (1983), revisited by Beckert (2010), illustrate how institutions generate safe processes to deal with events such as regulatory requirements. In this context, these take the form of meta-formal processes.

Champions

The steady-state stage of the model is reached when businesses seek to establish themselves as self-regulating paragons based on high levels of transparency, trust and reputation. This is the ultimate end to the regulatory process, that which is aspired to by the Better Regulation Delivery Office.

To reach this stage, the costs of regulation should largely be met by the body being regulated; the regulator can adopt a removed, supervisory relationship that is only rarely visible. Again, formality and informality offer an interesting perspective on this process. We return to the duality of regulatory process: transparency requires formal systems and structures that are open to inspection by the regulator; at the same time, these will be based upon the experiences and dialogue of individual business operators, driven by informal practices. The question for further analysis is whether trans-informal systems based upon individual expertise can ever be formal enough, can achieve transparency and legitimacy, can be cost effective to generate and police, and, ultimately, whether occupational experts actually are as competent as they might claim. Whilst building control (case study 1) developed out of individual expertise, it was a quite specific case, and professional associations undoubtedly have taken over control even if they still accept individual solutions. Case study 2 suggests that associations regard self-regulation as an important source of authority for themselves, and that governments, when pushed, prefer the perceived safety of a validating authority over trust in an individual.

Conclusions

Regulation requires the methodical provision of benchmarks that those being regulated are expected to achieve, of systems to measure this and of jurisdictional competence to carry out the regulatory process. Since the latter quarter of the last century, serious questions have been raised about the efficacy of such "command and control" regulatory processes. They are seen as expensive, cumbersome and no guarantee of an acceptable level of compliance. Whilst it is tempting to regard this deregulation as either an informalisation of the process or an entangling of the formal and the informal, the removal of a single determining regulatory authority should not be seen as informalising the process; the same basic principles are still being carried forward. I prefer to view the process as one of "trans-informality", in which the same rational-legal approach is adopted into certain informal systems that are deemed appropriate to the requirements of regulating competence in terms of skills (means) and output (ends). In the short-run at least, these informal heuristic systems are simply re-analysed rather than actually transformed. Even the ideology behind the deregulation of building control did not go so far as to promote purely ad hoc provision of service, but it was keen to allow for many interpretations of how to achieve the finished result of a safe building. Over time, these informal systems were either discarded or increasingly formalised into competing systems for assuring competency. However, I think there are conceptual differences, too. These systems are not, or do not start out, as rational-legal within a Weberian tradition, rather they rely on what might be viewed as "traditional" forms of occupational control and provision. They are also not transparent or visible to others, and are not externally measured. Instead, the regulation of process gives way to the regulation of ends. Socially constructed norms based upon empiricism, pragmatism and "rule of thumb" expertise replace scientifically rationalised ones. But, these are then given permanence and are held up to scrutiny by other experts and (especially) clients and customers. Hence, the "trans-informality" process.

I adopt the term "meta-formality" to distinguish where competing dominating authorities are given agency to regulate using quasi-command and control procedures. The most obvious area for this is where professional associations are allowed to determine regulatory protocols. Here, professional associations rely upon the same formalised characteristics of the state, but are accorded less control. In my example of the provision of valuation services, there is an insistence on the methodical provision of a single approach (the means) and of an agreed universal accord on meaning. However, competition between two UK institutions, and also across Europe, leaves different terms, different indicators of competency and only a weakly coupled mutual recognition of other, equally methodical, positions. It is this looseness of the coupling between independent systems that generates "meta-formality" since it removes the universality of meaning and recognition by creating competing and different constructions of jurisdictional competence and its provision. The loss of a determining authority and its replacement with competing authorities is central to this idea of generating very strong formality, but at the same time, it is also somewhat incomplete.

The use of the term "meta" in this case reflects both conceptions of the term. First, my concept of meta-formality is very much adjacent to formality. In some ways, it is a regression to a more adolescent, weaker, form. In others, it may be an evolutionary one. Either way, the one follows the other. At the same time, epistemologically, I adopt the prefix meta- in referring to the idea that what I am studying concerns who has produced the formality, when, why, in what format etc.

Whether either "meta-formality" or "trans-informality" will remain as valid labels is another area of work-in-progress, just as my examination of smart regulation will continue through observing additional systems and spheres of regulation. However, whilst these particulars will undoubtedly evolve, I think the construction and destruction of formality remains a useful approach to analysing the issue of regulation.

References

Abbott, Andrew. 1988. *The System of Professions*. London: University of Chicago Press.

Barnard, Chester. 1938. *The Function of the Executive*. Cambridge (MA): Harvard University Press.

Baum, Andrew, Neil Crosby, Paul Gallimore, Patrick McAllister, and Adelaide Gray. 2000. *The Influence of Valuers and Valuations on the Workings of the Commercial Property Investment Market*. London: RICS.

Beckert, Jens. 2010. "Institutional Isomorphism Revisited: Convergence and Divergence in Institutional Change." *Sociological Theory* 28: 150–167.

BIS. 2011. "Web Statement on Better Regulative Executive." Accessed 26 February 2014. http://webarchive.nationalarchives.gov.uk/+/http://www.bis.gov.uk/policies/better-regulation/better-regulation-executive.

Black, Julia. 2009. "Legitimacy and the Competition for Regulatory Share." *LSE Law, Society and Economy Working Papers* 14. London: LSE.

DETR. 1997. *Building Regulations 1991: A Consultation Paper on Proposals for Competent Persons*. London: DETR Building Regulations Division.

DETR. 1999a. *Building Control Performance Standards*. London: DETR.

DETR. 1999b. *The Building Act 1984 – Taking Forward Self-Certification Under the Building Regulations*. London: DETR.

DiMaggio, Paul J., and Walter W. Powell. 1983. "The Iron Cage Revisited: Institutional Isomorphism and Collective Rationality in Organizational Fields." *American Sociological Review* 48: 147–160.

Directive 98/32/EC of the European Parliament and of the Council of 22 June 1998.

Eade, Christine. 1999. "Raven Flies Free." *Property Week*, June 11.

Eccles, Timothy, and John Pointing. 2013. "Smart Regulation, Shifting Architectures and Changes in Governance." *International Journal of Law in the Built Environment* 5: 71–88.

Eccles, Timothy. 1993a. "Some Concerns with the Vocabulary and Culture of the Valuation of Land and Buildings." Paper presented at the European Real Estate Conference "European Real Estate: An Agenda for Research", Amsterdam, Netherlands, 15–16 July.

Eccles, Timothy. 1993b. "The Vocabulary of Development." Paper presented at the Conference of European Faculty of Lane Use and Development "The Theory and Practice of the Evaluation of Land and Buildings", Kingston, UK, 15–17 April.

Eccles, Timothy. 1995. "The Vocabulary of Development." In *Theorie und Praxis der Wertermittlung von Gebäuden, Grund und Boden. Travaux scientifiques de la Faculté Européenne des Sciences du Foncier Strasbourg*, edited by Aimé de Leeuw and Sarah L. Sayce, vol. 20, 89–102. Frankfurt am Main: Peter Lang.

Eccles, Timothy. 2009. *The English Building Industry in Late Modernity: An Empirical Investigation of the Definition, Construction and Meaning of Profession*. London: London School of Economics (unpublished).

European Commission. 2013. "What is Better Regulation?" Accessed May 7, 2013. http://ec.europa.eu/governance/better_regulation/br_what_en.htm.

Finn, M. 2007. "In the Beginning." *Building Control*, March.

Fisher, Elizabeth. 2007. *Risk Regulation and Administrative Constitutionalism*. Oxford: Hart Publishing.

Fournier, Valerie. 1999. "The Appeal to »Professionalism« as a Disciplinary Mechanism." *Sociological Review* 42: 280–307.

Giddens, Anthony. 1990. *The Consequences of Modernity*. Cambridge: Polity Press.

Giddens, Anthony. 1991. *Modernity and Self-Identity*. Stanford (CA): Stanford University Press.

Giddens, Anthony. 1994. "Living in a Post-traditional Society." In *Reflexive Modernization*, edited by Ulrich Beck, Anthony Giddens and Scott Lash. 56–109. Cambridge: Polity Press.

Gunningham, Neil. 2009. "Environmental Law, Regulation and Governance: Shifting Architectures." *Journal of Environmental Law* 21: 179–212.

Hobbes, Thomas. 1968 [1651]. *Leviathan*. Harmondsworth: Penguin.

IASB. 2011. *IFRS13 Fair Value Measurement*. London: IASB.

IVSC. 2005. *International Valuation Standards*. Chicago: Appraisal Institute.

Johnson, Tony. 1972. *Professions and Power*. London: Macmillan.

Larson, Magali. 1977. *The Rise of Professionalism*. London: University of California Press.

Loosemore, M., and N. Adonakis. 2007. "Barriers to Implementing OHS Reforms – The Experiences of Small Subcontractors in the Australian Construction Industry." *International Journal of Project Management* 25: 579–588.

Peters, Richard. 1956. *Hobbes*. London: Pelican.

Raz, Joseph. 1990. *Authority*. London: Wiley-Blackwell.

RICS. 1994. *Commercial Property Valuations*. London: RICS.

RICS. 2002. *Property Valuation: The Carsberg Report*. London: RICS.

RICS. 2009. *RICS Valuer Accreditation and Regulation*. London: RICS.

RICS. 2010. *Valuation – Professional Standards*. London: RICS.

RICS. 2014. *Valuation – Professional Standards*. London: RICS.

RICS Valuation Faculty. 2003. *Valuation and Sale Price Variance Report 2003*. London: RICS.

Stevenson, Tom. 1993. *Surveyors Uneasy in the Spotlight*, Accessed May 5, 2013. http://www.independent.co.uk/news/business/surveyors-uneasy-in-the-spotlight-tom-stevenson-considers-the-plight-of-property-valuers-deeply-embarrassed-by-the-queens-moat-houses-debacle-1507533.html.

Stinchcombe, Arthur. L. 2001. *When Formality Works: Authority and Abstraction in Law and Organizations*. Chicago: University of Chicago Press.

Wikipedia. Undated. "Real Estate Appraisal." Accessed April 7, 2013. http://en.wikipedia.org/wiki/Real_estate_appraisal.

Aleksandra Herman

The Reconfiguration of Power as a Legitimization of Informal Political Actions in Local-Level Politics in Contemporary Poland

Introduction: Objectives and analytical assumptions

This paper considers local politics viewed from the angle of self-governance in contemporary Poland, based on the concept of the reconfiguration of state power against the backdrop of Polish historical and social conditions. Within this perspective, it presents separate social institutions operating on the bottom level of self-governance and considers their political potential in the local environment.

The assumed key hypothesis is that in the conditions of power reconfiguration and self-governance, the range of local politics broadens to include the previously non-politicized, new areas of the public sphere. At the same time, broadening the range of politics within the system of reconfigured power leads to mechanisms arising that work to merge the informal and formal political influences, as well as to formalize informal influences. As a consequence of this, informal political actors and initiatives are pulled into the area of local politics through legal instruments. The broad perspective on local politics in developed countries in this text is analyzed using theoretical models that come from literature on classic political anthropology. Such an approach is rendered possible by the fact that a broad range of politics in the local environment, according to concepts of reconfiguration, matches its scope in the primitive communities. Thus, the paper attempts to compare the applicability of two separate theoretical frameworks in contemporary anthropological research: the classic political anthropology models, and contemporary models of power reconfiguration.

The core analysis conducted in this text covers selected informal mechanisms of local politics within the conditions of the reconfiguration of power in contemporary Poland. However, as this analysis is not a monograph on a specific, defined local community, it also includes historical and cultural determinants of the wider national (Polish) arena. Within this perspective, specific institutions that are common throughout Poland (villages and FWAs) are analyzed as potentially useful tools in local politics. Both of these types of solutions have a bottom-top nature, balancing on the borderline between formality and informality.

Auxiliary research questions emerge from both the assumed hypothesis and from the reasons behind the employment of the classic analytical model. Does the division between formal and informal political influences become blurred as a result of the reconfiguration of power at the local level? Can informality be locally empowered as an important factor of bottom-top political influence? Is the politicization of the private domain, connected with public participation, a phenomenon working to open the local communities to activity and self-governance, or is it working to close it?

The final objective of the analysis is to attempt an evaluation of realistic possibilities for the attainment and use of informal influence on local power through local communities in contemporary Poland, and to evaluate informality in local politics under conditions of the reconfiguration of power.

Classic models of local-level politics in cultural anthropology and the question of informality

The development of political anthropology could only begin after the breakdown in ethnocentrism, previously manifested through the fusion of the theory of politics with the theory of state. The first ventures into anthropological fieldwork that focused on politics took place in Africa during the beginning of decolonization (Lewellen 2003, 7–8). In studying what constituted "politics" in segmentary societies, researchers focused on the systems of kinship and what influenced them, because these systems of kinship gave rise to political decisions. Thus, already in its early days, political anthropology identified the specificity of the political sub-system, which, in fact, rests on other sub-systems from which it is inseparable. As such, it may be classified as both formal and informal. The separation of politics from the state has led to the conclusion that each community shapes politics as a sphere of the realization of power. This conclusion also applies to societies with minimum power.

At the same time, the issue of the "inseparability" of the political sub-system caused problems with both the operationalization of the scope of research and with the theoretical assumptions. The problems occurred whenever attempts were made to transfer the cognitive grid, elaborated through research on simple societies, onto state systems that had already built a structure of formal power, administration and coercion mechanisms within the processes of Weber's (2003 [1905]) rationalization.

An accurate definition of politics in anthropological fieldwork has been provided by Turner, Swartz and Tuden (2006, 7): politics refers to "the processes involved in determining and implementing public goals, and in the differential achievement and use of power by the members of the group concerned with these goals"[1]. Therefore, the adjective "political" will apply to everything that is at once public and goal-oriented, and that involves a differential of power (in the sense of control) among the individuals of the group in question (Turner 2006, 7). An approximate method for the inclusion of informality into politics was proposed by Swartz (1969, 8–10) through concentrating on political actors (political field) and influences by varied local conditions (arena).

The inseparability of the public and the private in anthropological research, and so of the formal and informal, is also noticeable whenever we attempt to find a definition of "informality" in cultural anthropology literature. It is absent from both the *Encyclopedia of Social and Cultural Anthropology* (Barnard and Spencer 2007) and from the back-of-the-book index in Lewellen's monograph, *Political Anthropology: An Introduction* (2003). It is not even defined in *Political Anthropology* (2006) by Swartz, Turner and Tuden.

However, for the purposes of the proposed approach to research on local politics, the definition of informality formulated by Misztal (2002, 8) might be of help: "a form of interaction among partners enjoying relative freedom in interpretation of their roles' requirements". In this, she makes a reference to Gluckman's (1955, 19; 1965, 256–257) concept of multiplex ties between actors. This is characteristic of both primary and contemporary small-size communities where the same actors partake in multiple interactions in striving for multiple objectives, because the same people play various roles in different subsystems. We find an analogous phenomenon in modern societies, where the significance of the dominant role decreases and society allows actors to be more "multi-role" performers (Misztal 2002, 43). The contemporary overlapping of the public and private domains or realms results in an increasing role for social networks. These networks encompass the general structure of informal relationships within defined associational structures, connected with the increasing variation of acceptance for departing from the rigid scheme of one's role (Misztal 2002, 22).

1 At first thought, the term "politics" brings up many associations suggested by the English language that differentiate between *polity*, *policy* and *politics*. In order to not mix them up, it is best to clarify to what they refer: (a) a form of organization of power in human societies; (b) types of activities taken up to deal with public issues; (c) strategies resulting from rivalry between people and groups (see, *inter alia*, Balandier 1967, 32).

Reconfiguration as the new framework in the anthropological research of contemporary local-level politics

Research on contemporary national political processes, that is, those not restricted to local communities, leads to the conclusion that fundamental changes in power structures, induced by globalization, are under way. The main point of reference in research on the local level has been the concept of the reconfiguration of power and state responsibility, introduced by Banaszak, Beckwith and Rucht (2003). These authors have constructed a typology of independent power relocation forced by structural shifts within the state system in the EU[2]. Reconfigured power is subject to radical and diffused transformations of which the society is not aware. The process takes place through the multifaceted delegation of the traditional tasks of state power and of state responsibilities. The authors distinguish four directions of the reconfiguration:

(1) uploading – the transfer of certain entitlements of state power to the level of supra-state bodies and treaties;
(2) downloading – the transfer of certain entitlements of state power to the lower level of self-governance of the local entity;
(3) lateral loading – the transfer of certain entitlements of the elected power to unelected bodies (like courts, quasi-NGOs, etc.);
(4) off-loading – the transfer of responsibility for care and related services to NGOs or businesses.

Pursuant to this division, the scope of politics remains identical to that defined by Turner, Swartz and Tuden (2006), but reconfiguration brings on the empowerment of new subjects and relationships of power, and the legitimization of their political influence. In reference to the above, it is important to note that the scope of reconfigured power is dynamically staked out by various actors exerting different types of influence on public goals[3]. The empowerment of informal initiatives and the opening up of formal authorities to the influences exerted by "second-line" actors is an additional, presumed feature of local politics in a reconfigured state, where informal institutions that have a significant impact on

2 However, it is also possible to apply this concept to the analysis of state or political systems of other countries or larger regions of the world.

3 One characteristic example is the striving to influence power devoid of any intention to take it over, which is a distinct feature of not only the so-called "new social movements" (della Porta and Diani Mario 2006, 36–37), but also of numerous local social initiatives.

the shaping of political objectives receive the possibility of legalization. Cooperation between the key realms – private and public – renders it possible to translate informality into formality, based on the process of informality encompassing formal structures.

In this article, I intend to focus on the downloading and off-loading directions of reconfiguration, as they are directly present and implemented in the political field of local communities. Quite unlike uploading and lateral loading, they rest on mobilisation and bottom-up political participation, and on the strengthening of self-governance of local authorities. In local communities we are dealing with the multiplexity of ties, a highly dynamic political field, a fluidity of delineations between the private and the public and, consequently, greater opportunities for influencing public goals by local actors. Therefore, in accordance with the main hypothesis, two factors lead to the broadening of the political field onto previously non-political areas: on the one hand, the formal and systemic reconfiguration of power, and on the other hand, the theoretically identified determinants of the local community. What is more, these new areas will not be pulled only into political processes (which is a well-known phenomenon); as a result of the tools available through the downloading and off-loading of power, even informal political influences will be integrated into the formal sphere through their being tied together with formal initiatives via legal solutions.

The noticeable liberalization of the EU "soft laws" contributes significantly to the strengthening of informal political mechanisms – it gives space for the temporary empowerment of so-called social initiatives, which may be used not only by grassroots movements striving to exert an influence on politics, but also by professional political actors in direct power struggles. However, the main idea behind such reinforcement of these initiatives is to support the grassroots informal organizations in making their activities more professional and to take a fuller presence on the power stage. This transposition, incidentally, fits right into the Polish conditions governing social participation presented in the next paragraph. However, the question remains whether this available mechanism is indeed employed.

Local-level self-organization in contemporary Poland: Cultural determinants of the global arena

The idea of a civil society calls for an appropriate balancing of not only the power that is exercised by democratically elected authorities, but also of the power that is held by citizens in the form of public participation. The assumption of cooperation imposes, upon both parties, certain models of functioning that are based on mutual communication. The role of individuals in a civil society, then, calls for the transgression of the private sphere, and a certain transposition of private needs and expectations onto the public sphere. By doing this, individuals can communicate those needs and expectations to the authorities and strive for their realization in the public domain. Seligman (1995, 5) states that if the sense of a shared public is constitutive of civil society – that is, a free and equal citizenry – so is the very existence of the private. Therefore, in order to form a strong public sphere, a clearly defined private domain – the area for the shaping of citizenry – is necessary. The specific "missing link" in the cooperation between the authorities and citizens is the sector of non-governmental organizations. An analogous mechanism functions in the micro-scale of the local environment. The three vertices of the considered triangular model of local self-governance are: citizenry, NGO organizations and local authorities.

Contemporary patterns of self-organization in Poland largely differ from the assumptions of a civic society. As a result of historical and social determinants, they continue to resist attempts at transposing patterns of mature citizenship. Contrary to the stereotypes, self-organization in contemporary Poland does not suffer from limitations brought on by the lack of a proper social structure (or at least, this is not its sole shortcoming). Polish civic participation is in fact conditioned by the experience of almost two centuries of national conspiracy against colonial rule.

The perspective of post-colonialism, understood as a collection of methodological approaches (Said 1979, 23, 110), concentrates on researching the effects of the colonial period. The principal point of such argumentation is that emancipation from colonial domination has never caused an automatic restoration of the social balance in formerly colonized societies, and the period of subjugation often left a permanent mark on them, both in the material and the social infrastructure. Primarily, there are still strong and significant remnants of the absence of Polish statehood following the partitions perpetrated by Prussia, Russia and Austria, which lasted from 1772 until 1918. Secondly, the consequences of the 20th century occupations are also at play: aggression by Germany and the USSR in World War II, and the post-war subjugation to the Soviet Union – introduced by the 1945 Yalta Conference – were experienced jointly with other nations of Eastern and Central Europe.

Wedel (1986) sums up her research conducted in Poland during the 1980s and 1990s by describing it as a country that, during the days of the communist regime, was characterized by widespread distrust towards bureaucracy, information, economics and state officials. These attitudes stem from the Polish experiences, which were, simultaneously, the experiences of individual citizens and their families. They form a specific family burden, imprinted in the patterns of everyday life and passed from one generation to the next, boiling down to the necessity of having to circumvent the law in order to survive. These hindrances may also be attributed to the post-colonial heritage of Soviet supremacy, which conditions both the absence of civic activity and the deficit of democracy in the decision-making processes in the public sphere. Therefore, a certain vein of "clientelism", the most basic patterns of which are still present in the society and in the system, was determined by the state's monopolization of each and every aspect of social life (activity) and of economic life (goods and services), while the state itself was suppressed by the domination and control of the USSR.

Present-day Polish local self-organization may have dwindled, but its intention is still to protect citizens from authorities by keeping them away from them. The organizations steer clear of formal power structures, rather than seeking access to them and therefore they do not exert influence by participating in these structures. The main problems of the non-governmental organizations are: insufficient social, financial and organizational potential,[4] absence of the perception of being able to influence the social environment; and the unwillingness of the organizations themselves to cooperate and participate (Bogacz-Wojtanowska, Dutkiewicz and Górniak 2011, 26–29).

To exemplify the tendencies described above: as many as 60% of commune offices have not kept registers of non-governmental organizations active in their territories, despite the fact that nearly 91% of them have appointed units or persons responsible for contacts with non-governmental organizations. This indicates that, even though the cooperation is limited, the authorities have recently started to appreciate the importance of the social sector (Sobiesiak-Penszko 2012, 138). An analysis of the structure of cooperation between the authorities and the NGO sector leads to the conclusions that in Poland, the role of non-governmental organizations as service providers has solidified, and that their operations concentrate mainly on the aspects of distributing public resources (Sobiesiak-Penszko 2012, 140–142). This, however, undermines the fundamental function of the NGO sector as a mediator in the dialogue between citizens and authorities.

With regard to the readiness of Polish society for taking up participation, most citizens are willing to participate as long as the activity is somehow connected to

4 Public resources continue to be the major source of funding for the sector (USAID 2012).

their personal interests, even if they have to bear certain burdens (the burden of, for example, changing plans). The current rate of involvement stands between 23.9% and 30.4% ("pragmatic participation"). 6.3% to 7.2% of people are willing to participate in matters not connected to their personal interests, but only as long as no burdens are generated for them as a result ("occasional participation"). And, finally, 22.3% to 27.1% of persons are willing to participate regardless of their own interests and burdens ("unconditional participation") (Olech and Kotnarowski 2012, 154).

Therefore, the "Polish paradox" might be based on the fact that in the same country in which a massive bottom-up movement like Solidarity had emerged in protest against the communist authorities, there is still an absence of awareness about self-governance as a grassroots activity, which may potentially prove a powerful political tool in the hands of citizens. It is, however, important to remember the historical context of the functioning of Solidarity, which unified numerous antagonistic movements in the overarching relation of "us versus them" (where "them" referred to the official authority structures). Currently, we are still observing a degree of distrust towards the authorities, although one much less pronounced, especially in the case of local authorities, to whom citizens feel "closer". However, the pluralism of values and ideologies fosters the dissipation of social activity, and not its unification in a fight against one common enemy. Contemporary self-governance, especially as an essential element of the reconfigured state, has as its objective its participation in power, rather than its combating against it.

Contemporarily it is emphasized that in Poland, provisions of law serve merely as a formal framework that enables the inclusion of inhabitants of self-governance units in the decision-making processes. The quality of this participation, however, is largely determined by factors which lie outside of the law itself (Krajewska 2011, 22, 42–43). The culture of cooperation remains far behind the legally regulated options, although it is continuously improving. Amendments made in 2010 to the Act of the 24th of April 2003 on Public Benefit Activity and Volunteer Work mandated some forms of cooperation between local authorities and CSOs, with the outcome that more representatives from the NGO sector now provide their opinions on laws and programs proposed by the government. Since 2003, individual citizens have been able to designate 1% of their tax liabilities to organizations with public benefit status. More organizations receive such status every year, while other organizations intensify their campaigns to also achieve the public benefit status. As a result, the percentage of citizens that designated 1% of their tax liabilities increased from less than 3% in 2004 to 38% in 2010 (USAID 2012, 153–154). Despite this, some difficulties still exist. They include a lengthy, complicated and bureaucratic registration process for NGOs; the registration's dependence on the will, knowledge and attitudes (even political) of particular officials from the registration courts; and the fact that organizations that are not convenient for authorities might be excluded from receiving government support, especially at the local level (USAID 2012). The observed politicization of the non-governmental sector negatively affects the functioning of NGOs, both in the lateral dimension (cooperation between organizations and citizens), as well as in the vertical (cooperation with authorities).

It is commonplace for the data from national NGO databases regarding local associations (formal and informal) to diverge from the actual state of affairs – many such organizations have ceased to exist, but this fact is not recorded in the national databases (USAID 2012). At the same time, there are many new, often informal associations that are able to survive without public funding. Moreover, within the structure of non-governmental organizations in Poland, it is not uncommon for the authorities themselves to encourage organizations that are already deeply embedded in the local structure to formalize their activities in order to gain a broader spectrum of options due to receiving support. In certain cases, this may clash with the very functions of these organizations, which, although not holders of the appropriate mandate, should above all represent the public interest instead of providing services to the organs of public authority (Makowski 2011, 13). However, there are legal mechanisms in place that facilitate the solicitation of funding for the pro-social activities of informal entities. Advocates of informal activities emphasize that due to the very form of their undertakings, their energy serves to fuel real accomplishments instead of being wasted on bureaucracy and documentation. Informal leaders also point to the value of the trust vested in them by public opinion, as their projects – completed either on a cash-free basis or for small amounts of money – are visible to all. It is difficult to evaluate this phenomenon as a positive one, as this is an informality of an escapist nature.

In the following paragraphs, two areas of local self-governance in Poland are presented: *solectwo* (as a territorial community of the lowest administrative level, the equivalent of "village") and Farmers' Wives Association (FWA, as a commonplace, rural, organized form of civic engagement for women). The outline of the customary and legal determinants of the functioning of these institutions is illustrated with detailed examples from the research conducted. The presented examples show not only the inseparability of formality and informality in an organized grassroots activity in local communities, but also the specificity of Polish informality in the light of the discussed determinants.

Law versus structure: The case of the village *[solectwo]*

"Law versus structure" in the above subtitle indicates that the legal framework provides a certain scheme to be used by the structure for civic initiatives, as an element of self-governance in the process of the downloading of power.

While analyzing the issues connected with civic and political activities, it is important to pay attention to the legal conditions and practices relevant to the functioning of bottom-level self-governing bodies. A territorial self-government is the most important form of self-governance, construed as the shifting of certain competences within the framework of public administration directly to territorial communities. The institution of *solectwo* could be construed as a special type of participation mechanism, characteristic of rural communities (Makowski 2010). As the smallest territorial units operating below the level of communes, they have no legal personality[5]. At the same time, the legal structuring of village authorities lends them some potential space for self-governance based on direct democracy, as well as certain channels for its accomplishment through political lobbying in the commune authorities. The dual character of village organization within Polish law is also attested to by the fact that even though their authorities are appointed through democratic and direct elections, the National Electoral Commission (the highest electoral body in Poland) does not collect any data regarding these elections.

Villages, as units auxiliary to communes, may only manage their financial affairs within the limits of commune budgets. However, the rules governing these finances (the so-called village fund) must be enacted by the commune and written in the village statute. Unfortunately for villages, commune authorities are not always necessarily interested in grassroots activities nor are they open to such initiatives. In practice, however, neither the villages nor the communes have so far developed this sort of bottom-up activity, due to the low participation of the village inhabitants and the lack of funding in the communes. The law allows villages to manage financial means from various sources (e.g. revenues from administering communal property, revenues generated by their own activities, such as the organization of events, and national and EU funding or donations). However, village councils and commune representatives *[voyts]* often leave these affairs unattended, as they feel they are not important. Due to this neglect, commune statutes often lack appropriate provisions regulating

5 Territorial self-government in Poland is made up of regional territorial units [*województwa* – voivodeships], then *poviats* – a lower level, and finally of *gminy* or communes, the lowest level of territorial self-governing units.

this issue or, alternatively, there is a mere reference to the village statute, which points nowhere (Iwanicka 2013).

The village's resolution-passing body is the rural meeting – that is, an assembly of inhabitants that reaches all the way down the power-downloading ladder. The executive authority is held by the village mayor, who enjoys legal protection as a public functionary. The role of a village mayor [Polish *sołtys*, German *Schultheis* or Latin *sculdasius*] in Poland is that of a public post, historically and socially legitimized since the medieval times, connected with the German town law foundations. The village council is an auxiliary body: it fulfils an advisory and consultative function for the village mayor[6].

In 2005, research was conducted on village mayors in three communes (listed herein as commune C, commune S, and commune O) located in various regions of Poland[7]. Despite the fact that nine years have passed, this research remains probably the most complex and up-to-date work regarding village mayors and their work. Moreover, the results of this research were elaborated neither within the perspective suggested by the theory of the reconfiguration of power, nor as data to evidence the mechanisms of the lowest-level self-governing territorial communities. In the conducted research, however, a number of patterns of cooperation between village mayors and their councils were encountered. In each of the cases, respondents highlighted that they were built around family ties or personal connections. Two frequent models of interactions at the borderline of formal and informal are especially worth presenting. The first one is the model of an active married couple, where one of the spouses holds the function of the village mayor and the other is a member of the village council. The model is built around meetings in private homes or gardens for a cup of tea or coffee (women), or for a couple of beers or vodka (men)[8].

---

6 Village authorities were approved by the Act of 8th March 1990 on Commune Self-Governments.

7 Jointly, I conducted 68 interviews, under the direction of Professor Małgorzata Fuszara, as part of the grant *Enlargement, Gender and Governance: The Civic and Political Participation and Representation of Women in the EU Candidate Countries,* financed in the EU Fifth Framework Programme. An unpublished report on this research has been previously put to use in, among others, Fuszara (2006; 2013), Czech (2007) and Kurczewski (2009).

8 The role of alcohol in "getting things done", as the factor most associated with "breaking the ice" of formality and entering into an informal relationship, was described in detail by Wedel (1986), a long-time researcher of contemporary Polish customs in the mid-1980s and early 1990s.

> My council is great. There are only men in it; they are all members of the Voluntary Fire Brigade[9]. My husband is in the council too. Whatever I tell them should be done, they do it. They have never refused. Partly to show my gratitude, I have been trying to obtain funds to renovate the fire brigade's watchtower. Obviously, this watchtower serves the whole village, but since I'm dealing with firemen, they are concerned with this all the more. It's not difficult to work things out with them. We often meet informally, talk about what else could be done. Sometimes I have to buy them vodka for this social work and drink with them at the end of the day. So I buy and I drink. The council works a lot, it also shows initiative. I think they like and respect me. I figure this from the fact that they never say 'no' to me (mayor of commune C, female, 46 years old, married, farmer, 2005).

> My council is enviable. There are only young men there. They are always brimming with ideas, always up to things, and in private life they are all friends. They also show a lot of initiative – for example, recently they got junior high students to help in hay making on one of the meadows. Then there was a big bonfire for everyone. I had to buy drinks for the working youth with my own money, but the sausages for the bonfire were bought with the village funds that were left over after the last investment. It was great fun, the kids also enjoyed themselves (mayor of commune O, female, 49 years old, married, farmer, 2005).

A pattern of conflict (one case) also had informal relationships at its roots:

> All the members of my council are men, who don't listen to me. They often meet on their own, drink vodka – I can't be expected to drink with them. And then they defy me. They don't want to work with me. Sure, another woman in the council would probably help. Obviously, a woman sees many things differently. Still, I don't think that their defiance comes from the fact that I'm a woman. There had been all kinds of reactions to the man before me (mayor of commune O, female, 46 years old, married, housewife, 2005).

Unfortunately, most often the village mayors stressed that cooperation with their councils was less than satisfactory (35 of the pool of 65) or even non-existent (11 of 65), and they emphasized the absence of any civic activity:

> Councillors only advise rather than do anything themselves, but my demands on them are not high (mayor of commune S, female, 53 years old, widow, farmer, 2005).

> The council does nothing, no one has time for unpaid work (mayor of commune S, male, 40 years old, married, farmer, 2005).

---

9 The history of Voluntary Fire Brigades in Poland goes back to the mid-19th century. Nowadays, the local units have the legal form of associations, but they operate also on the basis of the Act of 24 September 1991 on Fire Protection. Each commune, on whose territory Volunteer Fire Brigades exist, cooperates with these units to ensure safety, and the brigades additionally carry out the commune's tasks related to culture, education and sports.

> There are problems with the council; I usually act alone (mayor of commune S, male, 60 years old, divorced, pensioner, 2005).

> [The] council lacks its own ideas – it's passive (mayor of commune O, female, 46 years old, married, farmer, 2005).

> [The] council does absolutely nothing – we never meet, there is a lack of consultations and decisions (mayor of commune C, female, 55 years old, married, pensioner, 2005).

> The council doesn't even help me in communicating announcements to the Agricultural Market Agency. If there's something pressing going on, it's me who has to drive around the village (mayor of commune C, female, 54 years old, married, farmer, 2005).

The village mayors would also complain about the lack of interest of the inhabitants, which rendered it impossible to carry out any social initiatives:

> People don't care; they give it all a miss. My council is the same (mayor of commune C, female, 46 years old, married, ward nurse, 2005).

> There's no one to work with, people are just not up to it. And yet, everyone brings their problems to the village mayor (mayor of commune O, male, 32 years old, married, farmer, 2005).

The same reasons were identified as the basis for the malfunctioning of the rural meetings, which the village mayors also complained about.

The examples confirm the accuracy of the classic anthropological models, according to which it is impossible to separate formality and informality within broadly construed political activity, and likewise inconceivable to separate the private and public realms (the meetings of social actors in private homes, drinking vodka, etc.). At the same time, the presence of these mechanisms results in the activity of the local authority structures, in a way lending political empowerment to the local community and thus fitting into the model of reconfigured power. For this reason, the institution of villages, along with its separated authorities for political activity, constitutes a significant example of informal, but legitimized, practices of political activity in bottom-level local communities. What is more, this is an example not from the NGO sector but from self-government, which makes it an interesting hybrid of civic and political activity. In theory at least, it is capable of fulfilling the idea of direct democracy due to its social micro-scale.

However, it is still a body of self-governing administrative structure, and as such it is not particularly popular among the locals. Despite the availability of a sufficient legal framework, in practice it is the structural conditions that determine the low rates of social activity. Global determinants hamper self-governance, which is a necessary precondition for a well-functioning reconfigured state. Generally, however, informality in the discussed institution favours its effectiveness, both with regard to models of cooperation and the use of multiplex ties. A certain blurring of the borders between the formal, informal, private and public realms is a factor conducive of social mobilization, both positive (examples of cooperation in the public sphere, which are based on personal connections) and negative (refusal to cooperate in the public sphere, based on personal connections).

Structure versus law: The case of the Farmers' Wives Association (FWA)

The second case illustrates an attitude that is structurally contrary to the case of village councils. It exemplifies a situation where active civic participation has been attained, but it refuses to be harnessed by legal regulations. It might sound like a paradox that such initiatives carried out without any public funding often become economically prosperous and successful, but they do. It is often due to the authentic human activity factor involved.

FWAs started to emerge in the second half of the 19th century. From the very beginning, their mission was to fulfil various social and organizational functions for women from rural areas. This approach has secured this form of local women's association a great deal of social legitimization and allowed them to exert a significant impact in the localized public sphere. Nowadays their activities focus on improving local social and health care facilities for children and families, and on strengthening female entrepreneurship. More recently, these associations have begun to put greater emphasis on promoting local culture, which has been very well received by young girls, thus resulting in a renewed bout of FWA's popularity.

Polish law reserves the name FWA for organizations with the status of social-professional associations of farmers (so-called agricultural co-operatives). However, the term is also commonly used by other rural and small-town female initiatives, such as associations (formal, commonly known as "registered") and spontaneous, informal social activities. Choosing an association's status is connected with the profile of activities and with the planned development perspectives, as well as with available options to raise funds. On the one hand, FWAs – functioning as agricultural co-operatives – may distribute any potential profits to their members, whereas an association is obliged to allocate any such profits to statutory objectives. At the same time, associations enjoy a greater wealth of options in raising funds for their activities from external sources. In both cases they may, however, apply to carry out tasks delegated by self-government authorities.

On a side note, agricultural co-operatives are voluntary organizations united within the National Union of Farmers, Co-operatives and Agricultural Organizations [*Krajowy Związek Rolników, Kółek I Organizacji Rolniczych, KZRKIOR*], which is the only operating Polish farmers' union. This status is imposed on the FWAs by the Act of October 8th 1982 on Socio-Occupational Organizations of Farmers. The inclusion of social activity into the framework of one federative-structure organization had an ideological background: it was used as a tool of state politics against non-partisan farmers. Nowadays, some registry courts refuse to register farmers' wives' associations, basing this decision on the provisions of the Act of 1982.

However, some of the FWAs are not aware of the formal and legal aspects of their functioning. The reason for this, among others, is that the FWAs are connected with a specific farmers' union that has always taken care of these matters for them. In the 1990s, many of these agricultural co-operatives ceased their operations, which also put a formal end to the functioning of FWAs. In the same period, FWAs themselves also became less active. At the same time, these organizations were not formally stricken from the register of co-operatives, a fact of which the current members of FWAs are not aware. It is not always clear who is keeping the registration information, nor is it clear where it is kept. In certain cases it is easier to simply form a new organization. Mass reactivation of FWAs (reactivation, not establishment of new ones) has been taking place over the last decade in step with the development of various initiatives in rural areas. However, self-governments and the animators of Regional Centres of European Social Funds [*Regionalne Ośrodki Europejskiego Funduszu Społecznego, ROEFS*] or NGO advisors usually lack knowledge regarding the laws regulating the functioning of agricultural co-operatives or FWAs[10].

But, in practice, voluntary and mass cooperation between *voyts* (commune representatives) and FWAs is as widespread as it is thanks to the informal ties between the two. FWAs are active leaders, traditionally interested in local affairs and thus are trustworthy. For example, some female FWA leaders from one Mazovian commune (commune C) emphasized the informal character of their activities despite the pressure exerted on them by the *voyt* to register. They refused because – as they were explaining – their activity would become just an obligation and not something to enjoy anymore. This argumentation clearly highlights the differentiation between the informal, identified as grassroots and spontaneous, and the formal, which is viewed as limiting and laborious. Despite the informal status, their activity has received financing from the *voyt* on numerous occasions: village

10 This information was obtained through e-mail interviews with Iwona Raszeja-Ossowska (2013), an independent expert on the development of organizations operating in rural areas.

fairs, folk costumes, renovation of premises in a municipal building, congresses and even a sightseeing tour. During the celebrations of the 70th anniversary of their establishment, the most active leaders received commemorative medals. The FWA reciprocates by, for example, helping to organize harvest festivals, during which they sing a song about the *voyt*. They have also won a culinary contest for the best local dish, which made the village famous in the region. Moreover, the FWAs cooperate with other organizations traditionally present in the Polish countryside that also enjoy high levels of social legitimacy, such as the Voluntary Fire Brigades or parishes. Therefore, they have managed to exert a relatively strong political impact by being deeply embedded in the local structure, whether formal or otherwise.

The analysis shows that informality does not lock local organizations out of political influence – it is quite the opposite, in fact. In the described examples, political influence is more a result of the deep-rootedness and multiplexity of ties in the local community, rather than of the legalization of operations in the form provided by the national legislation. Perhaps this also leads to the shaping of another feature of informality, i.e. the tendency to operate informally due to the greater possibility for flexibility and to obviate bureaucracy, which is an inherent burden of formalization. Based on the presented examples, it is possible to conclude that, as a result of the effective operations of informal organizations and of the influence they exert on the local authorities, the borderline between the formal and the informal becomes blurred – as also perceived by the local community – but this inference refers to the discussed examples only.

Summary: Informal political actions in local-level politics within the framework of power reconfiguration

The introduction of this article puts forward the hypothesis that under conditions of power reconfiguration and the development of self-governance, the local political field expands to include the previously non-political public sphere, extending over new actors, resources and values. As such, the broad definition of modern local politics in the perspective of power reconfiguration converges with the now-classic approaches to research from the first half of the 20th century. This expanded conception employs a particular emphasis on both anthropological political models (the classic political anthropology model and the contemporary model of downloading of reconfigurated power) for eliminating the dichotomy between the formal and informal spheres. Thus the broad definition of politics applied in cultural anthropology is further reinforced by the processes of seeking power and exercising power in contemporary local communities.

The examples presented in the article, although selected on the basis of subjective judgment, show the self-governance of local communities from two perspectives: territorial self-governance, and social self-organization. Moreover, the examples of institutions were purposefully chosen due to the significance of informality in their operations. The analysis has confirmed that both types of considered institutions exert political influence and, as such, are included in the local politics through defined legal mechanisms. The main determinant of this influence and of the potential inclusion is not the feature of formality or informality. Rather, it is the extent to which these institutions are deep-seated in the local community, and the strength of the multiple ties between the actors involved. Reconfiguration (in the directions of the downloading and off-loading of power), then, leads to the expansion of the political sphere, and thereby also to the politicization of resources that were previously non-political. Power reconfiguration creates opportunities for real self-governance through grassroots activities.

Interestingly, the Polish escapism reflected in the tendency to cling to informal action – i.e. in circumstances where the law is getting "softer" and bottom-up structures gain power – may actually become a valuable socio-political resource. The historical and social determinants of the wider national (Polish) arena have a hampering effect on civic activity. However, the already "classical" anthropological determinants of the local arena (multiplexity of ties, acceptance of informality) may positively influence the self-governance initiatives of local residents, as these factors help to break through the inhibiting restraints of the national cultural legacy. Informal initiatives still exhibit a strong proclivity towards maintaining their independence, which unveils the the local's perception of the formal power (even the local-level governance) as being alien and unfriendly. Moreover, supporters of such actions emphasize that informality is a way for them not to waste energy on bureaucratic documentation. It might sound like a paradox that such initiatives carried out without any public funding often become economically prosperous and successful, thanks precisely to the authentic human activity factor involved. For this reason, villages are able to control their direct authorities and to lobby for their interests, while FWAs are strong local actors with an important mission of promoting their villages, which often brings them tangible profits. Naturally, the presence and the significance of informal influences on power may occur regardless of the degree of state power reconfiguration. It is, however, noteworthy that in the traditional analytical approach to state power (e.g. according to the categories offered by political science) informality is omitted and does not constitute a significant factor subject to systemic inclusion, and is not acknowledged as politically substantial. Meanwhile, from the anthropological perspective, informal mechanisms are an integral – and potentially empowered – element of the local structures of power, regardless of the framework of analysis.

Regarding the auxiliary research questions that stemmed from the hypothesis put forward, it is possible – based on the analyzed examples – to draw the conclusion that the reconfiguration of power on the local level, and especially the legal instruments for including informal political forces, favour the blurring of boundaries between the formal and the informal in the political field. This is attested to by the models of cooperation (or lack thereof) seen between village councils and village mayors, as well as the patterns of intersection of different social ties between the same actors. As the civil society shapes and solidifies, the borderlines between the private and public realms also become increasingly hazy – that is, members of the local communities more frequently step into the public sphere in order to voice their private interests in an organized manner. Due to the observed effectiveness of the informal activities, and also thanks to their real political clout, informality may be locally acknowledged as a substantial factor of bottom-top political influence. The respect enjoyed by the FWAs in their respective local communities is an example of this. This respect is manifested by locals supporting their initiatives, and by participating in what they do (attending the events they organize; taking part in charitable actions, whether by giving or receiving; speaking of them proudly, presenting them as "local achievements"). Finally, the politicization of the private realm, construed as public participation and the empowerment of the local community, seems to be a phenomenon that may potentially work to open the local communities to civic activity. This is evidenced by the tendency to evaluate informal activity in a positive way, as spontaneous and unrestricted by bureaucracy.

In these final words, besides indicating the possibilities opening up for informal initiatives under a system of reconfigured power, it is also worthwhile to point out their shortcomings. One shortcoming is the very limited capacity to absorb financial resources, and the consequent dependence on financing by local authorities. The second shortcoming is the lack of the professionalization of the informal NGO operations in Poland, as well as problems with maintaining the continuity of their activities resulting from insufficient financing options. These make it difficult also for the formal organizations to keep up their professional activities. Moreover, lingering in informality does not contribute to overcoming the negative historical conditions, but rather conserves them. While the liberalization of the Polish law on associations seems inevitable (Krajewska 2011, 26), subsequent amendment drafts are being rejected. Perhaps this is because the process of power and responsibility reconfiguration also has a feedback aspect to it: most importantly, it significantly curtails the range of state competence. All the while, large central power structures must ensure their own legitimacy and they do so by imposing limits on self-governance in their territories. The process of power reconfiguration, then, has one more aspect – that of an internal conflict consuming the state administration. Therefore, even if the specific appreciation of informality is an inherent element of contemporary institutional and custom-related transformations, and even if it is manifested through a distrustful attitude towards bureaucracy, its local impact might, with time, become a subject of bureaucratic control by the central authorities.

References

Act of 8 October 1982 on Socio-Occupational Organizations of Farmers [Ustawa z dnia 8 października 1982 r. o społeczno-zawodowych organizacjach rolników]. Dz.U. 1982 nr 32 poz. 217. Accessed April 20, 2014. http://isap.sejm.gov.pl/DetailsServlet?id=WDU19820320217.

Act of 8 March 1990 on Commune Self-Governments [Ustawa z dnia 8 marca 1990 r. o samorządzie terytorialnym]. Dz.U. 1990 no, 16 item 95. Accessed April 20, 2014. Available also: http://isap.sejm.gov.pl/DetailsServlet?id=WDU19900160095.

Act of 24 April 2003 on Public Benefit Activity and Volunteer Work [Ustawa z dnia 24 kwietnia 2003 r. o działalności pożytku publicznego i o wolontariacie]. Dz.U. 2003 nr 96 poz. 873. Accessed April 20, 2014. http://isap.sejm.gov.pl/DetailsServlet?id=WDU20030960873.

Act of 24 September 1991 on Fire Protection [Ustawa z dnia 24 sierpnia 1991 r. o ochronie przeciwpożarowej]. Dz.U. 1991 nr 81 poz. 351. Accessed April 20, 2014. http://isap.sejm.gov.pl/DetailsServlet?id=WDU19910810351.

Balandier, Georges. 1967. *Anthropologie politique.* Paris: Presses Universitaires de France.

Banaszak, Lee Ann, Karen Beckwith, and Dieter Rucht. 2003. "Interactive Changes in Movements and States." In *Women's Movements Facing the Reconfigured State*, edited by Lee Ann Banaszak, Karen Beckwith and Dieter Rucht. 1–29. Cambridge: Cambridge University Press.

Barnard, Alan, and Jonathan Spencer, eds. 2007. *Encyclopedia of Social and Cultural Anthropology*. London and New York: Routledge.

Bogacz-Wojtanowska, Ewa, Marek Dutkiewicz, and Katarzyna Górniak. 2011. "Problem jakości współpracy między organizacjami pozarządowymi i administracją publiczną w kontekście dostępnych badań. Analiza materiałów zastanych." In *Jakość współpracy między organizacjami pozarządowymi i administracją publiczną. Raport końcowy z badań*, edited by Grzegorz Makowski, 13–40.

Warsaw: Institute of Public Affairs: 2011. Accessed August 18, 2013. http://www.isp.org.pl/uploads/filemanager/Program%20Spoleczenstwa%20Obywatelskiego/ISPjakowsppracyraportkocowy.pdf.

Czech, Henryk. 2007. "Czynniki kształtujące aktywność polityczną na wsi oleskiej." In *Lokalne wzory kultury politycznej*, edited by Jacek Kurczewski, 387–426. Warsaw: Trio.

della Porta, Donatella, and Mario Diani. 2006. *Social Movements. An Introduction*. Oxford: Blackwell.

Fuszara, Małgorzata. 2006. *Kobiety w Polityce*. Warsaw: Trio.

Fuszara, Małgorzata. 2013. "Sołtyski w Polsce." Accessed June 3, 2013. http://www.feminoteka.pl/downloads/soltyski_w_polsce.pdf.

Gluckman, Max. 1955. The Judicial Process among the Barotse of Northern Rhodesia. Manchester: Manchester University Press.

Gluckman, Max. 1965. *Politics, Law and Ritual in Tribal Society*. Oxford: Basil Blackwell.

Iwanicka, Joanna. 2013. "Porady prawne Fundacji Wspomagania Wsi. Od Piaseczna do Wiązowny." Accessed April 23, 2013. http://www.gazetasolecka.pl/fww3.html.

Krajewska, Anna. 2011. "Prawne aspekty partycypacji na poziomie lokalnym." In: *Prawo a partycypacja publiczna. Bilans monitoringu 2010*, edited by Grzegorz Makowski and Paulina Sobiesiak, 47–58. Warsaw: Institute of Public Affairs. Accessed August 18, 2013. http://isp.org.pl/uploads/pdf/1110451964.pdf.

Kurczewski, Jacek. 2009. *Ścieżki emancypacji. Osobista teoria transformacji ustrojowej w Polsce*. Warsaw: Trio.

Lewellen, Ted C. 2003. *Political Anthropology: An Introduction* (3rd edition). Westport, Conn: Praeger.

Makowski, Grzegorz. 2010. "Fundusz sołecki nie będzie już tylko sołecki?" Accessed April 23, 2013. http://www.isp.org.pl/kwartalnik/aktualnosci,1,278.html.

Makowski, Grzegorz. 2011. "Nowe ramy uczestnictwa obywateli w procesach decyzyjnych." In *Prawo a partycypacja publiczna. Bilans monitoringu 2010*, edited by Grzegorz Makowski and Paulina Sobiesiak, 9–21. Warsaw: Institute of Public Affairs. Accessed August 18, 2013. http://isp.org.pl/uploads/pdf/1110451964.pdf.

Misztal, Barbara. 2002. *Informality: Social Theory and Contemporary Practice* (2nd edition). London and New York: Routledge.

Olech, Anna, and Michał Kotnarowski. 2012. "Jednostki pomocnicze samorządu: między obywatelem a władzą." In *Dyktat czy uczestnictwo? Diagnoza partycypacji publicznej w Polsce*, edited by Anna Olech, vol. 1, 211–233. Warsaw: Institute of Public Affairs. Accessed August 18, 2013. http://www.isp.org.pl/uploads/pdf/1408475792.pdf.

Said, Edward. 1979. *Orientalism*. New York: Random House.

Seligman, Adam. 1995. *The Idea of Civil Society* (2nd edition). Princeton: Princeton University Press.

Sobiesiak-Penszko, Paulina. 2012. "Organizacje pozarządowe a partycypacja publiczna." In *Dyktat czy uczestnictwo? Diagnoza partycypacji publicznej w Polsce*, edited by Anna Olech, vol. 1, 137–148. Warsaw: Institute of Public Affairs. Accessed August 18, 2013. http://www.isp.org.pl/uploads/pdf/1408475792.pdf.

Swartz, Marc J. 1969. "Introduction." In *Local-Level Politics. Social and Cultural Perspectives*, edited by Marc J. Swartz. 1–46. London: University of London Press Ltd.

Swartz, Marc J., Victor Turner, and Arthur Tuden. 2006. "Introduction." In *Political Anthropology*, edited by Marc J. Swartz, Victor Turner and Arthur Tuden. 1–41. New Brunswick: Aldine Transaction.

USAID (United States Agency for International Development). 2012. *The 2011 Sustainability Index For Central and Eastern Europe and Eurasia. 15th Anniversary Edition*. Bureau for Europe and Eurasia, Office of Democracy, Governance and Social Transition. Accessed August 18, 2013. http://transition.usaid.gov/locations/europe_eurasia/dem_gov/ngoindex/reports/2011/2011CSOSI_Index_complete.pdf#page=160.

Weber, Max. 2003 [1905]. *The Protestant Ethic and the Spirit of Capitalism*. New York: Courier Dover Publications.

Wedel, Janine. 1986. *The Private Poland: An Anthropologist's Look at Everyday Life*. New York: Facts on File Incorporated Publications.

Francisco Linares

Social Networks, Social Norms and Workers' Resistance: A Computational Simulation Analysis

Introduction

The aim of this chapter is to account for certain consequences of the embeddedness of workers' behavior in informal social relations which are developed within a firm's formal organization, thus extending Granovetter's (1985) famous thesis of the embeddedness of economic actors in social structure. Granovetter (1985, 495) himself recognized in his seminal paper that "it seems plausible [...] that the network of social relations within the firm might be more dense and more long-standing on the average than that existing between". A nice empirical example of this issue is provided by Leon's (2011) game theoretical analysis of informal norms of resistance, which emerged within teams working in a firm of the automotive sector located in Barcelona's industrial area. The goal of these norms was to lower the production quota, that is, the number of pieces to be produced per day established by the firm's managers. According to Leon (2011, 78–79) the set of informal norms ruling the behavior of workers included: (a) the duty to hide skills while being timed, (b) the duty to work at the same rhythm set by other colleagues on the production line, and (c) the duty to hide from management the means of resistance and tricks that colleagues employ in managing the production rhythm. Leon convincingly shows how the incentive system among team mates works to back these prescriptions.

The topic of Leon's empirical research has a long tradition in sociology, going back at least to the famous Hawthorne experiments conducted by Mayo (1945), who opened new venues of research in social psychology, the science of management, and the sociology of organizations. From a theoretical point of view, Leon's approach is in debt to Coleman's (1990) theory of the corporate actor. Coleman's (1990, 423) theory made a strong criticism of Weber's conception of organizations, "where only the central authority is treated as a purposive actor" and "the fact that the persons who are employed to fill the positions in the organization are purposive actors as well is overlooked".

Following these guidelines, my goal is to show how different network topologies of informal relations (links which run parallel to the firm's formal positions) can foster, or not, the emergence of social norms of resistance (in the sense of Leon's article). In doing so, I will use a relatively infrequent tool of analysis among sociologists: an agent-based simulation model, which attempts to capture some of the main features of workers' social interactions.

Recent revisions to the literature of organizations, social networks and worker resistance (Vaughan 1999; Brass et al. 2004; Roscigno and Hodson 2004) show that this topic has been largely overlooked by scholars. While literature on networks and organizations, as well as on organization and worker resistance, is quite large, I have been unable to find evidence related to the effect of network topology on worker resistance within organizations. It seems as if it is just assumed that, quite obviously, some social relations among peers must exist if workers are to produce any resistance action; but the details of how network topology should be to promote these relations are largely unexplored. For instance, in their rigorous examination of 82 workplace organizational ethnographies, Roscigno and Hodson (2004) conclude that strike action and other individual resistance strategies (such as work avoidance or absenteeism) are more likely to occur in workplaces characterized by systematic and ongoing interpersonal conflict with management, union presence and, sometimes, bureaucratic structure. Adding that "social relations on the shop floor play a meaningful role in prompting both collective and individual manifestations of class resistance" (Roscigno and Hodson 2004, 33). However, with the term, "social relations," they are mainly referring to conflict (or the absence of conflict) between managers and workers. In a previous work (Dixon and Roscigno 2003), the authors specifically attempted to show the impact of social networks on strike behavior. But, again, the details of the network topology are missing since what is measured is the amount of union activity with variables such as the number of card-carrying individuals.

The chapter will proceed as follows: First, because these resistance norms cannot be taken as a given (that is, they emerge at some point as a product of agents' interactions) and because there is not yet a widely accepted sociological theory of the emergence of norms, a comprehensive definition of social norms and an account of how they are supposed to emerge are provided in this paper. Secondly, some basic notions of "complex adaptive systems" and the "new science of networks" are introduced. These notions are required to understand the agent-based simulation model and how it works. A brief exposition of the model is then provided, but details are left to the appendix. Next, the main results of the model are elucidated. The chapter ends with a concluding section, which discusses the results on which the simulations shed light.

The puzzle of social norms

In the analytical tradition, norms are thought of as devices that allow social dilemmas to be overcome (Ullmann-Margalit 1977; Coleman 1990). A social dilemma exists when the behavior of individuals, even if rational, may produce an undesired, suboptimal social outcome. Rousseau's parable of the stag hunt, where two men must cooperate to hunt a stag, but either of them may be easily tempted to trap a hare, is the classical illustration. Extending this reasoning to the topic of this chapter, all workers of a production plant may have an incentive to follow some norm of resistance; however, workers may be tempted to defect from this common interest either to avoid possible punishments from a supervisor or to fulfill upward mobility expectations.

In order to preclude these temptations, at least three possible mechanisms could operate in real societies. First, individuals may hold moral preferences towards the collective good; meaning that "moral norms" guide their behaviors. Second, individuals may be frightened by an external authority (the state), which holds the capacity to punish them if they do not pursue the collective good; meaning that individuals are ruled through "legal norms". Third, individuals may mutually encourage or discourage each other's behavior by promoting the collective good; meaning that they are ruled by a "social norm". Of course, these three mechanisms actually operate concurrently, making it difficult to empirically ascertain whether the observed behavior is due to one, two or all three of them. Nevertheless, by analytically isolating one of them, we can better understand how it works. Obviously, it is impossible to do so in an empirical setting (you cannot, for instance, "switch morality off" in real individuals), but it can be done (clearly, under simplifying assumptions) in an artificial society (see below).

The first task to accomplish is to answer a simple, but not easy, question: what exactly is a social norm? The question is not an obvious one because, turning to sociological literature, we find definitions of social norms with very different implications: punishment-based definitions, expectation-based definitions and emotion-based definitions.

Thus, for the political scientist Axelrod (1986, 1097 – emphasis added): "A norm exists in a given social setting to the extent that individuals usually act in a certain way and are often *punished* when seen not to be acting in this way". This definition assumes that the existence of a certain norm can be empirically tested because it involves two kinds of behavior: the usual behavior performed by individuals, and the sanctioning behavior also performed by them. On other grounds, for the analytical philosopher Bicchieri (2006, 2 – emphasis added), "the existence of a social norm depends on a sufficient number of people *believing* that it exists and pertains to a given situation, and *expecting* that enough other people are following it in those kind of situations". Like the preceding definition, this one assumes that a social norm is something shared by a number of people, but stresses a component which is more difficult to test empirically: beliefs and expectations. For Bicchieri, sanctions may or may not be a component of "social norms," which is to say that they are not a necessary condition for them to exist. Finally, for the economist, Ostrom (2005, 121–122 – emphasis added), "Norms are prescriptions held by an individual that an action or outcome in a situation must, must not, or may be permitted [...] The changes may occur as a result of *intrinsic motivation* such as pride when keeping a norm or guilt when breaking a norm". In this case, as in Axelrod's definition, the emphasis is placed on sanctions; but the nature of the sanctioning process is somehow different, since it is assumed to be an internal (and emotional) one. Of course this process may be triggered (and usually will be triggered) by the awareness of not fulfilling others' expectations, and therefore implies that the behavior is performed under the possible scrutiny of others (Elster 2009).

What of these components (external sanctions, expectations, internal emotions), if any, are necessary conditions to account for the existence of a social norm? In this chapter I will support a "combined definition" (Opp 2001; see also Horne 2001), and assume that:

(1) A social norm is a non-intended process, whose outcome is the emergence of a certain pattern of behavior held by a large part of a population,
(2) where the behavior of individuals is controlled in a decentralized way (meaning, by other individuals),
(3) by means of an influence process based on the progressive spreading of the belief of what is expected in that population, and
(4) where that process is backed by informal sanctions which may (but not necessarily) produce certain emotions such as pride or guilt.

For most sociologists, the existence of "informal sanctions" is a necessary condition for a social norm to exist. Other kinds of regular behavior, such as conventions or social routines produced by mutual expectations are not, in fact, social norms because they do not require social sanctions in order to be maintained over time. The role played by sanctions is, however, mainly as a deterrent. Both in case studies (Ostrom 2005) and experimental settings (Fehr and Gätcher 2000), it is a well-established empirical fact that sanctions are rarely performed. Any understanding of social norms must therefore account for this fact, and provide an explanation for why so little sanctioning behavior is required to maintain prosocial behavior. I will assume (Linares 2012) that the mechanisms which produce this fact work as shown in *diagram 1*, which represents a cycle where:

(1) The initial triggering of sanctions pushes the generalization of beliefs about expected behavior (influence-spreading mechanism).
(2) The more generalized the expected behavior, the less triggering of sanctions is required (sanctioning-reduction mechanism).

Diagram 1: Mechanisms producing a social norm

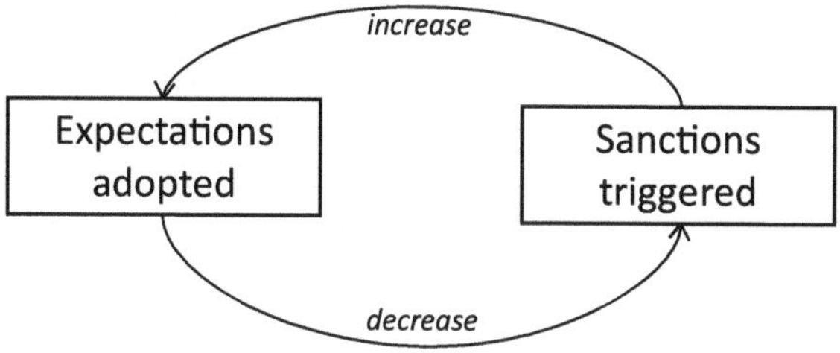

Complex adaptive systems, social networks and agent-based models

"Complex Adaptive Systems" is the term which has been coined to refer to systems, whether social, biological, or of another kind, whose aggregate behavior is the result of actions performed by different individuals who continuously adapt to an environment, which itself is constituted by others individuals who also engage in an adaptive behavior (Miller and Page 2007). These multiple behaviors, which mutually adapt to each other, may combine in complex ways to produce self-reinforcing dynamics, which are difficult to understand. This is so because the behavior that emerges from individuals' mutual adaptations, *aggregate behavior*, may in fact be the product of a long chain of interactions and show patterns that are quite distant from individuals' initial motivations to act.

These individuals, as eruditely claimed by Granovetter (1986), interact neither in an infrasocialized state of nature nor in an oversocialized social structure. Rather they are sensitive to the particular network in which they are embedded, with a corollary of this argument being that different topologies of networks will produce different interaction patterns. The "science" of networks, to which Granovetter himself is a recognized contributor, however, was almost dormant until the late 1990s when new technological innovations made possible new progress, which has been called "the new science of networks" (Barabási 2003; Watts 2003). The central concepts of this "new" science are: "small-world" and scale-free networks. The term "small-world" refers to a fact known at least since Milgram's (1967) famous study showing that, on average, two stranger American citizens are separated, on average, by just six intermediaries. Watts and Strogatz (1998) formally defined small-world networks as networks with high local clustering and short path lengths. In these networks, which are quite characteristic of social communities, every agent is just a few "handshakes" away from any other.

Not all social networks are like that, however. Hierarchical communities have a characteristic topology which has been called "scale-free" (Barabási and Albert 1999). This kind of network is assumed to be the product of a "preferential attachment" mechanism: a new agent who enters the community will have a strong incentive to create links with the most linked agents. As a result of this, a "rich get richer" or "Mathew effect" occurs so that a minority of agents holds the vast majority of links, becoming "hubs". In this kind of topology, the path between any two nodes is also short, but this is not so because of the more or less even distribution of links, but due to the opposite phenomenon: a highly skewed links distribution.

The difference between "small-world" and "scale-free" networks is clearly shown by comparing *bar graph 1.a*, which represents the distribution of ties typical of a "small-world" network, to *bar graph 1.b*, which represents a typical distribution of a scale-free network. In both cases the x-axis represents the number of ties and the y-axis represents the number of individuals in a network of 20 nodes holding 0, 1, 2 … n ties.

Bar Graph 1.a: Example of "small world" topology

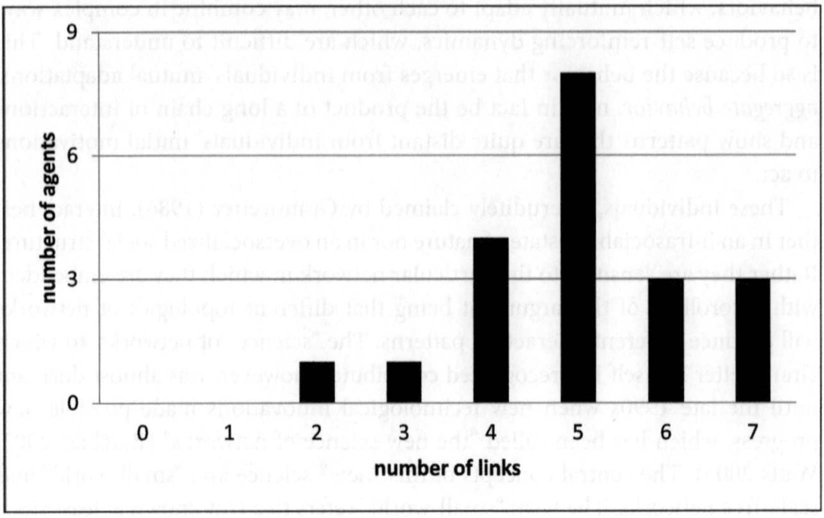

Bar Graph 1.b: Example of scale-free topology

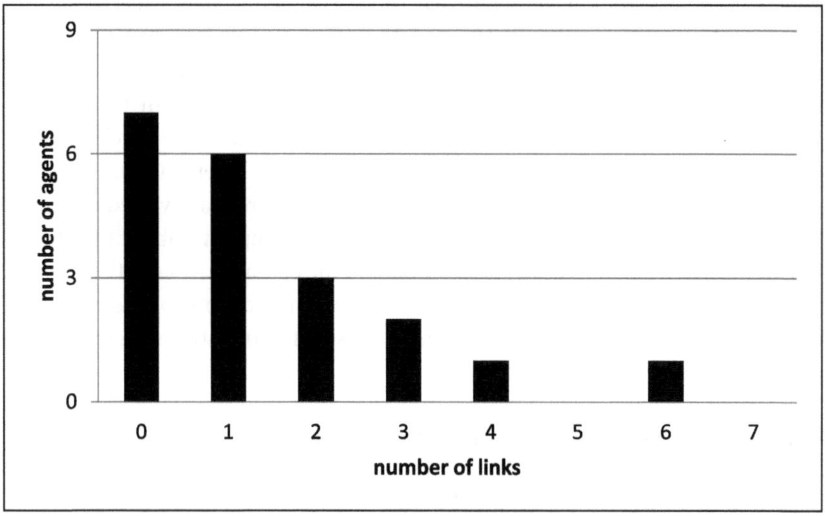

A group of workers that jointly produce a certain good while adjusting their behaviors under each other's monitoring is indeed an example of a complex adaptive social system. These individuals will likely differ in at least two parameters: their willingness to engage in resistance and their tolerance to social pressure. Will those more willing to resist be able to successfully establish a production rhythm below the formal target established by the firm's managers? As individuals interact and push new adjustments onto other individuals, one of the three following aggregate outcomes will emerge: (a) no individuals or just a few will cooperate in a resistance norm; (b) some individuals will cooperate; (c) most or all individuals will cooperate. The question is, which outcome will be reached? And, furthermore, what elements result in a certain outcome being reached? Is it a matter of the original number of highly motivated workers? Does it depend on the number of people displaying a low degree of tolerance toward social pressure? Or is it dependent on the properties of the network structure?

Of course these questions are not easy to answer in a rigorous, deductive way. Fortunately, there is a tool that may help: agent-based models (ABMs). An ABM is a formal and simple representation of the reality which, unlike other formal and simple representations (such as differential equations), can easily deal with heterogeneity in a population of individuals (that is, individuals may differ in many traits, such as "tolerance") as well as with decision rules other than rationality. ABMs are thus nicely suited for analyzing complex adaptive systems.

These models have several applications. By means of empirically calibrating their parameters, they have been used to explain actual data, such as fertility trends in France (González-Bailón and Murphy 2013), local youth unemployment rates in Stockholm (Hedström 2005), or educational achievement in France (Manzo 2013). Notwithstanding these empirical applications, in the realm of sociology ABMs are, at the moment, mainly a formal tool for developing and exploring the implications of middle-range theories (Gilbert 2008). According to this aim, which is also the aim of this paper, ABMs are mainly used to explore the logical consequences following from a set of assumptions about the characteristics of agents, their rules of interaction and the characteristics of the environment. The main theoretical and methodological implications of these kinds of models are bottom-up explanations, the analysis of cumulative systems and the production of artificial experiments.

Explanations, generative social science and mechanisms

As Epstein and Axtell (1996) nicely put it at the end of their path-breaking *Growing Artificial Societies*, where the now well-known Sugarscape model is analyzed:

> From an epistemological stand point, what "sort of science" are we doing when we build artificial societies like Sugarscape? Clearly, agent-based social science does not seem to be either deductive or inductive in the usual senses. But then what is it? We think generative is an appropriate term. The aim is to provide initial microspecifications (initial agents, environments and rules) that are sufficient to generate the macrostructure of interest. We consider a given macrostructure to be "explained" by a given microspecification when the latter's generative sufficiency has been established (Epstein and Axtell 1996, 177).

This "generative" approach, implying that a given social pattern is explained from the bottom-up, provides a sensible answer to the micro-macro problem, masterfully traced by Coleman (1986) ten years earlier:

> The major theoretical obstacle to social theory built on a theory of action is not the proper refinement of the action theory itself, but the means by which purposive actions of individuals *combine* to produce a social outcome" (1986, 1321 – emphasis in the original).

Epstein and Axtell's generative approach is also very close to Elster (1989) and others' (e.g., Hedström and Swedberg 1998) defense of "mechanisms" as the building blocks of sociological explanations. According to Hedström and Bearman (2009, 5), a mechanism "refers to a constellation of entities and activities that are organized such that they regularly bring about a particular type of outcome". Although a common unit of analysis in sociology is the individual, nothing in the concept of "mechanism" precludes the unit either being a "supra-individual" entity, such as a collective, or a "sub-individual" entity, such as the components of an individual decision-making process (e.g. attitudes, values, emotions, etc.). The concept of "mechanism" does not exclude a rational conception of action either. What the concept of "mechanism" does imply, is that whatever the entities and their rules of behavior, it has to be shown that they must regularly produce the outcome that is to be explained. Artificial society modeling is a nicely suited tool to accomplish this task, and it is a key component of the analytical sociology agenda (Hedström and Bearman 2009).

Emergence and cumulative systems

One of the most intriguing characteristics of society is the strong stability of many social patterns. Despite the fact that we all have the experience of living in an era of change, certain characteristics of society seem either to change very slowly or not change at all. The distribution of wealth among different social classes, rules of domestic labor assignment and school achievement rates of students of different backgrounds are just a few examples. Assuming that all that happens in society is a result of individual actions, the question to answer is, how is it that individuals act in ways that produce such aggregate patterns, which are often unintended, undesired, and even detrimental to many of them?

The answer lies in the fact that the relations individuals produce when interacting with one other often produce a new reality that, so to speak, "traps" individuals. As in the case of undergraduate students living in a residence hall who develop a stable system of informal rules concerning the use of the common kitchen in a few days, once a given distribution of rights and resources is established in any realm of society it will likely show a self-perpetuating trend, since agents are now forced to mutually adjust their behavior under the new conditions, eventually reaching an equilibrium (though possibly "unfair"). These complex adaptive systems, where the emergent outcome feeds back on the original system of action, are known as "cumulative systems" (Boudon 1979) and are a common object of sociological analysis.

Research methods and simulation experiments

Sociologists do not deal with "real" as opposed to "artificial" agents. The numbers that statistical software analyzes (and the sociologist interprets) are real only in the sense that they have been collected and introduced in the computer by someone, as well as in the sense that they represent properties of "real" individuals. However, it is a basic assumption of science that we simply cannot know what these individuals "are" in reality (whatever the "true" reality is). The same is equally true for the qualitative data produced by recording speech and action: we simply do not hear or see the records with the omniscient eyes of a divinity.

Both quantitative and qualitative data are essential to produce simulation models in so far as these models are not built to reflect someone's fantasies about society. The representation of reality – the model – has to be grounded in empirical knowledge of the world if it is to provide an explanation of that world at all. In any given theory (whatever the theoretical style), we will find concepts (e.g., properties of individuals such as "sensitivity to the influence of others") with difficult, or even impossible, empirical measure. When agent-based modelers are faced with this problem, some solution must be found in order to make the simulation run. The solution consists of substituting unknown empirical data for random numbers which are extracted from a theoretical distribution. This is why the results of ABMs must be accounted as averages of a sufficiently large number of simulation runs. While this procedure may be considered an artifact, notice that it is quite honest: the modeler explicitly recognizes the lack of knowledge that in a narrative style of theorizing often goes unnoticed (and sometimes hidden under a prose whose eventual literary beauty is not an essential element of a proper explanation).

This "artificial" way of proceeding has a further advantage, which is key for the analytical agenda in social sciences: the possibility of carrying out "artificial experiments". When conducting field social research, it is almost impossible to answer "what if" questions that may be relevant for increasing the understanding of a social phenomenon. When there are *competing theoretical understandings* of an issue, as in the case of "social norms," relevant questions arise, such as, "What if the topology of the social network were different?" "What if people were not sensitive to others' expectations?" However, by artificially manipulating parameters, it is possible to show whether a given prerequisite (e.g. network closure) is actually a necessary condition to "grow up" the social pattern.

A simple model of the emergence of resistance norms

Brief description

Briefly, the model used in this chapter (which is fully described in the appendix) represents a landscape where "patches" hold a certain reserve of a valuable resource that, for the sake of illustration, I will assume is grain. The population of 20 artificial agents represents workers who have the job collecting a certain amount of the valuable resource per minute (the quota formally established by the firm). Agents interact for a working period of 480 minutes. As suggested in the introductory section, a social norm may prescribe a working rhythm lower than that formally established by the managers. However, for that social norm to emerge two conditions have to be met:

(a) Agents have to be sensitive to others' expectations. It is assumed that this happens through an "influence process": agents will change their behavior if it differs from the behavior displayed by a number of agents in their local environments (see details in the appendix).
(b) Agents have to be sensitive to social sanctions provided by peers. In the model, a number of "unconditional agents" (see details in the appendix) will be introduced. These agents are invested with the capacity to sanction, but no assumption is made about the nature of the sanction or the nature of the internal psychological process driving the behavior of the target of the sanction[1].

Unconditional agents are assumed to be the "leaders" of the resistance behavior. They are unconditional in the sense that, whatever the behavior of other agents, they always produce below the manager's quota. Other agents have a binary choice between collecting the quota or an amount below it (which, because of the exposition convenience, is fixed and the same for all agents). Thus, in the initial ("set up") stage these agents are programmed to meet the quota. However, as the simulation runs, their behaviors will evolve according to the interaction rules mentioned above; this is why they are "conditional" agents.

Because the central interest of the paper is on the emergence of social norms of resistance among peer workers, other important features of the firm's internal activity, such as the procedures for supervising workers' activity or the effects of workers' formal organization (i.e. unions) have been obviated in the model; which is not to say that they are not important at all in reality. Nonetheless, some conclusions related to unions' activity will be derived based on the realization of the simulation results.

Output variables, parameters and simulation experiments

The main output of the model is the amount of resources collected by agents at the end of the simulation run. Concerning the emergence of resistance norms,

1 Public goods experiments have shown that certain individuals have the capacity to impose sanctions regardless of the cost; a trait that has been called "strong reciprocity" (Gintis 2000). It has been estimated that roughly one-third of the population holds this trait.

three outputs will be recorded: the number of defectors (that is, the number of agents that produce the quota); the number of cooperators (that is, the number of agents that produce below the quota); and the number of sanctions performed during a day's work.

It is assumed that all these variables will change in accordance with two parameters:

(a) The number of "unconditional agents" initially introduced in the simulation.
(b) The distribution of links among agents.

The first parameter has an obvious effect: the larger the number of unconditional agents (i.e. agents which unconditionally produce an amount below the quota) the lower the total amount of resources collected. This does not necessarily imply, however, that a social norm of resistance will emerge, since the distribution of links among agents (either the "small-world" type or "scale-free" type) may facilitate, or preclude, the social influence process that is required for such a norm to exist.

Since the solution to this puzzle is not straightforward, a number of "simulation experiments" have been conducted. In this way, while holding constant the type of network topology, the effect of increasing or decreasing the number of unconditional agents can be shown. And conversely, while holding constant the number of unconditional agents, the effect of changing the type of network topology can be shown. For every experimental condition, 50 repetitions were run. *Figures 1* to *3* (below) show the means of the results for every output.

Main results

Figure 1 shows the amount of grain collected at the end of the simulation as a function of the number of unconditional cooperators at the beginning of the simulation. The lines show an obvious decreasing trend, meaning that the larger the number of initial unconditional cooperators (i.e. agents who unconditionally follow the norm of resistance), the smaller the amount collected. There are, however, strong differences among *experimental conditions*. The baseline model (short-dashed line) shows the outcomes produced by a *random rule of behavior* for conditional cooperators, so these agents have a 50% chance of either cooperating or defecting, regardless of the behavior of other agents. Because of this feature (besides the fact that there are no links among actors), this experimental condition is called "asocial". Interesting enough, the outcomes produced by the *scale-free experimental condition* (long-dashed line) do not sharply differ from the asocial model. On the other hand, the outcome produced by the *small-world condition* (continuous line) is initially higher than the baseline model; however, once a small critical mass of unconditional cooperators exists (4 agents), the collection of grain drops from 750 units to 630.

Figure 1: Production of grain

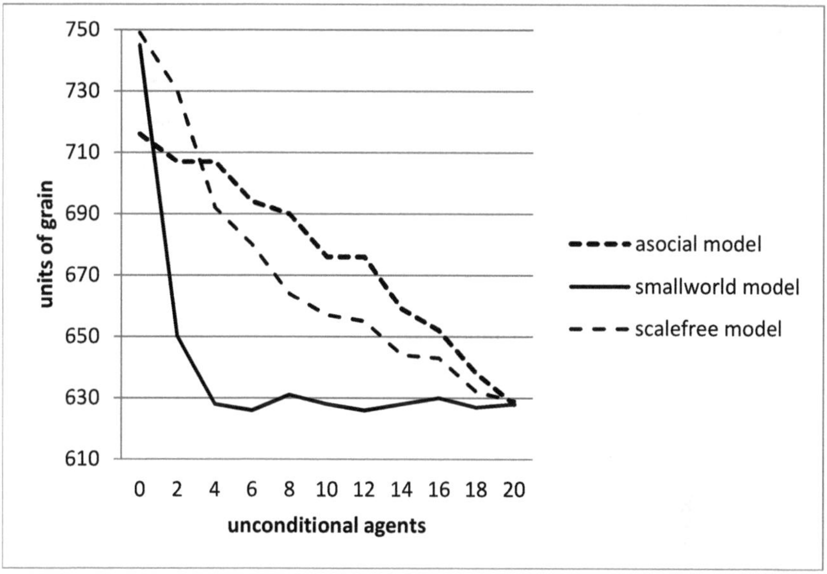

Figures 2.a and *2.b* show the number of agents cooperating and defecting, respectively, for every experimental condition. The resistance of conditional agents is increasingly fostered as the number of initial unconditional cooperators increases. But in the small-world condition cooperation grows much faster, so only four initial unconditional cooperators are able to induce the cooperation of almost all conditional cooperators. In the scale-free condition the same number of unconditional cooperators, however, induce the resistance of half of the conditional cooperators.

Figure 2.a. Amount of cooperation

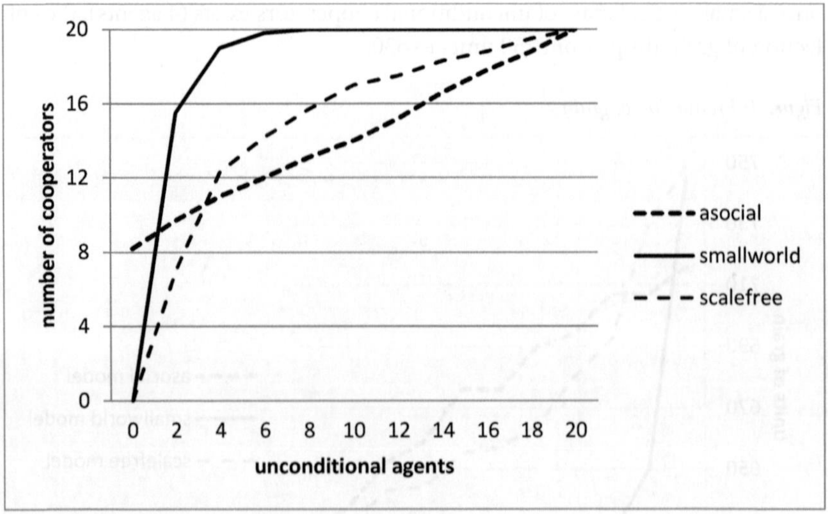

Figure 2.b: Amount of defection

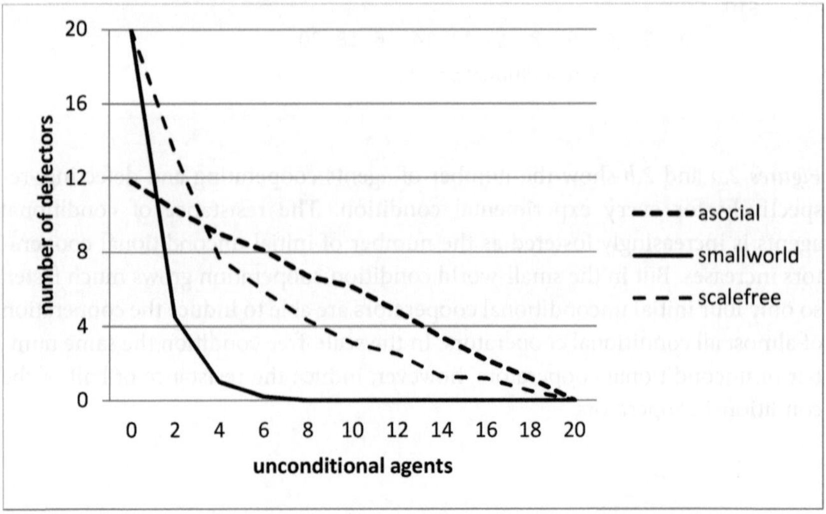

Why should it be so? The evidence about the number of executed sanctions is shown in *Figure 3*, which displays the number of sanctions at the end of the simulation relative to the initial number of unconditional cooperators in both experimental settings. Even if sanctioning activity follows a similar pattern in both experimental conditions (peaking when the number of unconditional agents equals four), it is clear that sanctioning activity is much larger in the small-world condition than in the scale-free; this variance being responsible for the above-mentioned difference in resistance behavior and, hence, collection of grain.

Figure 3: Sanctions at the end of the simulation run

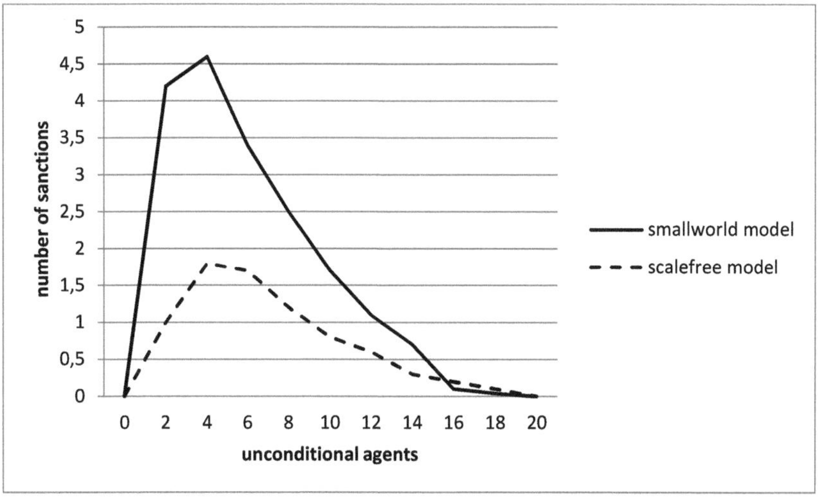

Figures 4.a and *4.b* show the results of two typical simulation runs (hence, the numbers displayed on the y-axis are not mean values) comparing the evolution of both models as the time step progresses from 0 to 30 (afterwards the results are stable), when the initial number of unconditional cooperators is set at 4. *Figure 4.a* shows the accumulated number of sanctions. *Figure 4.b* shows the corresponding evolution of defecting behavior in these typical simulations. In both models all the sanctions are produced in the earliest steps of the simulation. However, it seems that an identical number of unconditional agents do not produce the same amount of sanctions under different topological conditions. It clearly suggests that "scale-free" networks preclude the execution of enough sanctions to control most of the agents, even though the required sanctioning activity to achieve this end is fairly low (on average, less than five).

Figure 4.a: Evolution of sanctions

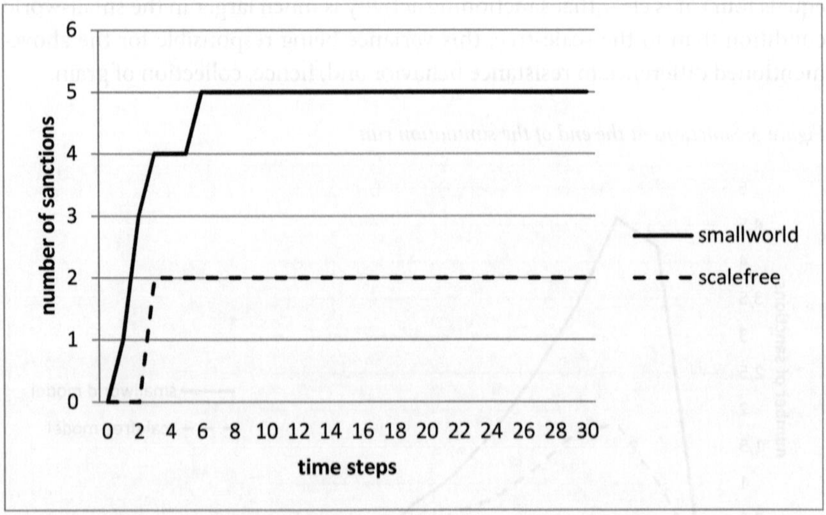

Figure 4.b: Evolution of defection

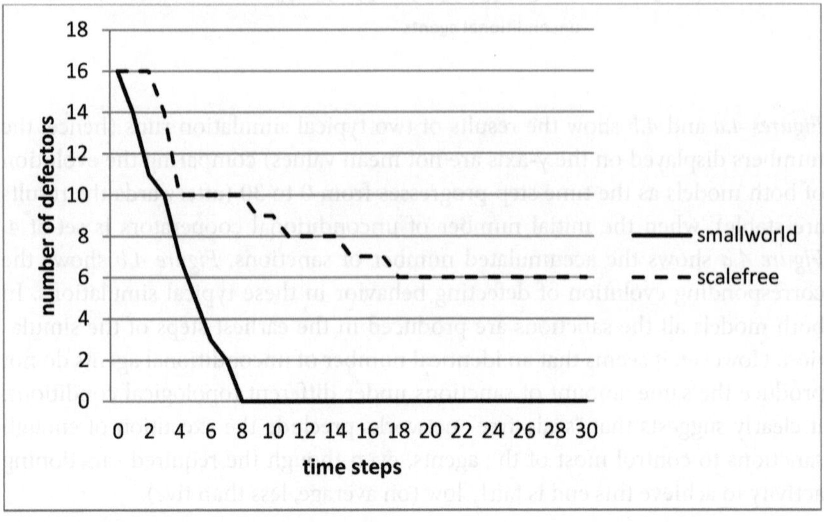

Discussion and conclusions

This chapter addresses the question of the emergence of resistance behavior among workers. This kind of behavior is assumed to be a "social norm of resistance" in so far that it results in a decrease in the output produced (a very typical form of resistance) that totals an amount below the formal goal established by the firm's manager, by means of a social influence process backed by social sanctions.

The specific problem the paper focuses on is the effect of network topology on this social process. The so-called, "new science of networks" has pointed at two different topologies as typical of many social settings: small-world networks (networks where the distribution of ties resembles a bell curve) and scale-free networks (networks with a heavily uneven distribution of ties). Do they have a differential impact on the emergence of these informal norms of resistance? As far as the author knows there is no empirical research attempting to answer this problem.

The approach utilized in this chapter consists of the building of a computational model (specifically an agent-based model) that attempts to capture some essential features of the issue. In this model a population of 20 agents "works" for one working day. They "must" collect a certain amount of grain per minute. The simulation shows how, when a few "unconditional" agents produce below this target, the abovementioned social mechanisms foster a pattern of behavior that leads to a decline in the amount of grain collected.

However, results are not the same under different topological conditions: this informal norm of resistance is weaker when a scale-free topology is assumed, since it produces very little sanctioning activity. This is so because of two reasons: (a) in the scale-free model some agents have no ties at all, so they cannot be influenced in any way; (b) most important from a sociological point of view, in scale-free networks the production outcome is highly sensitive to the position of "unconditional" agents in the network. Thus, if unconditional agents were the ones to occupy the few positions with the largest number of links, the influence process will very likely spread to almost the whole population. However, in the model, nothing guarantees this outcome since positions are assigned randomly.

Some considerations on the role played by workers' formal organizations within firms can be derived from these results. The existence of union activity implies a scale-free topology, since union leaders become "hubs" (nodes with a disproportionately large number of ties). But unionization also guarantees that the more connected positions are occupied by the agents who are also more likely to initiate resistance actions. Nevertheless, if unions are not present, informal resistance in the workplace is still possible as long as, as shown by the simulation model, relations among peer workers have a small-world structure.

These conclusions push towards two directions: on the one hand, from a methodological point of view, agent-based models can be fruitfully used not only to replicate actual empirical patterns, but also to suggest venues of further research when empirical information is missing. On the other hand, according to the conclusions obtained in this paper, more detailed information about network topologies is required if we are to improve our knowledge of worker resistance.

Acknowledgements

I am grateful for informal comments about a previous version of this paper from other participants in the workshop, "Formality and Informality: From Decoupling to Entanglement – Workshop on the Formal and the Informal in Law, Institutions and Economy" (Warsaw, Poland, May 2013), as well as the formal and detailed comments of two anonymous reviewers. Financial support from the Spanish MICNN funds for research projects is acknowledged (reference CSO2010-00034).

Appendix: Description of the model

Concerning the explanation of the emergence of social norms, three families of models are of interest to this paper:

(1) "Social influence" models (e.g., Gould 1993; Axelrod 1997), which show how the behavior of agent *i* depends on the number of other agents in his environment displaying the same/opposite kind of behavior.
(2) "Decentralized sanctions" models (e.g., Axelrod 1986; Heckathorn 1989), which show how the behavior of agent *i* depends on the behavior of a number of other agents in his environment with the capability to provide punishments/rewards. A major achievement of these models was finding out that "hypocrisy" (encouraging others to do what one is not willing to do) is a good strategy to promote the public good.
(3) "Critical Mass" models (e.g., Marwell et al. 1988; Kim and Bearman 1997), which show how the generalization of a certain behavior depends on the existence of a certain (small) number of agents displaying that behavior, and the possibility of transmission (either by means of an influence mechanism or a sanctioning mechanism) of that behavior through social networks.

There is a number of models in the literature addressing similar issues (e.g. Bravo 2010; Gächter and Thöni 2011), with Kitts' (2006) simulation of "Norms Amid Ties of Amity and Enmity" bearing the closest resemblance to ours. In this model, agents are subject to a "mimetic influence" (which is the label Kitts uses to refer to the process of social influence) as well as to an "inductive influence" (which is the label used to refer to the sanctioning process). Some specific features that characterize the model set forth in this chapter include: (a) agents are not embedded in a fixed network, but move around; (b) agents are not influenced by all other agents but only by those in their local environments; and (c) agents are not sanctioned by all other agents but only by those with the capacity to sanction in their local environments.

Several platforms are available for programming agent-based models. NetLogo, by Wilenski (1998), has proven to be a powerful tool which has a large community of users worldwide. It is the one used to develop the model presented in this chapter. Because computer programs can be difficult to understand (and because replication is a key component in the quest for scientific knowledge), a standardized protocol for describing models has been developed. The so-called "Overview, Design concepts and Details" (ODD) protocol (Railsback and Grimm 2012) contains the basic information required to understand the architecture of the model as well as how it works. Following this protocol, the main features of the model are presented below:

Overview

(1) *Purpose*: The main aim of the model is to improve our theoretical understanding of how social norms emerge. The specific problem the model addresses is: under what conditions will a pro-social behavior spread among a population of individuals in the absence of moral and/or legal enforcement? In the context of the model, the term pro-social behavior means for workers to follow a social norm of resistance which dictates to produce below the quota established by the firm.

(2) *Entities, state variables, and scales*: The model has three kinds of entities: the environment, agents and links among agents. The environment consists of a torus of 33x33 patches which have only one state variable: "grain-store" (numerical). All agents have the following state variables: "cooperated?" (boolean), "influence-sanction-threshold" (numerical), "vision-scope" (numerical), "degree" (numerical), "influence-rate" (numerical), "sanctioned?" (boolean), "initial-number-of-links" (numerical), "initial-links" (numerical) and, a key variable, "unconditional?" (boolean).
Global variables are: "number-of-sanctions" (numerical), "grain" (numerical), "population" (numerical), "number-of-unconditionals" (numerical), "average-node-degree" (numerical) and "small-world" (boolean).
There is no spatial scale, since real environment is not simulated.
Every tick simulates one minute of real time, but this convention is not grounded in any empirical data.

(3) *Process overview and scheduling:* The model includes the following actions executed every time step in the same order:
   (a) The variable "grain-store" of every patch is updated according to two processes that are independent from agents' behavior. These are called "get-rainfall" and "get-dry".
   (b) Agents are asked either to "take-grain" (if the patch's grain-store is large enough) or to "move" (in order to search for grain).
   (c) If the agent's characteristic of "unconditional?" is true, he takes grain cooperatively (i.e. he takes less grain than the target). If this condition is not met, he is asked to "being-influenced?", which may push the agent to take grain cooperatively.
   (d) If the influence process has not produced cooperation, the agent is asked to "be-sanctioned?", which may push the agent to take grain cooperatively.
   (e) If the agent defects, he meets the target established by the firm.
   (f) Outputs are displayed.
   (g) The simulations stop at the end of the "working day" (480 ticks).

Design concepts

(4) *Design concepts*
Basic Principles: The model attempts to capture three related theoretical propositions: (a) for a social norm to exist there must be a social influence process; (b) for a social norm to exist there must also be a decentralized sanctioning process; (c) social ties may foster its emergence. The model also attempts to capture two empirical facts: (a) sanctioning is rare; (b) some agents have an intrinsic motivation to execute sanctions. The model does not, however, specifically represent the psychological mechanism that drives changes in the behavior of sanctioned agents.
Emergence: The expected results are that, when "conditional cooperators" are subject to influence and sanctioning processes, a small number of "unconditional cooperators" will trigger the spreading of resistance behavior. Thus, as the model runs, it should be observed that: (a) cooperative behavior increases, and (b) the cumulative number of executed sanctions decreases.
Adaptation: The adaptive behavior of agents is not assumed to follow any empirical rule. All agents move around following a random path. Unconditional cooperators perform a non-social behavior in the sense that their behavior does not respond to the presence of other agents (that is why they are "unconditional"). Conditional cooperators perform a social

behavior in the sense that their behavior responds to the presence of other agents.

Objectives: Agents are not programmed to pursue any goal (although they behave as if they had the goal of collecting grain). They just implement a rule of taking either a "small amount" or a "large amount" of grain under certain conditions (in the case of unconditional cooperators they always take a small amount), but the behavior rules are not intended to reproduce rationality. There are neither prediction nor learning processes.

Sensing: All agents are assumed to know the "grain-store" of the patch they are on. Conditional cooperators are assumed to know the number of agents which are actually displaying a pro-social behavior in their local environments (that is, within a given radius) and calculate its rate to the total agents in their local environments. They are also assumed to know whether there is an unconditional cooperator in their Moore neighborhoods (the eight patches surrounding their own patch).

Interaction: Agents interact with patches, subtracting either a "small amount" or a "large amount" from the patches' "grain-store". Agents interact among themselves by means of two processes: influence and sanctioning (see details below).

Stochasticity: Stochastic processes are used in the initialization to set the patches' grain-store, agents' position and conditional cooperators' influence-sanction-threshold, as well as in the initialization of networks. Random numbers are also used in the "take grain" and sanctioning processes (see details below).

Collectives: There are no agent-sets, but a main distinction is traced between unconditional and conditional agents, since they behave in very different ways.

Observation: At the end of every simulation, the required outputs are: (a) the units of grain collected; (b) the number of cooperators (i.e. agents who followed the norm of resistance); (c) the number of defectors (i.e. agents who did not follow the norm of resistance); (d) the accumulated number of sanctions. The evolution of these indicators through the time steps is shown in plots.

Details

(5) *Initialization*: The "population" (i.e. number of agents) is set at 20, in order to resemble middle size teams. The number of initial "unconditional cooperators" is a parameter which can be manipulated in every simulation

(min=0 and max=20). When the global variable "small world" is set true, links among agents are randomly created to match the value of the average node degree, which is set at 5. When "small world" is set false the links are randomly created to fit a gamma distribution with parameters 1.2 and 0.5, in order to simulate a scale-free distribution of nodes.

For all agents, allocation is randomly assigned. For unconditional cooperators, the variable "cooperated?" is set as "true", while for conditional cooperators it is set as "false". For conditional cooperators, the "influence-sanction-threshold" is randomly assigned between 0.0 and 1.0 (uniform distribution), and "vision-scope" is set at 4. For unconditional cooperators, "degree" is calculated as the number of links to other agents.

Other global variables (grain and number-of-sanctions) are set at zero.

The value of the state variable of patches, grain-store, is randomly assigned between 0.0 and 6.0 (uniform distribution)

(6) *Input data*: No input data are required.

(7) *Submodels*:

Patch submodels: Every patch "gets rainfall" (that is, they increase their grain-store by +0.1) at every time step if a random number between 0 and 100 is below 3. Every patch "gets dry" (that is, they decrease their grain-store by −0.1) if a random-float number between 0 and 100 is below 0.1.

Agent submodels:

Influence: At every time step, the conditional agents calculate the rate of agents within their local environment with the attribute "cooperated?" set as "true" to the total agents within their scope of vision. If this number is larger than their influence-sanction-threshold, then the agents take-grain-cooperatively (which means they take a "small amount", subtracting 0.1 from the patch grain-store); otherwise the agents "take grain" (see details below). Local environments are determined either by the scope of vision (when the average node degree equals 0) or by linked neighbors (when the average node degree is larger than 0).

Sanctioning: At every time step, the conditional agents check if there is an unconditional cooperator in one of their eight neighboring patches (when the average node degree equals 0) or in their set of linked neighbors (when the average node degree is larger than 0). If this happens to be the case, it is assumed that the conditional cooperator is exposed to some external reward or punishment that will cause either pride or guilt if he cooperates or does not cooperate, respectively. As in the case of influence, it is assumed that all agents are not equally sensitive to this process. Therefore, if a random number between 0.0 and 1.0 is above the influence-sanction-threshold, the agent is sensitive to this, and takes a small amount of grain. The reason for restating the local environment as the Moore neighborhood in this process (when the average node degree equals 0) is to reduce the chance of executing sanctions (since, as stated before, sanctions rarely occur).

Taking grain procedures: When an agent takes grain in a cooperative setting the grain-store of the patch is reduced -0.1, but when an agent "defects" it is reduced -0.2

Move: An agent remains in his patch as long as there is grain to be taken from it. When this condition is not met, the agent randomly moves until he finds a new patch with enough grain-store.

References

Axelrod, Robert. 1986. "An Evolutionary Approach to Norms." *American Political Science Review* 80: 1095–1111.

Axelrod, Robert. 1997. "The Dissemination of Culture: A Model with Local Convergence and Global Polarization." *Journal of Conflict Resolution* 41: 203–226.

Barabási, Albert-Lászl6. 2003. *Linked*. Cambridge (MA): Plume.

Barabási, Albert-Lászl6, and Reka Albert. 1999. "Emergence of Scaling in Random Networks." *Science*: 286: 509–512.

Bicchieri, Cristina. 2006. *The Grammar of Society*. Cambridge: Cambridge University Press.

Boudon, R. 1979. *La logique du social. Introduction à l'analyse sociologique*. Paris: Hachete.

Brass, Daniel J, Joseph Galaskiewicz, Henrich R. Greve, and Wenpin Tsai. 2004. "Taking Account of Networks and Organizations: A Multilevel Perspective." *Academy of Management Journal* 47: 795–817.

Bravo, Giangiacomo. 2010. "Agent's Beliefs and the Evolution of Institutions for Common-Pool Resource Management." *Rationality and Society* 23: 117–152.

Coleman, James S. 1986. "Social Theory, Social Research and a Theory of Action." *American Journal of Sociology* 91: 1309–35.

Coleman, James S. 1990. *Foundations of Social Theory*. Harvard (MA): Harvard University Press.

Dixon, Marc, and Vincent J. Roscigno. 2003. "Status, Networks and Social Movement Participation: The Case of Striking Workers." *American Journal of Sociology* 108: 1309–1335.

Elster, Jon. 1989. *Nuts and Bolts for the Social Sciences*. Cambridge: Cambridge University Press.

Elster, Jon. 2009. "Norms." In *The Oxford Handbook of Analytical Sociology*, edited by Peter Hedström and Peter Bearman, 195–217. Oxford: Oxford University Press.

Epstein, Joshua M., and Robert Axtell. 1996. *Growing Artificial Societies: Social Science from the Bottom Up*. Washington (DC): The Brookings Institution.

Fehr, Ernst, and Simon Gächter. 2000. "Cooperation and Punishment in Public Goods Experiments." *American Economic Review* 90: 980–994.

Gächter, Simon, and Christian. Thöni. 2011. "Micromotives, Microstructure, and Macrobehavior: The Case of Voluntary Cooperation." *Journal of Mathematical Sociology* 35: 26–65.

Gilbert, Nigel. 2008. *Agent-based Models*. London: Sage Publications.

Gintis, Herbert. 2000. "Strong Reciprocity and Human Sociality." *Journal of Theoretical Biology* 206: 169–179.

González-Bailón, Sandra, and Tommy E. Murphy. 2013. "The Effects of Social Interactions on Fertility Decline in Nineteenth-Century France: An Agent-Based Simulation Experiment." *Population Studies: A Journal of Demography* 67: 135–155.

Gould, Roger. 1993. "Collective Action and Network Structure." *American Sociological Review* 58: 182–196.

Granovetter, Mark. 1985. "Economic Action and Social Structure: The Problem of Embeddedness." *American Journal of Sociology* 91: 481–510.

Heckathorn, Douglas. 1989. "Collective Action and the Second Order Free Rider Problem." *Rationality and Society* 1: 78–100.

Hedström, Peter. 2005. *Dissecting the Social: On the Principles of Analytical Sociology*. Cambridge: Cambridge University Press.

Hedström, Peter, and Peter Bearman. 2009. "What is Analytical Sociology All About? An Introductory Essay." In *The Oxford Handbook of Analytical Sociology*, edited by Peter Hedström and Peter Bearman, 3–24. Oxford: Oxford University Press.

Hedström, Peter, and Robert Swedberg. 1998. *Social Mechanisms. An Analytical Approach to Social Theory*. Cambridge: Cambridge University Press.

Horne, Christine. 2001. "Sociological Perspectives on the Emergence of Social Norms." In *Social Norms*, edited by Michael Hechter and Karl D. Opp, 1–30. New York: Russell Sage Foundation.

Kim, Hyojoung, and Peter S. Bearman. 1997. "The Structure and Dynamics of Movement Participation." *American Sociological Review* 62: 70–93.

Kitts, James A. 2006. "Social Influence and the Emergence of Norms Amid Ties of Amity and Enmity." *Simulation Modelling Practice and Theory* 14: 407–422.

Leon, Francisco J. 2011. "Peer Loyalty and Quota Restriction as Social Norms: A Case Study of Their Emergence." *Rationality and Society* 23: 75–115.

Linares, Francisco. 2012. "A Multi-Agent Simulation of a Social Norm Spreading Mechanism." *Revista Española de Investigaciones Sociológicas* 138: 19–40.

Manzo, Gianluca. 2013. "Educational Choices and Social Interactions: A Formal Model and a Computational Test." *Comparative Social Research* 30: 47–100.

Mayo, Elton. 1945. *The Social Problems of an Industrial Civilization*. Cambridge (MA): Harvard University Press.

Marwell, Gerald, Pamela Oliver, and Ralph Prahl. 1988. "Social Networks and Collective Action: A Theory of Critical Mass III." *American Journal of Sociology* 94: 502–534.

Milgram, S. 1967. "The Small-World Problem." *Psychology Today*, 2: 60–67.

Miller, John H., and Scott E. Page. 2007. *Complex Adaptive Systems*. Princeton (NJ): Princeton University Press.

Opp, Karl-Dieter. 2001. "How Do Norms Emerge? An Outline of a Theory." *Mind & Society* 3: 101–128.

Ostrom, Elinor. 2005. *Understanding Institutional Diversity*. Princeton (NJ): Princeton University Press.

Railsback, Steven F., and Volker Grimm. 2012. *Agent-Based and Individual-Based Modeling*. Princeton (NJ): Princeton University Press.

Roscigno, Vincent J., and Randy Hodson. 2004. "The Organizational and Social Foundations of Worker Resistance." *American Sociological Review* 69: 14–39.

Ullmann-Margalit, Edna. 1977. *The Emergence of Norms*. Oxford: Oxford University Press.

Vaughan, Diana. 1999. "The Dark Side of Organizations: Mistake, Misconduct, and Disaster." *Annual Review of Sociology* 25: 271–305.

Watts, Duncan, and Steven H. Strogatz. 1998. *Collective Dynamics of "Small-World" Networks*. Nature 393: 440–442.

Watts, Duncan J. 2003. *Six Degrees. The Science of a Connected Age*. New York: W.W. Norton & Company.

Wilenski, Uri. 1998. NetLogo. Center for Connected Learning and Computer-Based Modeling, Northwestern University, Evanston, IL. Accessed January 16, 2014. http://ccl.northwestern.edu/netlogo.

Contributors

Stef Adriaenssens is an assistant professor in economic sociology and in culture and economies at the Faculty of Economics and Business of KU Leuven, Brussels campus. His recent publications deal with underground and informal activities, such as begging, informal work, and prostitution.

Grzegorz Bryda is an assistant professor at the Jagiellonian University in Kraków, and Rector's Proxy for the Evaluation of the Quality of the Education System. He has experience in the implementation of quantitative and qualitative research, statistical techniques for data analysis, data and text mining and computer assisted qualitative data analysis (Computer-assisted qualitative data analysis software) for commercial companies and public institutions.

Robert Dingwall is a consulting sociologist and professor at Nottingham Trent University. He has published widely on aspects of work, organization and interaction in the fields of health care, law, and science and technology.

Timothy Eccles is a course leader at Nottingham Trent University. His research publications concern the embeddedness of professional gestalts within occupations and how this develops into particular technical templates of "best" practice. From this theoretical work he was part of a research project that directly led to improvements in developing, benchmarking and enforcing "professional" practice within the commercial real estate sector. He has also co-written one student textbook, and is currently working on a second.

Liela Groenewald is a senior lecturer at the University of Johannesburg, where she teaches Urban Sociology and Social Science Research Methodology. Her research and publications focus on informal settlement in post-apartheid South African cities and in particular on the construction of citizenship for the residents of informal settlements by the post-apartheid local state, as well as the social construction of identity. She is also Secretary of the South African Sociological Association and convenor of its working group on Urban Sociology.

Jef Hendrickx is a member of MEES (Mathematical Economics, Econometrics and Statistics) at KU Leuven, campus Brussels. He teaches statistics and econometrics at the Faculty of Economics and Business. He specializes in research on hard-to-reach populations. Technically he is mainly interested in applications of multilevel methods.

Aleksandra Herman is an assistant professor at the University of Warsaw. She deals with issues related to retaining cultural diversity in multicultural local communities and the anthropology of corporality. She is an active researcher in the field.

Mikko Lagerspetz is a professor of sociology at the Åbo Akademi University in Turku. Before that, he worked at Tallinn University, Estonia, as a professor and director of the Centre for Civil Society Study and Development. He has written around 100 scientific publications mainly on civil society, cultural policies, minorities and social problems. In 2004, he received the Estonian Annual State Prize for Research in the Social Sciences.

Hans-Joachim Lauth is a professor for comparative politics at the University of Würzburg. His publications cover several studies about democratization and measuring democracy. He is also author and editor of four different textbooks on comparative politics. In the last decade his research interests have been focused on informal rules and their relevance to politics and polities.

Francisco Linares is an assistant professor at the University of La Laguna, where he teaches social theory and statistics for the social sciences. He is the author of a number of articles on the topics of unintended consequences, collective action and social norms. His current interests include the applications of agent-based models in sociological research.

Adriana Mica is an assistant professor at the University of Warsaw. Her interests concern the diffusion of paraformality in the field of short-term loans in Poland, and the types of organizational failure in dog population management in Romania.

Barbara A. Misztal is a professor at the University of Leicester. She is interested in sociological theory, political and cultural sociology and specifically the issues of trust, memory and the role of public intellectuals. Her new book project aims to enhance sociological understanding of the spectrum of normality by illustrating it with the help of literature.

Grażyna Skąpska is a professor at the Jagiellonian University in Kraków. Her interests are focused on the application of contemporary sociological theories to socio-legal phenomena, transitional justice, constitutionalism and the rule of law. She is president of the Polish Sociological Association, and is a member of the International Sociological Association Executive Committee.

Dieter Verhaest is an assistant professor in labour economics and macroeconomics at KU Leuven, campus Brussels, where he is a member of the Human Relations Research Group. He is also affiliated to SHERPPA, Ghent University. His research interests include, amongst others, the match between education and work, participation in post-school training activities and the youth labour market.

Jan Winczorek is an assistant professor at the University of Warsaw. He has a vivid interest in the sociology of law and theoretical sociology.

Rafał Wiśniewski is an assistant professor at Cardinal Stefan Wyszyński University in Warsaw. He is vice-president of the Polish Sociological Association and specializes in the sociology of culture and intercultural communication.

Studies in Social Sciences, Philosophy and History of Ideas

Edited by Andrzej Rychard

Vol. 1 Józef Niżnik: Twentieth Century Wars in European Memory. 2013.

Vol. 2 Szymon Wróbel: Deferring the Self. 2013.

Vol. 3 Cain Elliott: Fire Backstage. Philip Rieff and the Monastery of Culture. 2013.

Vol. 4 Seweryn Blandzi: Platon und das Problem der Letztbegründung der Metaphysik. Eine historische Einführung. 2014.

Vol. 5 Maria Gołębiewska / Andrzej Leder/Paul Zawadzki (éds.): L'homme démocratique. Perspectives de recherche. 2014.

Vol. 6 Zeynep Talay-Turner: Philosophy, Literature, and the Dissolution of the Subject. Nietzsche, Musil, Atay. 2014.

Vol. 7 Saidbek Goziev: Mahalla – Traditional Institution in Tajikistan and Civil Society in the West. 2015.

Vol. 8 Andrzej Rychard / Gabriel Motzkin (eds.): The Legacy of Polish Solidarity. Social Activism, Regime Collapse, and the Building of a New Society. 2015.

Vol. 9 Wojciech Klimczyk / Agata Świerzowska (eds.): Music and Genocide. 2015.

Vol. 10 Paweł B. Sztabiński / Henryk Domański / Franciszek Sztabiński (eds.): Hopes and Anxieties in Europe. Six Waves of the European Social Survey. 2015.

Vol. 11 Gavin Rae: Privatising Capital. The Commodification of Poland´s Welfare State. 2015.

Vol. 12 Adriana Mica / Jan Winczorek / Rafał Wiśniewski (eds.): Sociologies of Formality and Informality. 2015.

www.peterlang.com